Fish and Diadromy in Europe (ecology, management, conservation)

Developments in Hydrobiology 200

Series editor

K. Martens

Fish and Diadromy in Europe (ecology, management, conservation)

Proceedings of the symposium held 29 March – 1 April 2005, Bordeaux, France

Edited by

Sylvie Dufour[1], Etienne Prévost[2], Eric Rochard[3] & Patrick Williot[3]

[1]*CNRS-National Museum of Natural History, Paris, France*
[2]*INRA, UMR Ecobiop, Saint Pée sur Nivelle, France*
[3]*Cemagref, Estuarine ecosystems and Diadromous fish research unit, Cestas, France*

Reprinted from Hydrobiologia, Volume 602 (2008)

Springer

Library of Congress Cataloging-in-Publication Data

A C.I.P. Catalogue record for this book is available from the Library of Congress.

ISBN-13: 978-1-4020-8547-5

Published by Springer,
P.O. Box 17, 3300 AA Dordrecht, The Netherlands

Cite this publication as Hydrobiologia vol. 602 (2008).

Cover illustration: Salmon spawning. Photo credit: D. Grimardias (INRA, UMR Ecobiop).

Printed on acid-free paper

Printed in the Netherlands

TABLE OF CONTENTS

Hydrobiologia (2008) 602:1–3
DOI 10.1007/s10750-008-9289-0

FISH AND DIADROMY IN EUROPE

Preface

Eric Rochard

This symposium was organized by Cemagref[1] and INRA[2] on behalf of the Diadfish network www.diadfish.org and was sponsored by the Région Aquitaine and the French Secrétariat d'état à la francophonie. It was the first one in Europe to gather scientists working on different diadromous species and addressed both theoretical and applied issues, whatever be the disciplines.

Sixty talks and fifty-eight posters were presented[3]. However, almost all the proposals were in the field of the life sciences; in spite of our efforts to promote social and human sciences, they were too rarely addressed in the communications. However I am sure that in the next symposium organized by the Diadfish network there will be much more interdisciplinarity. Selected talks and posters covered a large number of species; salmonids and eel were the most frequently mentioned and that probably represents their current importance in the scientific communities in Europe. Both are used partly as models of anadromous and catadromous life histories, respectively. Salmonids are still the most studied group: the recent increasing concern about the biological consequences of global change has renewed the high interest in these cold-water species. The drastic decrease in the European eel and the imminence of European regulations provoked numerous studies on several aspects of the ecology, physiology, and population dynamics of this species. Assessment of the effective impacts of different human activities on this species is also a very urgent question. Talks were thematically very diverse: biogeography, diadromous fish in the ecosystems, life history strategies, tactics and adaptations, and human pressures on diadromous fish.

It has been a very enthusiastic experience; we expect it will be fruitful and will help our community to initiate new researches at larger scales and to improve our expertise.

This special issue does not reflect all the diversity of what was presented during the symposium, but it does represent a nice sample of what this scientific community produces. The selection process was

Guest editors: S. Dufour, E. Prévost, E. Rochard & P. Williot
Fish and diadromy in Europe (ecology, management, conservation)

Electronic supplementary material The online version of this article (doi:10.1007/s10750-008-9289-0) contains supplementary material, which is available to authorized users.

E. Rochard (✉)
Estuarine Ecosystems and Diadromous Fish Research Unit, Cemagref, 50 Avenue de Verdun, 33612 Cestas, France
e-mail: eric.rochard@cemagref.fr

[1] French research institute for agriculture and environment engineering www.cemagref.fr

[2] National institute for research in agronomy www.inra.fr

[3] The complete program of the symposium is available as Electronic Supplementary Material at http://dx.doi.org/10.1007/s10750-008-9289-0

difficult and the organizers express their sincere gratitude to all the scientists who helped them in the preparation of the meeting and in the selection and revision of the papers. I especially want to thank the guest editors of this volume for their very valuable efforts.

The Diadfish network

The aim of this network, composed of 21 member institutes and 13 associated institutes, is to strengthen exchanges and to facilitate cooperation between European research institutes dealing with diadromous fish.

Member institutes

Centre for Environment, Fisheries and Aquaculture Science CEFAS (United Kingdom).

Centre national du machinisme agricole, du génie rural, des eaux et des forêts Cemagref (France).

Consejo Superior de Investigaciones Científicas CSIC (Spain).

Danube Delta National Institute for Research and Development (Romania).

Finnish Game and Fisheries Research Institute (Finland).

Fisheries Research Services, Freshwater Laboratory (United Kingdom).

German Oceanographic Museum (Germany).

Institute of Biology, Leiden University (The Netherlands).

Institute of Ecology of Vilnius University (Lithuania).

Institute of Oceanography, Lisbon (Portugal).

Institute of Nature Conservation (Belgium).

Leibniz Institute for Freshwater Ecology and Inland Fisheries (Germany).

National board of fisheries (Sweden).

National Institute of Agronomical Research INRA (France).

National Museum of Natural History MNHN (France).

Norvegian Institute for nature Research NINA (Norway).

Polish Inland Fisheries Institute (Poland).

Research Center in Biodiversity and Genetic Resources, ICETA - University of Porto (Portugal).

Swedish University of Agricultural Sciences SLU (Sweden).

Universita di Roma “Tor Vergata” (Italy).

Veterinary Medical Research Institute, Hungarian Academy of Sciences (Hungary).

Associated institutes

Centre for Ecology and Hydrology CEH (United Kingdom).

Netherlands Institute for Fisheries Research (The Netherlands).

Veszprem University, Department of Animal Physiology and Nutrition, Veszprem (Hungary).

Research Institute of Fisheries, South branch Federal Centre for Genetics and Selection in Fish Culture, Krasnodar (Russia).

Society to Save the Sturgeon, Rostock (Germany).

Université de Liège, laboratoire de démographie des poissons, Liège (Belgium).

Electricité de France EDF, Laboratoire National d’Hydraulique et Environnement LNHE, Chatou (France).

Institute for Inland Water Management and Waste Water Treatment, Lelystad (The Netherlands).

National Polytechnic Institute INP, Toulouse (France).

University of La Rochelle, La Rochelle (France).

National University of Ireland NUI, Galway (Ireland).

Institut National des Sciences et Techniques de la Mer INSTM, Salammbo (Tunisia).

Environment Agency, National Salmon and Trout Fisheries Centre NSTFC (Great Britain).

To know more: www.diadfish.org

Hydrobiologia (2008) 602:5–14
DOI 10.1007/s10750-008-9290-7

Diadromy, history and ecology: a question of scale

R. M. McDowall

Abstract The existence of diadromous migrations has significant implications for understanding a broad series of biogeographical and ecological questions and for doing so across a broad range of spatial and temporal scales. Understanding these implications is important for interpretation of patterns in historical and ecological biogeography as well as in community ecology and conservation. This article explores these implications.

Keywords Diadromy · Dispersal · Diversity · Migration · Scale

Introduction

Diadromous species are those that undertake regular, seasonal and life-stage-consistent migrations between marine and freshwater environments. There are around 250 diadromous fish species and, in addition, there is a probably small and as yet undocumented number of diadromous decapod crustaceans and gastropod molluscs (Carpenter, 1982; Ford & Kinzie, 1982; McDowall, 1988; Koike, 1996; Hodges & Allendorf, 1998; Myers et al., 2000; Fievet & Eppe, 2002; Fossati et al., 2002; McDowall, 2004a). There is an old saying that What you see depends on where you are looking from, and for diadromous fishes the implications of being diadromous varies according to the scale at which one is addressing the phenomenon.

Diadromy is a geographically widespread attribute that is found in nearly all geographical areas, and from the cold-temperate to the tropics—though few diadromous species are recognised from South America and southern Africa (McDowall, 1988). They are characteristic of both continental lands and oceanic islands. The relatively small number of diadromous species creates interesting patterns of distribution, whether at the large temporal and spatial scales of global historical biogeography, or at the narrower scales of ecological biogeography, community assembly and ecology and macroecology, as well as within individual species.

Diadromous species have caused considerable confusion, historically, as they are very often regarded as 'freshwater species', with their sea-going habits often being ignored by, being unknown to, or are simply disbelieved by some commentators (Rosen, 1974). On the other hand, many fish biologists regard the broad ranges of some diadromous species as being due to they having a marine ancestry, but this is generally untrue—most of them share a common ancestry with other diadromous species (McDowall, 1993, 2004a)—though this does beg the broader

Guest editors: S. Dufour, E. Prévost, E. Rochard & P. Williot
Fish and diadromy in Europe (ecology, management, conservation)

R. M. McDowall (✉)
National Institute of Water and Atmospheric Research, PO Box 8602, Christchurch, New Zealand
e-mail: r.mcdowall@niwa.co.nz

question of the marine or freshwater origins of these diadromous groups. Diadromy is a seemingly ancient behaviour in some substantial groups, such as lampreys, anguillid eels, sturgeons, salmonids, osmerids, galaxiids, alosid shads, cottids and sicydiine gobies, and their ancestral biome is uncertain (see discussion of this for salmonids in McDowall, 2002a). Nevertheless there are a few known instances where isolated species clearly of marine ancestry have become diadromous, as in the Tasmanian congolli, *Pseudaphritis urvilii*, the New Zealand torrentfish, *Cheimarrichthys fosteri*, and a species of flounder, *Rhombosolea retiaria*. These are, however, very much minorities across the diversity of diadromous fishes (McDowall, 1988, 2004a).

Different types of diadromy

It is important to recognise that diadromy, as defined by Myers (1949), comes in three distinct forms: anadromy, catadromy and amphidromy. Anadromy and catadromy are recognised much more widely than amphidromy, which is sometimes mistaken for a form of anadromy (Nelson, 1976; Bell et al., 1995). Therefore, I want to emphasise that amphidromy is very different from anadromy, particularly with regards to the return migration to freshwater:

- In anadromous species, the fish returning to freshwater have completed their somatic growth, are fully mature, and are about to spawn in freshwater, whereas:
- Amphidromous fishes return to freshwater as small juveniles, usually less than ca. 50 mm long, and they will undertake prolonged feeding and undergo most of their somatic growth also in freshwater, prior to maturing and spawning there.

In a functional, freshwater, community-ecology sense, amphidromy is actually more similar to catadromy than to anadromy, especially as regards the early juvenile-to-adult growth to maturity.

There is very little amphidromy on continental land masses, but those who have worked across the tropical to temperate islands of the globe tend to be well familiar with amphidromy, especially in the sicydiine gobies, but also in some galaxiids (McDowall, 1988).

As far as I am aware, no new anadromous or catadromous species of fish has been described in at least the last 50, perhaps 100 years. This is untrue of amphidromous species, many of which are relatively newly described, especially among the sicydiine gobies (Watson, 1992, 1995a, b, 1996; Watson & Kottelat, 1994, 1995; Allen, 1997; Watson et al., 1998, 2002; Watson & Allen, 1999). Not surprisingly, their life histories remain poorly elucidated, though many of these gobies are described as amphidromous (Keith et al., 1999, 2002; Marquet et al., 2003). This is due, in large measure, to the fact that the gobiid fishes are typically small, poorly known and are found on remote oceanic islands. Often they are of little or no fisheries/economic importance, and many more may await discovery. However, some amphidromous species are important in fisheries, as is true of sicydiine gobies across the tropics [Philippines, Manacop (1953); in the Caribbean, Erdman (1961), Bell et al. (1995); La Reunion, Aboussouan (1969); Sri Lanka, Deraniyagala (1937)]; the osmerid *Plecoglossus altivelis* in Japan (Okada, 1960); and galaxiids in New Zealand (McDowall, 1990).

Diadromy at the global, oceanic or continental scale

Many diadromous species are very widespread geographically. Owing often to their common treatment as ‘freshwater species’ (which they are not), their wide ranges have caused considerable confusion and debate. Thus, the sicydiine goby *Sicyopterus lagocephalus* is found from the central Pacific across the Indo-Pacific as far west as the Comoro Islands (Balon & Bruton, 1994; Keith et al., 2002). Other diadromous species, too, are widespread across broad oceanic gaps—Atlantic salmon, various osmerid smelts, and sturgeon on both sides of the Atlantic Ocean; Pacific salmons of the genus *Oncorhynchus* and other osmerid smelts on both sides of the northern Pacific Ocean.

In the cool-temperate southern hemisphere, the small diadromous galaxiid, *Galaxias maculatus*, is known from western and eastern Australia, Tasmania, Lord Howe Island, New Zealand, the Chatham Islands, Patagonian South America and the Falkland Islands; and the diadromous southern pouched lamprey, *Geotria australis*, is known from Australia in the west, through New Zealand across the Pacific Ocean to Patagonian South America.

Certainly, not all diadromous species are widespread but some are; moreover the ostensibly freshwater species, that are reportedly widespread, are often diadromous. Were these fish regarded as 'marine' rather than 'freshwater' fishes, their distributions would probably create no more interest or contention than the distributions of plenty of widespread marine fishes.

The role of diadromy in promoting broad-scale distributions of fish species can be seen in action in occasional instances: the American shad, *Alosa sapidissima*, was translocated from the east coast of North America to the west coast, where it became established; it was not long before the species had become widespread along the Pacific coast of North America, and then across the North Pacific to invade river systems of the eastern, Pacific coasts of Eurasia (Whitehead, 1985). Another instance involves the recent arrival in New Zealand of the Australian longfin spotted eel, *Anguilla reinhardtii*. This species began to turn up in New Zealand rivers about 25 years ago owing, presumably, to some changes in the ocean currents in the south-western Pacific Ocean and the Tasman Sea (Jellyman et al., 1996; McDowall et al., 1998). This is probably the only observed natural, long-distance dispersion/range extension of any diadromous fish species. Thus, what we have very occasionally observed happening in the recent past, is probably indicative of what has happened throughout the evolutionary histories of diadromous fishes. The fact that these species spend regular parts of their life histories at sea has to be highly influential in their distributions—diadromy is a 'driver' of broad-scale distributions of fish species, whether at the oceanic scale, or at finer scales.

Thus, overall, the presence of diadromy may have been a crucial element in the broad-scale distributions of some significant freshwater fishes. Its role in promoting species' dispersals and broad distributions is widely demonstrated in the cool southern hemisphere. Darwin (1872) long ago suggested that the broad range of the small southern cool temperate *Galaxias maculatus*, discussed above, could best be explained by transoceanic dispersal, as did Boulenger (1902), over a century ago. More recent expositions on the contribution of this range (McDowall, 1964, 1969) have been treated with scorn or ridicule by some commentators (Croizat et al., 1974; Rosen, 1974; Craw, 1979), but molecular studies (Waters et al., 2000), and other evidence (McDowall, 2002b), support a dispersal scenario.

Even the Australian/New Zealand mudfishes of the genus *Neochanna* (f. Galaxiidae) are now seen as having dispersed across the Tasman Sea. Flannery (1984) predicted that if it proves to be the case that these New Zealand mudfish are most closely related to the Tasmanian freshwater mudfish *N. cleaveri*, even dispersalists such as McDowall will concede their probable Gondwanan origin (McDowall, 1984), but this is not so! Morphological and molecular data both do support a shared common ancestry of the Tasmanian and New Zealand mudfishes (McDowall, 1997, 2004b; Waters & White, 1997). However, what neither Flannery nor I did not know until the late 1980s, was that the Tasmanian species has a diadromous whitebait juvenile (Fulton, 1986). Moreover, molecular data suggest that a lineage split between the Tasmanian and New Zealand species of *Neochanna* took place only around 10 million years ago, and this date is far too recent to implicate involvement in Gondwana (Waters & McDowall, 2005)—New Zealand became separated from Gondwana (and thus lost its land connection with Australia) around 82 million years ago (Cooper & Millener, 1993).

A role for diadromy in restoration after perturbation at the catchment scale

Fluvial habitats are well known to be seriously vulnerable to diverse natural and anthropogenic impacts from a wide range of influences over medium to short time scales. Habitats were undoubtedly seriously perturbed, often destroyed completely, during the Pleistocene glaciations, this being especially well understood in North America (Pielou, 1991). Fish biogeographers have discussed restoration of fish populations to the extensive areas of North America that were defaunated by the advance of glacial ice. As the ice retreated, fish were able to re-invade the forming river systems and lakes (Legendre & Legendre, 1984; Bernatchez & Wilson, 1998). To some extent this reinvasion was probably through fluvial connections and river captures across the re-emerging landscape. However, there can be little doubt that much of the re-invasion was by diadromous fishes spreading northwards through coastal seas as the glacial ice sheets retreated. This re-invasion is still continuing in Alaska, where newly accessible coastal stream habitats continue to be invaded exclusively by diadromous fishes (Milner, 1987; Milner & Bailey, 1989).

Similar impacts on stream fishes have been described in New Zealand, with residual effects from Pleistocene ice sheets being still evident in fish distributions—non-diadromous species are absent there, but diadromous species are widespread in areas most severely affected (Main, 1989; McDowall, 1996).

Also in New Zealand, large volcanic eruptions in recent times in mountains of the central and western North Island have had impacts on stream biotas that persist to the present day (McDowall, 1996); again, diadromous species are widespread in areas across which there was massive deposition of volcanic ash and tephra, whereas non-diadromous species are widely missing from the fish communities.

Implications for genetic structuring at the species scale

A series of papers has demonstrated that there is little genetic structuring across the geographical ranges of diadromous fishes. Comparisons of levels of genetic diversity among marine, freshwater and diadromous fish species, have shown that diadromous species have diversity comparable with marine species (Gyllensten, 1985; Ward et al., 1994). Diadromous species seem to exhibit little genetic structuring across their widespread populations (Zink et al., 1996; Dijkstra & Jellyman, 1999). Bernatchez & Wilson (1998) explored the phylogeography of boreal fishes in northeastern North America in relation to Pleistocene glaciation. My re-analysis of their data (McDowall, 1999) demonstrated, again, a distinct contrast between diadromous and non-diadromous species, with the latter tending to exhibit much greater nucleotide diversity.

Zink et al. (1996), Chubb et al. (1998) and Fitzsimons et al. (1990) explored genetic diversity among the sparse Hawaiian freshwater fish fauna. The five species present there (all of them gobioids) are diadromous, and they exhibit no genetic diversification across the Hawaiian Islands, and there is clearly continuing gene flow. Allibone & Wallis (1996) contrasted genetic diversity of diadromous and non-diadromous freshwater fishes in New Zealand. Dijkstra & Jellyman (1999) searched for genetic diversification in the Australian/New Zealand anguillid eel, *Anguilla australis*, and found none, concluding that the populations are panmictic across the Tasman Sea (around 2,000 km). Thus, there seems a general pattern of low genetic diversification across the ranges of diverse, widespread diadromous fish species. Fievet & Eppe (2002) explored genetics in a diadromous atyid shrimp, *Atya innocuous*, and similarly found little divergence across the species' range in the Caribbean. There seems to be a general lack of divergence across disjunct often widespread populations in a broad range of diadromous taxa.

Taxonomic differences—comparisons at the within species-group scale

These absences of within-species genetic structuring are distinct contrasts with what is found when diadromous species are compared with closely related non-diadromous species. Allibone & Wallis (1996), for instance, drew a strong distinction between the low genetic diversification in New Zealand's diadromous galaxiids and much greater diversification in non-diadromous ones. Goto & Andoh (1990) similarly found little divergence in a diadromous Japanese species of *Cottus amblystomopsis* but substantially greater diversity in multiply derived non-diadromous populations recognised as *C. nozawae*. Page et al. (2005) found little genetic structuring across the range of the amphidromous shrimp species genus *Paratya*.

Comparisons at the within-species scale

Zattara & Premoli (2005) have examined genetic structuring in Andean populations of *Galaxias maculatus*, reporting that lake-limited (non-diadromous) populations exhibit far greater genetic differences from each other than is seen across the ranges of diadromous populations. The obvious interpretation is that, when populations become isolated in lakes, they have diverged individually, owing to a variety of possible effects: local adaptation, genetic drift and/or founder effects deriving from establishment as small founder populations. Gene flow among diadromous populations, on the contrary, is likely to have inhibited such diversification. Thus, an historic-to-contemporary role of diadromy in facilitating dispersion across wide ranges can be seen in the lack of genetic structuring among populations of species across broad scales and in diverse regions and taxonomic groups.

Landlocking and the implications for speciation

Many diadromous species are capable of establishing populations that are restricted to fresh water, a process known as landlocking, though the habitats in which this takes place do not necessarily have barriers preventing fish from migrating to and from the sea. Instances are known in which 'landlocked' and diadromous stock (i.e. non-migratory populations) are sympatric in lakes, usually low elevation, sometimes brackish coastal lagoons (McDowall, 1972; Northcote & Ward, 1985; Katayama et al., 2000). In these instances cross-breeding between diadromous and non-diadromous individuals seems likely.

With time, isolated non-diadromous populations of diadromous species may diverge and become reproductively isolated from their ancestral diadromous populations, leading to distinct, non-diadromous species. This is widely known in lampreys, salmonids, osmerids, retropinnids and galaxiids (McDowall, 1988; Waters & Wallis, 2001).

The implications of homing and straying

Homing, the return of fish to their natal stream, is widely reported among diadromous fishes, especially salmonids (Quinn, 1984; Quinn & Dittman, 1990). Homing has implications for understanding the processes of genetic divergence in diadromous fishes. It is regarded as a mechanism that contributes to local adaptation of populations, especially in salmonids, and so may counteract the tendency for diadromy to foster gene flow among isolated populations of diadromous fishes. Individual stocks are widely recognised in some salmonids, for instance (reviewed in McDowall, 2001).

Implications for distributions at the regional/local scales

At regional and local scales there is a strong tendency for diadromous species to be much more widely distributed than non-diadromous species. I'll illustrate this from the New Zealand freshwater fish fauna, which comprises 17 extant diadromous species and ca. 24 non-diadromous ones (decisions about the taxonomic status of some lineages are yet to be made). Plots of species' ranges of New Zealand's fauna (Fig. 1) show that diadromous species are typically present along the entire 14° latitudinal range of the archipelago, or nearly so, whereas the distributions of the diadromous species bear no relationships to, and provide no hints of associations with, New Zealand's turbulent geological history. In contrast, the non-diadromous species have much narrower distributions, and their geographical patterns very often exhibit associations with that geological history. Species are present on the main islands of New Zealand across existing marine straits—but these were bridged by land in the Pleistocene. Sister species have distributions separated by mountain ranges. Distributions that seem enigmatic become comprehensible when an understanding of changes in river flow patterns is obtained (Waters et al., 2001). In addition, as discussed above, historical events like volcanism and glaciation have had impacts that are still reflected in the distributions of non-diadromous species, whereas reinvasions of river systems by diadromous species have restored their populations to these areas.

Exclusion from upstream habitats—impacts at the catchment scale

At the river catchment scale the requirement of these migratory species to occupy habitats by migrating upstream from the sea means that where there are impediments to migration, the upstream habitats will lack diadromous species, diadromous species can be excluded from upstream habitats in situations where there are high falls or natural dams on rivers. The extent of impacts of such impediments varies from species to species, since some are poor climbers, whereas other are very skilled and aggressive upstream migrants, the ultimate instance being

Fig. 1 Presence of diadromous species (upper panel) and non-diadromous species (lower panel) in half-degree latitudinal bands across New Zealand (note the generally widespread presence of the diadromous species and much narrower latitudinal bands in the non-diadromous species)

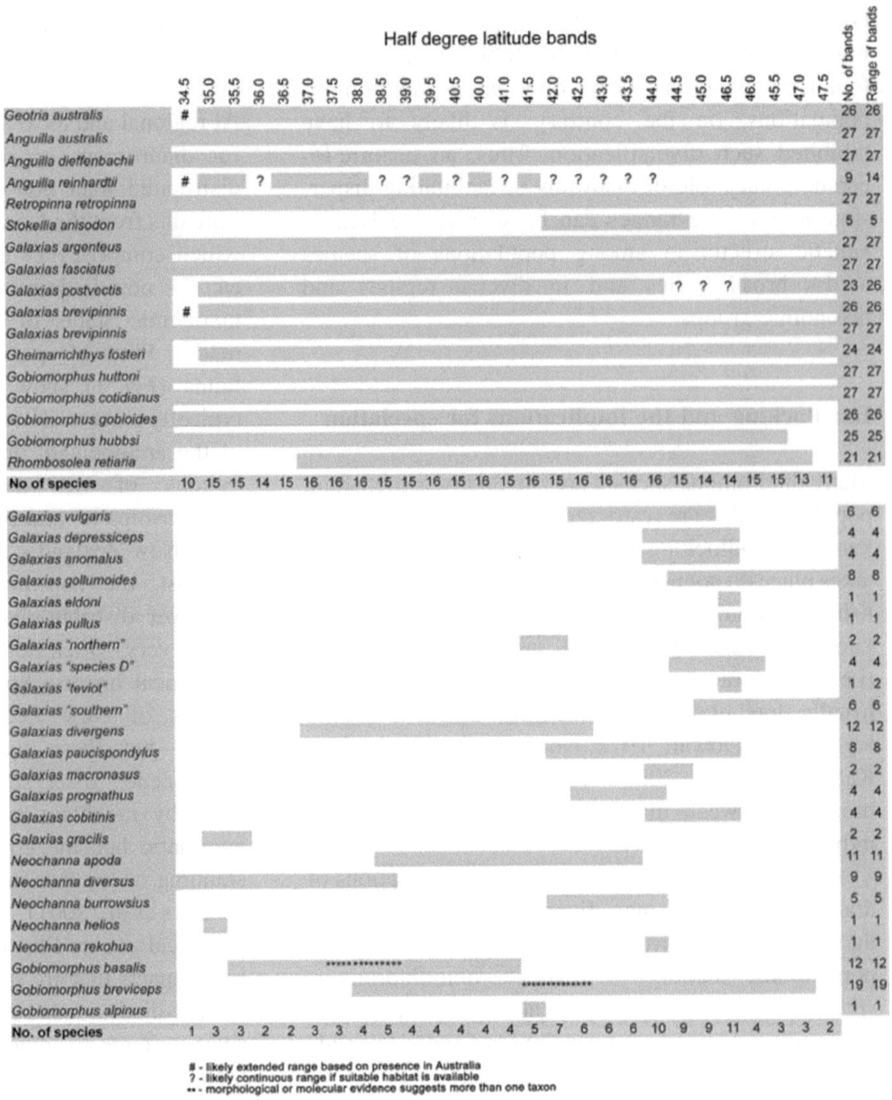

Hawaii's gobiid, *Lentipes concolor*, which has been found upstream of cumulative fall heights of over 600 m (Englund & Filbert, 1997). Other taxa, though perhaps not as aggressive as this, nevertheless can reach impressive heights, as in some galaxiids and anguillid eels (McDowall, 1990).

Local species richness—implications for catchment-level impacts

The fact that diadromous fishes invade fish communities in streams and lakes by migrating upstream from the sea, combined with the fact that there is wide variation among species in the extent to which they penetrate upstream, results in a downstream–upstream decline in species richness—there are almost always more diadromous species in downstream localities than in those further upstream (Figs. 2, 3).

Parallel to this decline in species richness is a change in the size and age structure of the populations of the species that migrate into river systems from the sea. Typically lower elevation sites, closer to the sea, have a broad ranges of ages and sizes of diadromous species, and there is an upstream tendency for the smaller, juveniles of the species to not be present. This can be attributed to the immigrants making a slow upstream movement, so that by the time they reach more upstream habitats, they have grown larger.

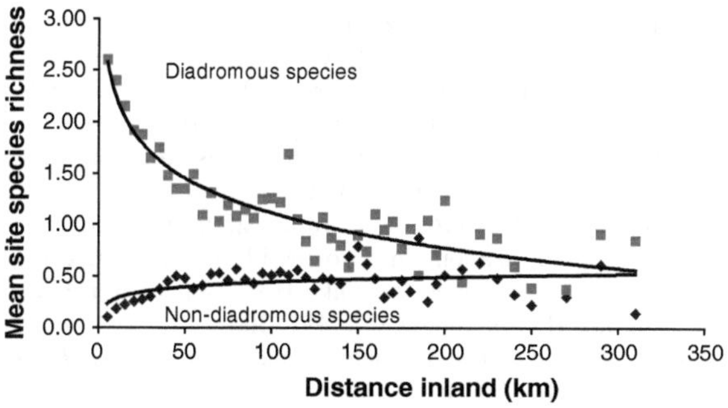

Fig. 2 Variation in species richness with increasing distance inland in 15,000 sites across the New Zealand landscape (upper curve is diadromous species and lower curve non-diadromous species)

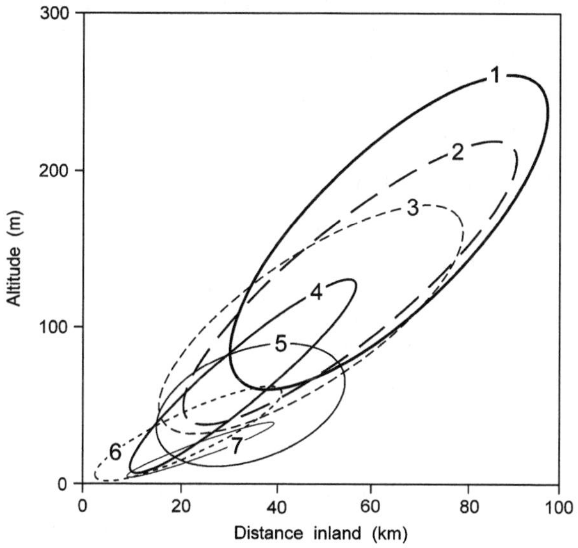

Fig. 3 Increasing species richness in the Grey River, New Zealand with rising elevation and distance inland (Gaussian ellipses enclose 50 of the distributions at each level of species richness)

Community assembly and structuring at the local/community scale

Diadromy can have a significant role in the assembly and structuring of fish communities. Stream and lake fish communities tend to be 'closed' owing to the geographically confined nature of individual streams and lakes, with aquatic access between waterways mostly possible only through coastal seas, so that community assembly processes are determined primarily by what takes place within these confined habitats. This is not true of diadromous fishes, as these invade river systems from the sea, making recruitment and community assembly open processes; as above, species richness of diadromous species tends to be higher at lower elevations (Figs. 2, 3).

A role in importing and exporting nutrients

Where there is massive migration of diadromous fishes between freshwater ecosystems and the sea, there is ecologically significant import and export of nutrients. It is especially true of Pacific salmons of the genus *Oncorhynchus*, whose vast immigrations from the sea are important sources of oceanic nutrients in often oligotrophic, mid-high elevation river systems in boreal forests (Mathison et al., 1988; Garman, 1992; Kline et al., 1993). There is also growing recognition that diadromy also involves significant export of terrestrial nutrients to sea (Jonsson & Jonsson, 2003; Moore & Schindler, 2005).

Summary

Thus, it seems that diadromy is highly influential in fish ecology and behaviour across a range of spatial scales relating to their:

- Distributions and biogeography;
- Genetic diversity across separate populations;
- Speciation processes;
- Community assembly and downstream upstream changes;
- Community species richness;
- Longitudinal distributions in river systems;
- Age and size structure of populations;
- Conservation; and
- Response to environmental perturbation.

It is thus profoundly important for good understanding of the ecology of fishes and better knowledge of the behavioural ecology of fishes is crucial to both their management and conservation.

References

Aboussouan, A., 1969. Note sur les Bichiques de l'ile de la Reunion. Revue des Travaux de la Station Marine d'Endoume h9: 25–31.

Allen, G. R., 1997. *Lentipes watsoni*, a new freshwater goby (Gobiidae) from Papua New Guinea. Ichthyological Exploration of Freshwaters 8: 33–40.

Allibone, R. M. & G. P. Wallis, 1996. Genetic variation and diadromy in some native New Zealand galaxiids (Teleostei: Galaxiidae). Biological Journal of the Linnean Society 50: 19–33.

Balon, E. K. & M. N. Bruton, 1994. Fishes of the Tatinga River, Comoros, with comments on freshwater amphidromy in the goby *Sicyopterus lagocephalus*. Ichthyological Exploration of Freshwaters 5: 25–40.

Bell, K. N. I., P. Pepin & J. A. Brown, 1995. Seasonal, inverse cycling of length- and age-at recruitment in the diadromous gobies *Sicydium punctatum* and *Sicydium antillarum* in Dominica, West Indies. Canadian Journal of Fisheries and Aquatic Sciences 52: 1535–1545.

Bernatchez, L. & C. C. Wilson, 1998. Comparative phylogeography of Nearctic and Palearctic fishes. Molecular Ecology 7: 431–452.

Boulenger, G. A., 1902. The explanation of a remarkable case of geographical distribution among fishes. Nature 67: 84.

Carpenter, A., 1982. Habitat and distribution of the freshwater shrimp *Paratya curvirostris* (Decapoda: Atyidae). Mauri Ora 10: 85–98.

Chubb, A. L., R. M. Zink & J. M. Fitzsimons, 1998. Patterns of mtDNA variation in Hawaiian freshwater fishes: phylogeographic consequences of amphidromy. Journal of Heredity 89: 8–16.

Cooper, R. A. & P. R. Millener, 1993. The New Zealand biota: historical background and new research. Trends in Ecology and Evolution 8: 429–433.

Craw, R. C., 1979. Generalized tracks and dispersal in biogeography: a response to R.M. McDowall. Systematic Zoology 28: 99–107.

Croizat, L., G. J. Nelson & D. E. Rosen, 1974. Centers of origin and related concepts. Systematic Zoology 23: 265–287.

Darwin, C., 1872. On the Origin of Species by Means of Natural Selection, or the Preservation of Favored Races in the Struggle for Live, 6th Ed. John Murray, London.

Deraniyagala, P. E. P., 1937. Two catadromous fishes new to Ceylon. Ceylon Journal of Science B20: 181–184.

Dijkstra, L. H. & D. J. Jellyman, 1999. Is the subspecies classification of the freshwater eels *Anguilla australis australis* Richardson and *A. australis schmidtii* Phillipps still valid? Marine and Freshwater Research 50: 261–263.

Englund, R. & R. Filbert, 1997. Discovery of the native stream goby, *Lentipes concolor*, above Hawaii's highest waterfall, Hi'ilawe Falls. Bishop Museum Occasional Papers 49: 62–64.

Erdman, D. S., 1961. Notes on the biology of the gobiid fish *Sicydium plumieri* in Puerto Rico. Bulletin of Marine Science, Gulf and Caribbean 11: 448–456.

Fievet, E. & R. Eppe, 2002. Genetic differentiation among populations of the amphidromous shrimp *Atya innocuous* (Herbst) and obstacles to their upstream migration. Archiv für Hydrobiologie 153: 287–300.

Fitzsimons, J. M., R. M. Zink & R. T. Nishimoto, 1990. Genetic variation in the Hawaiian goby, *Lentipes concolor*. Biochemical Systematics and Ecology 18: 81–83.

Flannery T., 1984. New Zealand: a curious zoogeographic history. In Archer M. & G. Clayton (eds), Vertebrate Zoogeography and Evolution in Australasia (animals in space and time). Hesperian, Carlisle, 1089–1094.

Ford, J. I. & R. A. Kinzie, 1982. Life crawls upstream. Natural History (New York) 91: 60–67.

Fossati O., M. Mosseron & P. Keith, 2002. Distribution and habitat utilization in two atyid shrimps (Crustacea: Decapoda) in rivers of Nuku-Hiva Island (French Polynesia). Hydrobiologia 472: 197–206.

Fulton, W., 1986. The Tasmanian mudfish, *Galaxias cleaveri*. Fishes of Sahul 4: 150–151.

Garman, G. C., 1992. Fate and potential significance of post-spawning anadromous fish carcasses in an Atlantic coastal river. Transactions of the American Fisheries Society 121: 390–394.

Goto, A. & T. Andoh, 1990. Genetic divergence between the sibling species of river-sculpin, *Cottus amblystomopsis* and *C. nozawae*, with special reference to speciation. Environmental Biology of Fishes 28: 101–112.

Gyllensten, U., 1985. The genetic structure of fish: differences in the intraspecific distribution of biochemical variation between marine, anadromous and freshwater species. Journal of Fish Biology 26: 691–699.

Hodges, M. H. & F. W. Allendorf, 1998. Population genetics and pattern of larval dispersal of the endemic amphidromous gastropod *Neritina granosa* (Prosobranchia: Neritidae) Pacific Science 52: 237–249.

Jellyman, D. J., B. L. Chisnall, L. H. Dijkstra & J. A. T. Boubée, 1996. First record of the Australian longfinned eel, *Anguilla reinhardtii*, in New Zealand. Marine and Freshwater Research 47: 1037–1040.

Jonsson, B. & N. Jonsson, 2003. Migratory Atlantic salmon as vectors for the transfer of energy and nutrients between freshwater and marine environments. Freshwater Biology 48: 21–27.

Katayama, S., R. L. Radtke, M. Omori & D. J. Shafer, 2000. Coexistence of anadromous and resident life history styles of pond smelt, *Hypomesus nipponensis*, in Lake Ogawara, Japan, as determined by analyses of otolith structure and strontium: calcium ratios. Environmental Biology of Fishes 58: 195–201.

Keith, P., E. Vigneux & P. Bosc, 1999. Atlas des poissons et des crustacés d'eau douce de la Réunion. Muséum National d'Histoire Naturelle, Paris, 136 pp.

Keith, P., E. Vigneux & G. Marquet, 2002. Atlas des poissons et des crustacés d'eau douce de Polynésie Francaise. Paris, 177 pp.

Kline, T. C., J. J. Goering, O. A. Mathison & P. H. Poe, 1993. Recycling of elements transported upstream by runs of Pacific salmon: II. delta 25 N and 13 C evidence in the Kvichak River watershed, Bristol Bay, southwestern Alaska. Canadian Journal of Fisheries and Aquatic Sciences 50: 2350–2365.

Koike, K., 1996. Embryonic development and larval dispersal in the Neritidae (Gastropoda: Prosobranchia). Galaxea 13: 129–130.

Legendre, P. & V. Legendre, 1984. Postglacial dispersal of freshwater fishes in the Quebec Peninsula. Canadian Journal of Fisheries and Aquatic Sciences 41: 1781–1802.

Main, M. R., 1989. Distribution and post-glacial dispersal of freshwater fishes in South Westland, New Zealand. Journal of the Royal Society of New Zealand 19: 161–169.
Manacop, P. R., 1953. The life history and habits of the goby *Sicyopterus extraneus* Herre (anga) Gobiidae, with an account of the goby fry fishery of Cagayan River, Oriental Misamis. Philippines Journal of Fisheries 2: 1–60.
Marquet, G., P. Keith & E. Vigneux, 2003. Atlas des poissons et des crustacés d'eau douce de Nouvelle Calédonie. Museum National d'Histoire Naturelle, Paris, 284 pp.
Mathison, O. A., P. L. Parker, J. J. Goering, T. C. Kline, P. H. Poe & R. S. Scalan, 1988. Recycling of marine elements transported into freshwater systems by anadromous salmon. Verhandlungen de Internationale Vereinigung fur theoretische und angewandte Limnologie 23: 2249–2258.
McDowall, R. M., 1964. The affinities and derivation of the New Zealand fresh-water fish fauna. Tuatara 12: 59–67.
McDowall, R. M., 1969. Relationships of galaxioid fishes, with a further discussion of salmoniform classification. Copeia 1969: 796–824.
McDowall, R. M., 1972. The taxonomy of estuarine and brackish-lake Retropinna from New Zealand (Galaxioidei: Retropinnidae). Journal of the Royal Society of New Zealand 2: 501–531.
McDowall, R. M., 1984. Southern hemisphere freshwater salmoniforms: development and relationships. American Society of Ichthyologists and Herpetologists Special Publication 1: 150–153.
McDowall, R. M., 1988. Diadromy in Fishes, Migrations between Freshwater and Marine Environments. Croom Helm Publ., London, 308 pp.
McDowall, R. M., 1990. New Zealand Freshwater Fishes: A Natural History and Guide. Heinemann Reed, Auckland.
McDowall, R. M., 1993. A recent marine ancestry for diadromous fishes? Sometimes yes, but mostly no. Environmental Biology of Fishes 37: 329–335.
McDowall, R. M., 1996. Volcanism and freshwater fish biogeography in the northeastern North Island of New Zealand. Journal of Biogeography 23: 139–148.
McDowall, R. M., 1997. Affinities, generic classification, and biogeography of the Australian and New Zealand mudfishes (Salmoniformes: Galaxiidae). Records of the Australian Museum 49: 121–137.
McDowall, R. M., 1999. Diadromy and genetic diversity in Nearctic and Palearctic fishes. Molecular Ecology 8: 527–528.
McDowall, R. M., 2001. Anadromy and homing: two life history traits with adaptive synergies in salmonid fishes? Fish and Fisheries 2: 78–85.
McDowall, R. M., 2002a. The origin of the salmonid fishes: marine, freshwater...or neither. Reviews in Fish Biology and Fisheries 11: 171–179.
McDowall, R. M., 2002b. Accumulating evidence for a dispersal biogeography of southern cool temperate freshwater fishes. Journal of Biogeography 29: 207–220.
McDowall, R. M., 2004a. Ancestry and amphidromy in island freshwater fish faunas. Fish and Fisheries 5: 75–85.
McDowall, R. M., 2004b. The Chatham Islands endemic galaxiid: a *Neochanna* mudfish (Teleostei: Galaxiidae). Journal of the Royal Society of New Zealand 34: 315–331.
McDowall, R. M., D. J. Jellyman & L. C. Dijkstra, 1998. Arrival of an Australian anguillid eel in New Zealand: an example of transoceanic dispersal. Environmental Biology of Fishes 51: 1–6.
Milner, A. M., 1987. Colonization and ecological development of new streams in Glacier Bay National Park, Alaska. Freshwater Biology 18: 53–70.
Milner, A. M. & R. G. Bailey, 1989. Salmonid colonization of new streams in Glacier Bay National Park, Alaska. Aquaculture and Fisheries Management 20: 179–192.
Moore, J. W. & Schindler D. E., 2005. Nutrient export from freshwater ecosystems by anadromous sockeye salmon (*Oncorhynchus nerka*). Canadian Journal of Fisheries and Aquatic Sciences 61: 1582–1589.
Myers, G. S., 1949. Usage of anadromous, catadromous and allied terms for migratory fishes. Copeia 1949: 89–97.
Myers, M. J., C. P. Meyer & V. H. Resh, 2000. Neritid and thiarid gastropods from French Polynesian streams: how reproduction (sexual, parthenogenetic) and dispersal (active/passive) affect population structure. Freshwater Biology 44: 535–545.
Nelson, J. S., 1976. Fishes of the World. Wiley, New York.
Northcote, T. G. & F. J. Ward, 1985. Lake resident and migratory smelt *Retropinna retropinna* (Richardson), of the lower Waikato River system. New Zealand Journal of Marine and Freshwater Research 17: 113–129.
Okada, Y., 1960. Studies on the Freshwater Fishes of Japan. Prefectural University of Mie, Tsu.
Page, T. J., A. M. Baker, B. D. Cook & J. M. Hughes, 2005. Historical transoceanic dispersal of a freshwater shrimp: the colonization of the South Pacific by the genus *Paratya* (Atyidae). Journal of Biogeography 32: 581–593.
Pielou, E. C., 1991. After the Ice Age: The Return of Life to Glaciated North America. University of Chicago, Chicago.
Quinn, T. P., 1984. Homing and straying in Pacific salmon. In: McCleave J. D., G. Arnold, J. Dodson & Neill W. (eds), Mechanisms of Migrations in Fishes. Plenum, New York, 357–361.
Quinn, T. P. & A. H. Dittman, 1990. Pacific salmon migrations and homing: mechanisms and adaptive significance. Trends in Ecology and Evolution 5: 174–177.
Rosen, D. E., 1974. Phylogeny and zoogeography of salmoniform fishes and relationships of *Lepidogalaxias salamandroides*. Bulletin of the American Museum of Natural History 153: 265–326.
Ward, R. D., M. Woodwark & D. O. F. Skibinski, 1994. A comparison of diversity levels in marine, freshwater, and anadromous fishes. Journal of Fish Biology 44: 213–232.
Waters, J. M., D. Craw, J. H. Youngson & G. P. Wallis, 2001. Genes meet geology: fish phylogeographic pattern reflects ancient rather than modern drainage patterns. Evolution 55: 1844–1855.
Waters, J. M., L. H. Dijkstra & G. P. Wallis, 2000. Biogeography of a southern hemisphere freshwater fish: how important is marine dispersal? Molecular Ecology 9: 1815–1821.
Waters, J. M., & R. M. McDowall, 2005. Phylogenetics of the Australian mudfishes: evolution of an eel-like body plan. Molecular Phylogenetics and Evolution 34: 417–425.

Waters, J. M. & G. P. Wallis, 2001. Cladogenesis and loss of the marine life-history phase in freshwater galaxiid fishes (Osmeriformes: Galaxiidae). Evolution 55: 587–597.

Waters, J. M. & R. W. G. White, 1997. Molecular phylogeny and biogeography of the Tasmanian and New Zealand mudfishes (Salmoniformes: Galaxiidae). Australian Journal of Zoology 45: 39–48.

Watson, R. E., 1992. A review of the gobiid fish genus *Awaous* from insular streams of the Pacific plate. Ichthyological Exploration of Freshwaters 3: 161–176.

Watson, R. E., 1995a. Gobies of the genus *Stiphodon* of French Polynesia, with descriptions of two new species (Teleostei: Gobiidae: Sicydiinae). Ichthyological Exploration of Freshwaters 6: 33–48.

Watson, R. E., 1995b. Review of the freshwater goby genus *Cotylopus*. Ichthyological Exploration of Freshwaters 6: 61–70.

Watson, R. E., 1996. A review of *Stiphodon* from New Guinea and adjacent regions, with descriptions of five new species (Teleostei: Gobiidae: Sicydiinae). Revue Francaise Aquariologie 23: 113–132.

Watson, R. E. & G. R. Allen, 1999. New species of freshwater gobies from Irian Jaya, Indonesia (Teleostei: Gobiidae: Sicydiinae). Aqua: Journal of Ichthyology and Aquatic Biology 3: 113–118.

Watson, R. E., G. R. Allen & M. Kottelat, 1998. A review of *Stiphodon* from Halmahera and Irian Jaya, Indonesia, with descriptions of two new species. Ichthyological Exploration of Freshwaters 9: 293–304.

Watson, R. E., P. Keith & G. Marquet, 2002. *Lentipes kaaea*, a new species of freshwater goby from New Caledonia (Teleostei: Gobioidei: Sicydiinae). Bulletin Francaise de la Peche et de la Pisciculture 364: 173–185.

Watson, R. E. & M. Kottelat, 1994. *Lentipes whittenorum* and *Sicyopus auxilimentus*, two new species of freshwater gobies from the western Pacific (Teleostei: Gobiidae: Sicydiinae). Ichthyological Exploration of Freshwaters 5: 351–364.

Watson, R. E. & M. Kottelat, 1995. *Stiphodon* from Leyte Island, Philippines with descriptions of two new species (Teleostei: Gobiidae: Sicydiinae). Ichthyological Exploration of Freshwaters 6: 1–16.

Whitehead, P. J. P., 1985. Clupeoid fishes of the world (suborder Clupeiodei): an annotated and illustrated catalogue of the herrings, sardines, pilchards, sprats, shads, anchovies and wolf-herrings. Part 1. Chirocentridae, Clupeidae and Pristigasteridae. FAO Fisheries Synopsis 125(7): 1–303.

Zattara, E. & A. Premoli, 2005. Genetic structuring in Andean landlocked populations of *Galaxias maculatus*: effects of biogeographic history. Journal of Biogeography 32: 5–14.

Zink, R. M., J. M. Fitzsimons, D. L. Dittman, D. R. Reynolds & R. T. Nishimoto, 1996. Evolutionary genetics of Hawaiian freshwater fish. Copeia 1996: 330–335.

Hydrobiologia (2008) 602:15–25
DOI 10.1007/s10750-008-9287-2

Life history correlates and extinction risk of capital-breeding fishes

Henriette I. Jager · Kenneth A. Rose · Anna Vila-Gispert

Abstract We consider a distinction for fishes, often made for birds and reptiles, between capital-breeding and income-breeding species. Species that follow a capital-breeding strategy tend to evolve longer intervals between reproductive events and tend to have characteristics that we associate with higher extinction risk. To examine whether these ideas are relevant for fishes, we assembled life history data for fish species, including an index of extinction risk, the interval between spawning events, the degree of parental care, and whether or not the species migrates to spawn. These data were used to evaluate two hypotheses: (1) fish species with a major accessory activity to spawning (migration or parental care) spawn less often and (2) fish species that spawn less often are at greater risk of extinction. We tested these hypotheses by applying two alternative statistical methods that account for phylogenetic correlation in cross-taxon comparisons. The two methods predicted average intervals between spawning events 0.13–0.20 years longer for fishes with a major accessory activity. Both accessories, above-average parental care and spawning migration, were individually associated with longer average spawning intervals. We conclude that the capital-breeding paradigm is relevant for fishes. We also confirmed the second hypothesis, that species in higher IUCN extinction risk categories had longer average spawning intervals. Further research is needed to understand the relationship between extinction risk and spawning interval, within the broader context of life history traits and aquatic habitats.

Guest editors: S. Dufour, E. Prévost, E. Rochard & P. Williot
Fish and diadromy in Europe (ecology, management, conservation)

Electronic supplementary material The online version of this article (doi:10.1007/s10750-008-9287-2) contains supplementary material, which is available to authorized users.

H. I. Jager (✉)
Environmental Sciences Division, Oak Ridge National Laboratory, Oak Ridge, TN 37831-6036, USA
e-mail: jagerhi@ornl.gov

K. A. Rose
Department of Oceanography & Coastal Sciences and Coastal Fisheries Institute, Louisiana State University, Baton Rouge, LA 70803, USA

A. Vila-Gispert
Institute of Aquatic Ecology and Department of Environmental Sciences, University of Girona, Girona, Spain

Keywords Extinction risk · Comparative analysis · Life history · Migration · Spawning interval

Introduction

Capital-breeding species use stored energy to make a large, fecundity-independent investment at each breeding opportunity (Bull & Shine, 1979). Species with high, fecundity-independent costs include those that make long migrations to breed and those that provide considerable parental care before birth (e.g., live-bearing) or after birth (e.g., nest-guarding). According to the capital-breeding theory, longer-than-annual intervals between breeding events evolved in association with capital breeding because more offspring were produced over a lifetime by skipping years than would be possible by spending energy to reproduce more frequently without either migrating or providing parental care. Bull & Shine (1979) give the following example,

> "Suppose a fish migrates every month to lay ten eggs; the migration uses stored energy which she gathers each month and could otherwise be converted into ten more eggs. Now, if she delays reproduction 1 month and simply stores the energy, at the time of the next reproduction she will have the capacity for 20 eggs and two migrations. However, she needs to migrate only once and can thus use the energy from the "extra" migration for an additional ten eggs. Her fecundity after 2 months is 30 eggs rather than 20."

Long spawning intervals occur both as a plastic life history response to energy stores or environmental conditions (skipped spawning) and as a "hardwired" life history feature. In this study, we focus on spawning interval at the evolutionary timescale, where the mechanism is natural selection. Jørgensen et al. (2006) used a dynamic programming model to predict that spawning should be skipped if the expected future gain in reproductive output, discounted by survival, more than balances the expected reproductive success the current year. According to their model, the incidence of skipped spawning increased linearly as the energetic or mortality costs of migration increased, thus supporting the capital-breeding hypothesis.

A distinguishing characteristic of capital breeders is that females reproduce in years when they have accumulated a threshold level of stored energy reserves; in contrast, income breeders spend energy on reproduction as it is gained (Jonsson, 1997). Murua & Saborido-Rey (2003) describe the breeding strategies of marine fishes. Among the strategies they describe for iteroparous fishes, the income-breeding strategy seems to correspond with that of fishes with asynchronous egg development and indeterminate fecundity who develop eggs more or less continuously. The capital-breeding strategy seems to correspond with group-synchronous marine fishes that have determinate fecundity. In the marine environment, income-breeding fishes tend to be small, pelagic species in temperate waters, whereas capital-breeding fishes tend to be demersal species in cold, marine waters (Murua & Saborida-Rey, 2003).

Capital breeding may be associated not only with less-frequent breeding, but also with a suite of correlated life history traits. Storing the energetic reserves and fecundity required for less-than-annual spawning favors a large body size. Species that spend reproductive capital on long breeding migrations tend to mature late and to have a large adult body size. It is less clear that large body size is correlated with parental care, particularly for oviparous species. Large body size also reduces predation risk and increases general resistance to stressors, which in turn, makes delayed maturation a feasible strategy. Jørgensen et al. (2006) found that the optimal age and length at maturation also increased in response to increased migration costs for their simulated cod populations.

In this paper, we consider whether the capital-breeding paradigm, which was developed for birds and reptiles (Johsson, 1997), is useful for understanding fish life histories. We assembled fish species data for three life history traits (interval between spawning events, degree of parental care, and migration strategy) and an index of extinction risk. We use these data to examine whether infrequent breeding is positively related to a species' migration strategy and degree of parental care. We also use the data to test whether extinction risk is higher for capital-breeding species.

Methods

Life history and extinction risk data

We assembled a database that included life history traits and the IUCN conservation status for a wide variety of fish species (see http://www.esd.ornl.gov/~zij/mypubs/). Three primary sources of data were: (1) Winemiller & Rose (1992) for life history traits of North American fishes, (2) Vila-Gispert & Moreno-Amich (2002) for life history traits of European freshwater fishes, and (3) FishBase (FishBase, 2004) for life history traits of additional marine and estuarine species and for the IUCN status and taxonomy (class, order, family, genus, species) for all fish species.

We conducted a literature search using terms "fishes" and {"spawning interval," "skipped breeding," "parental care," and "migration"}. We also searched for publications on each taxonomic fish order represented in our database or in FishBase with the key word "reproduction". We included those species in our analysis for which the needed life history information was available. The life history data in the Winemiller & Rose (1992) dataset were recorded by species, whereas the other two sources were organized by stock. For numeric variables, we averaged stocks in FishBase to aggregate to the species level. For IUCN status, we used the middle designation unless there were only two values, in which case the higher level of risk was assigned to the species. We excluded species known to us to be semelparous, such as Pacific salmon, lampreys, and some herrings.

We associated IUCN status, assigned by the World Conservation Union (Hilton-Taylor, 2000), to each species in the database. We converted these codes to numeric values: 4 = critically endangered, 3 = endangered, 2 = vulnerable, and 1 = lower risk. The lower risk category also included the subcategories of conservation-dependent, near-threatened, and least-concern. Species not found in IUCN Red List were assigned a code value of zero. We excluded species listed by the IUCN as data-deficient or not evaluated.

We used our data sources to quantify three life history traits: (1) the interval between spawning events (in years), (2) whether the species makes spawning migrations or not, and (3) an index that measures parental care. The interval between spawning events was reported by Winemiller & Rose (1992) and Vila-Gispert & Moreno-Amich (2002); we estimated values from the literature for species that we added from FishBase. Intervals for batch spawners were recorded as annual if oogenesis had one seasonal peak.

Migration status was indicated for a subset of fishes in all three databases. We changed the status values reported to non-migratory for some species if literature sources revealed that spawning migrations were shorter than $\sim$50 km. Distance migrated was not known for the majority of species. We defined an indicator variable, Imig, and assigned it a value of 1 for species that make spawning migration and a value of 0 for those that do not. For species listed as amphidromous in FishBase, we independently confirmed that the purpose of migration was spawning.

We adopted the index of parental care used by Winemiller & Rose (1992) and Vila-Gispert & Moreno-Amich (2002). We computed this parental care index for species that we added from FishBase for which the required information was available in the literature or the "AddInfos" field of FishBase. The parental care index (denoted as Cindex) was computed as a sum of three variables, $x_1 + x_2 + x_3$, that characterize effort in providing spawning habitat, effort to provide parental protection, and nutritional contribution, respectively. Index $x_1 = 0$ if there is no special placement of zygotes, 1 if zygotes are placed in a special habitat, and 2 if both zygotes and embryos are protected by a nest. Index $x_2 = 0$ if no parental protection of zygotes, embryos, or larvae; 1 = brief period of protection by one parent (<1 month); 2 = extended protection (>1 month) by one parent or brief protection by both; and 4 = extended period of protection by both parents. Index $x_3 = 0$ if parents extend no post-egg nutritional contribution to larvae, 2 = brief period of nutritional contribution to embryos or larvae, 4 = long period of nutritional contribution to embryos or larvae (1–2 months' gestation), 6 = 2–10 months' gestation, and 8 = >10 months' gestation.

We derived two additional indicator values from variables Imig and Cindex. To reduce our reliance on Cindex as a quantitative measure of parental investment, we defined binary indicator variable, Icare, which we defined as zero for Cindex < 4 and one for Cindex ≥ 4. We also defined an indicator variable, Major = 1 for those fish species for which reproduction involves a "major accessory activity" such as a long migration (Icare = 1) or extensive parental care (Imig = 1) and assigned Major = 0 for all other species.

We examined relationships between life history traits and two response variables (spawning interval and extinction risk), using a database of 353 species that represented 2 classes, 35 orders, 102 families, and 202 genuses.

Phylogenetic correlation in life history traits

One complication when studying relationships among life history traits across species is the effect of phylogeny on these relationships (Pagel & Harvey,

1988; Ives & Zhu, 2006). For example, perhaps two ancestral species radiated, one with a long spawning interval and migration and the other with a short spawning interval and no migration. These two ancestral species radiated, each evolving into ten new species with the same traits. Assume also that ancestral species with the other two combinations of traits (short spawning interval, migration and long spawning interval, no migration) did not radiate and evolved only one modern species each. Together, these data, with two species-rich clusters, would produce a spurious correlation between the two traits simply due to historical accident, and not necessarily due to adaptations that arose independently as distinct evolutionary events.

Comparative methods are used to account for phylogenetic constraints (see review by Gittleman & Luh, 1992). These methods reduce the power to detect significant relationships between or among life history variables by accounting for non-independence among related species. We attempted to account for phylogenetic constraints using two comparative methods: repeated-measures' regression and phylogenetic regression using independent contrasts. Both approaches can be described by the same basic model (Eq. 1) that relates the dependent life history trait, y, to the independent life history variables, $\mathbf{X}$ (lower case variables represent vectors and upper case, bold variables represent matrices).

$$y = \mathbf{X}\beta + \varepsilon, \tag{1}$$

where β is a vector of regression coefficients. The vector of errors, ε, follows a multivariate normal distribution with mean vector, 0, and covariance matrix, $\mathbf{R}$. A regression that does not consider phylogenetic correlation would treat species (or averages of higher taxonomic nodes) as independent ($\mathbf{R} = \sigma^2\ \mathbf{I}$). Instead, we estimate covariances among traits due to phylogeny and use these as off-diagonal elements in $\mathbf{R}$.

Repeated-measures' regression

We used repeated-measures' regression to model covariance among species sharing the same order. We implemented this approach using SAS™ Proc Mixed (Littell et al., 1996; See Electronic supplementary material—Appendix I). Covariance estimates were constrained to be positive, and species in different classes were assumed to be independent. The covariance matrix, $\mathbf{R}$, was constructed using compound symmetry, with variances and covariances estimated for each taxonomic order. This approach did not consider correlations between taxonomic levels lower than order (i.e., family and genus), and orders represented by fewer than three species had to be removed from the analysis to obtain estimates. Although some predictors had more missing values than others, this left about 334 species representing 19 orders. Estimates were obtained by using restricted maximum likelihood.

Phylogenetic regression

We used Grafen's (1989) method of phylogenetic regression to account for phylogeny by forming independent contrasts in which each radiation provides an independent data point. The basic idea of phylogenetic regression is not to use species as independent data points, but instead to use the higher nodes in the phylogeny (Grafen, 2006). Values at each radiation are weighted averages of species' life history traits at lower levels in its taxonomic subtree. Obtaining the weighted averages for each independent radiation involves solving a generalized least squares problem with phylogeny incorporated into the structure of the error term (Martins & Hansen, 1997). Under the model of Brownian motion evolution, the values of a species' trait follow a multivariate normal distribution with mean equal to the mean of the base of the tree and covariance matrix whose off-diagonal elements for species i and j (Eq. 2) decrease in proportion to the branch length, h_{ij}, of their shared lineage in the tree (Grafen, 1989). We assigned path lengths of 1, 2, 3, and 4 for species sharing genus, family, order, and class, respectively. Diagonal elements of $\mathbf{R}$ are given by variance, σ^2.

$$\mathbf{R} = \mathrm{E}\left(\varepsilon_i \varepsilon_j\right) = \sigma^2\left(1 - h_{ij}^{\rho}\right) \tag{2}$$

The idea is that if the same relationship is observed in two radiations, it cannot result from phylogenetic similarity (Grafen, 1989). Parameter ρ is the power to which heights are raised before computing path segment lengths (phylogenetic distance). The parameter ρ measures the strength of phylogenetic correlation. Parameters β and ρ are estimated, and an F-test is provided for the overall model or for specified terms of the regression model. Phylogenetic degrees of freedom (denominator) are the number of higher phylogenetic nodes that provide useful information (e.g., nodes for which all species below a node share the same value do not provide information).

For repeated measure and phylogenetic regression models with multiple predictors, we evaluated the importance of each predictor by testing the significance of each term, while controlling for the other predictors in the model. We report the one-sided P-value associated with the F-test because we postulate positive effects of all predictors including their interactions as the alternative hypothesis.

Power and hypothesis testing

We report the importance of each predictor by testing whether each estimated coefficient is significantly greater than zero. We report the one-sided P-value associated with the statistical tests because, in all cases, we postulate positive effects, including interactions, as the alternative hypothesis. Setting a critical Type-I error, $\alpha = 0.05$, is considered very conservative because it does not provide a good balance between avoiding wrong conclusions of significance and having the power to detect relationships (Toft & Shea, 1983; Gotelli & Ellison, 2004). We therefore specify $\alpha = 0.1$ and report the actual P-values to allow readers to make their own determinations of significance.

Accounting for phylogenetic correlation appropriately reduces the power of statistical tests, increasing P-values. Repeated-measures' regression is more powerful than phylogenetic regression because it uses more species-level data and does not consider correlations at the family or genus level. To the extent that the reduction in power is caused by properly considering correlations, the results of phylogenetic regression are more correct. To the extent that the reduction in power is caused by excluding relevant species-level information (for example, if we are interested in relationships among extant species regardless of their evolutionary histories), the results of repeated-measures' regression are more correct. Both methods could be improved on by obtaining better estimates of phylogenetic "distances" (evolutionary branching times) separating all pairs of species.

Are spawning intervals longer for fish species that migrate to spawn or provide parental care?

Capital breeding is thought to evolve in species for which breeding involves a "major accessory activity," such as a long migration or extensive parental care (Bull & Shine, 1979). We used repeated-measures' regression and phylogenetic regression to examine relationships between spawning interval (SI) and major accessory activity (Major). Next, we examined relationships between SI and the two constituents of Major: migration (Imig) and parental care (either Cindex, the original 9-value index, or indicator variable, Icare).

Are capital-breeding species considered at greater risk of extinction?

In our second analysis, we addressed the hypothesis that fish species that spawn less frequently tend to experience higher risk of extinction, as indicated by IUCN designation. We used box-whisker diagrams to graphically compare the probability distribution of spawning interval among IUCN risk categories. We also used phylogenetic regression to test for a positive association between our ordered index of IUCN risk status and spawning interval. This analysis included 41 independent contrasts. We are interested in a one-sided test to compare zero effect against the alternative hypothesis that SI has a positive effect on the ordered index of IUCN risk status.

Results

Species with long breeding intervals

Sixty-four species of the total 353 species in our sample had longer-than-annual breeding cycles (See Electronic supplementary material—Appendix II). The longest breeding intervals were found among the sturgeons. Like other sturgeons, female white sturgeon (*Acipenser transmontanus*) are physiologically capable of breeding at 2-year intervals, with a

1.8 year maturation cycle (Doroshov et al., 1997). In the wild, however, at least 1 year of resting has been observed for this species, and typical spawning intervals range from 3 to 9 years (Paragamian & Wakkinen, 2002). Larger shark species also exhibit longer-than-annual breeding cycles. Females of several shark species breed every other year (Frisk et al., 2001; Hueter et al., 2004), necessitated by a very long gestation period. Sturgeons and sharks accounted for a large proportion of the species in our database with longer-than-annual spawning intervals (See Electronic supplementary material—Appendix II). Other taxa included cavefishes, a sucker, several groundfishes, the coelanth, striped bass, and several salmonids.

Are spawning intervals longer for fish species that migrate and provide parental care?

Figure 1 summarizes the mean spawning intervals for classes and orders included in our analyses. Without accounting for phylogeny, average spawning interval was clearly higher for species with a major accessory activity than for those without (Fig. 2). Repeated-measures' regression found that the difference in

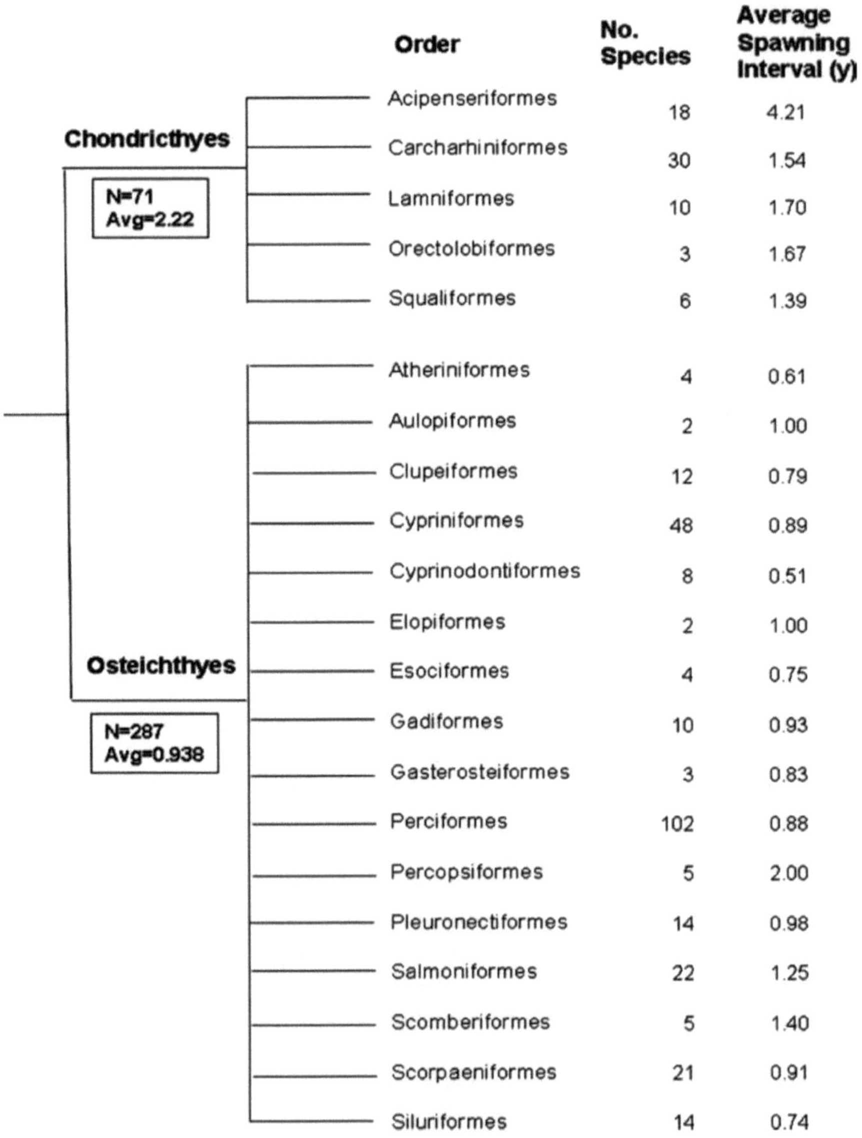

Fig. 1 Species counts and average spawning intervals for classes and orders of fishes in our data with more than one representative per order

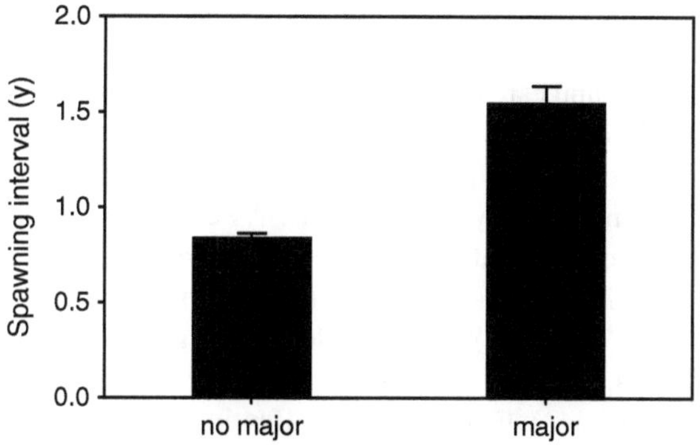

Fig. 2 Comparison of average spawning interval (±1 SE) for fish species with no major versus a major accessory activity to spawning

average spawning interval (SI) for species with Major = 1 and species with Major = 0, after taking between-order correlation into account, was highly significant ($\chi^2 = 563$, df = 31, $P < 0.0001$). Estimated coefficients are shown in Eq. 3.

$$SI = 0.8335 + 0.1313\ \text{Major} \tag{3}$$

The effect of Major was significantly greater than zero (one-sided $P = 0.0007$). Phylogenetic regression produced different estimates for this relationship (Eq. 4).

$$SI = 1.280 + 0.2032\ \text{Major} \tag{4}$$

An F-test with 1,85 degrees of freedom is reported, where the numerator degrees of freedom is the number of predictor variables and the denominator is the phylogenetic degrees of freedom (higher taxonomic nodes included in the analysis). The F-test value of 1.481 had an associated probability of $P = 0.227$. This corresponds to a one-sided T-test with a P-value of 0.1135, which is not significant at $\alpha = 0.1$. The exponent used in calculating heights was estimated to be $\rho = 0.4577$, indicating a moderate influence of phylogeny.

Next, we confirmed that spawning interval increased with each of the accessory activities that comprised the Major index variable. Collinearity between the two major accessories was evident. Without adjusting for phylogenetic effects, average spawning interval varied among the nine values of Cindex with a suggestion of longer spawning intervals with higher values of parental care (Fig. 3). Repeated-measures' regression found a significant positive relationship between SI and the parental care index ($\chi^2 = 552$, df = 28, $P < 0.0001$), as shown in Eq. 5.

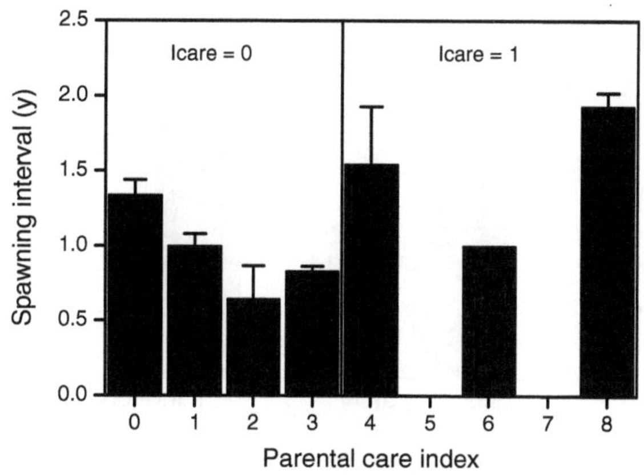

Fig. 3 Comparison of average (±1 SE) spawning interval for fish species with different levels of parental care. The vertical line separates species with Icare = 0 and those with Icare = 1

$$SI = 0.8266 + 0.0327\ \text{Imig} + 0.0239\ \text{Cindex} + 0.0685\ \text{Cindex*Imig} \tag{5}$$

The positive effects of Cindex (one-sided $P = 0.0239$) and the interaction (one-sided $P = 0.0075$) were significant, but that of Imig was not (one-sided $P = 0.2644$). Phylogenetic regression estimated the relationship shown in Eq. 6 (F-test with 3,82 degrees of freedom = 2.022, two-sided $P = 0.1171$).

$$SI = 1.196 + 0.1456\ \text{Imig} + 0.0379\ \text{Cindex} + 0.0288\ \text{Cindex*Imig} \tag{6}$$

The estimated exponent used in calculating heights was $\rho = 0.6907$. Only the parental care index had a significant positive influence (one-sided $P = 0.065$). However, a model with Imig by itself showed a clearly significant positive effect (F-test with 1,80 degrees of freedom = 5.209, one-sided $P = 0.0126$), suggesting the presence of collinearity between the two predictor variables, Imig and Cindex.

An alternative way to measure the effect of parental care used the binary variable, Icare, rather than the original parental care index, Cindex. These analyses supported the idea that average spawning intervals are longer for species that provide considerable amounts of parental care and those that migrate to spawn. Without accounting for phylogenetic correlation, average spawning interval was higher for species with Icare = 1 than for species with Icare = 0 (Fig. 4). Repeated-measures regression estimated the model given by Eq. 7 ($\chi^2 = 540$, df = 32, two-sided $P < 0.0001$).

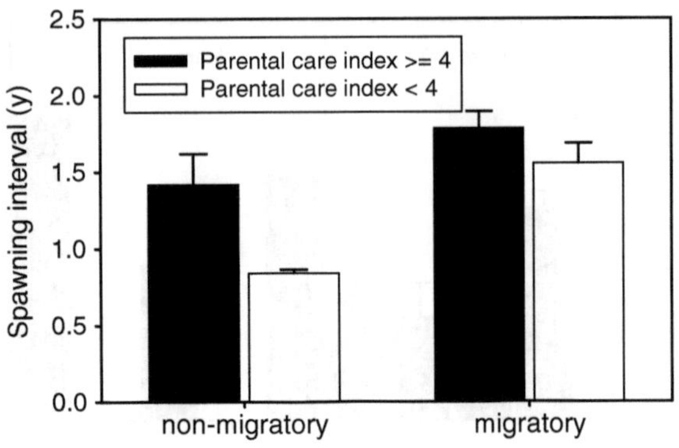

Fig. 4 Comparison of average spawning interval (±1 SE) for fish species with all four combinations of spawning migration versus no spawning migration and higher versus lower parental care

$$SI = 0.8221 + 0.1070\ \text{Imig} + 0.2215\ \text{Icare} + 0.0326\ \text{Imig*Icare} \quad (7)$$

The positive effects of Imig ($P = 0.0462$) and Icare ($P = 0.0040$) were both significant, but not that of the interaction term ($P = 0.3377$). Phylogenetic regression estimated different coefficients for the same model shown in Eq. 8 (F-test with 3,82 degrees of freedom = 1.922, two-sided $P = 0.1325$).

$$SI = 1.214 + 0.1770\ \text{Imig} + 0.2338\ \text{Icare} + 0.1837\ \text{Imig*Icare} \quad (8)$$

As with repeated-measures' regression, the positive effects of Imig (one-sided $P= 0.0628$) and Icare (one-sided $P= 0.0708$) were significant. The interaction was also significantly greater than zero (one-sided $P= 0.0982$). The estimated exponent used in calculating heights was $\rho = 0.4884$. These results suggest that the effect of Icare on spawning interval is stronger among migratory fishes than among non-migratory fishes.

The results of using two alternative statistical methods applied to three combinations of predictor variables suggested that both parental care and migration are associated with longer spawning intervals. The statistical relationship between spawning interval and migration was weaker when Cindex, the original parental care index, was included in the model rather than binary variable, Icare. This may be because the correlation between Icare and Imig is less than the correlation between Cindex and Imig, reducing the collinearity between the two variables.

Without accounting for phylogenetic correlation, we saw little visual evidence for a monotonic increase in spawning interval with increasing values of the original parental care index (Fig. 3). At the extreme of no parental care (Cindex = 0), we would expect a low average spawning interval for species. However, sturgeons, which do not provide parental care, had the longest spawning intervals. As expected, species assigned a parental care code of "4" had longer mean spawning intervals than those providing less parental care. This category included nest guarders (e.g., basses, catfish, rockfishes, and sticklebacks) and cavefishes, which incubate offspring in their gills. However, these nest-guarding species had longer average spawning intervals than species assigned a parental code of "6," which consisted of sharks with gestation periods less than 10 months. At the other extreme, species investing the most in parental care conformed to the expectations of the capital-breeding strategy, with a high average spawning interval (Fig. 3). This category included sharks with gestation periods longer than ten months and the coelacanth (See Electronic supplementary material—Appendix II).

Are capital-breeding species considered at greater risk of extinction?

The average spawning interval increased for successively higher categories of extinction risk, but there was considerable overlap of spawning intervals among the extinction risk categories (Fig. 5). Phylogenetic

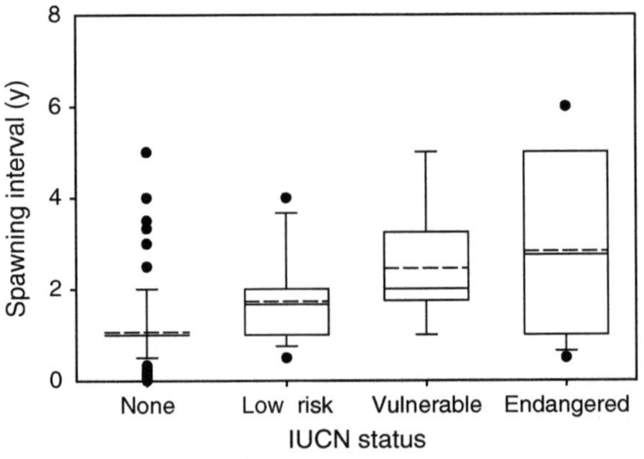

Fig. 5 Box-whisker diagrams showing the distribution of spawning intervals for species with different IUCN status. IUCN value "endangered" combines endangered and critically endangered species. The box defines 25th and 75th percentiles, the horizontal lines (usually) in each box define the mean (dashed) and median (solid). Outliers are shown as dots, and the 5th and 95th percentiles are shown by "whiskers"

regression suggested a significant positive association between extinction risk and spawning interval (F-test with 1,41 degrees of freedom = 11.745, one-sided $P = 0.0007$), as shown in Eq. 9.

$$\text{Risk} = -0.00002 + 0.3863\ \text{SI}. \qquad (9)$$

The estimated exponent used in calculating heights was $\rho = 0.3133$.

Discussion

Our analysis confirmed both hypotheses concerning capital-breeding fishes in all but one statistical test, despite the fact that the power of our tests was lowered by accounting for phylogenetic correlation. We found, as Villa-Gispert et al. (2002) did, that phylogenetic effects on life history traits were important, with the majority of variation in spawning interval occurring at the level of class and order. Below, we discuss results for each of the two hypotheses tested by this study.

Are spawning intervals longer for fish species that migrate to spawn or provide parental care?

Our life history analysis demonstrated that the capital-breeding hypothesis, originally developed for birds and reptiles, shows promise for fishes. Fishes with a major accessory activity (migration or parental care) were found to have longer intervals between breeding, even when phylogenetic correlations were accounted for. Species represented in our data confirmed the expected relationships predicted by the capital-breeding paradigm. On average, fish species that migrate to spawn had longer spawning intervals than non-migratory species, and within each migratory category, fish with higher parental care had higher spawning intervals (Fig. 4). Understanding the relationships among these traits can help us to understand the evolutionary constraints that capital-breeding fishes operate under when faced with new environmental challenges.

This analysis could be refined in future by using quantitative measures of the fitness cost of each major accessory activity. For migration, a measure of migration distance or the energetic cost of making a spawning migration could be used. For parental care, a measure of the risk of starvation incurred by the female parent due to the energy she expends on reproduction could be used. Although parental care in fish has been a popular subject for research, few studies provide quantitative information on both parental expenditure and investment (Smith & Wootton, 1995). Including this type of quantitative information would account for several things that our index did not. For example, large fishes can sustain longer periods of starvation than smaller fishes. As another example, although sturgeons do not provide parental care, their eggs can take up to 30% of female body weight (Wei et al., 1997), making it difficult for them to feed.

Our analyses did not consider within-species variation or the influence of environmental factors on spawning interval. Life history traits such as spawning interval exhibit phenotypic plasticity in some taxa. Individual variation in spawning interval is found among many fishes, including some capital-breeding taxa. Rideout et al. (2005) propose that reproductive cycles can be interrupted due to poor nutrition or inadequate environmental conditions, overcrowding, and a shortage of mates. Understanding what factors influence spawning frequency and understanding the consequences of infrequent spawning can be important to the management of capital-breeding stocks with plastic reproductive timing. Uncertainties in predictions of sustainable harvest or population recovery can be reduced if efforts are made to determine and include realistic estimates of spawning interval, rather than assuming either annual spawning or adopting the minimum interval that is physiologically possible.

Are capital-breeding species considered at greater risk of extinction?

We confirmed that fishes with longer spawning intervals tended to belong to higher IUCN extinction-risk categories (Fig. 5), and that a positive association exists between our IUCN index and spawning interval. Three caveats accompany this result. First, we note that variation in IUCN status among stocks was not considered in our analysis. Second, as in any regression, the direction of causality might be reversed. IUCN status is justified, in some cases, by population modeling of species' life history traits (Musick, 1999), as well as the actual status and trends of its stocks. However, most marine fishes listed in the IUCN Red List of Threatened Animals were placed there because of observed population declines.

Third, risks associated with spawning interval and correlated life history traits could be confounded, and parsing out the roles of individual traits on extinction risk is difficult. It is well known that migratory fishes tend to be at higher risk (e.g., Angermeier, 1995). Among migratory species large-bodied fishes tended to be at higher risk, whereas among non-migratory species small-bodied freshwater fishes with restricted geographic ranges tended to be at greater risk of extinction (Parent & Schriml, 1995; Angermeier, 1995; Reynolds et al., 2005). We recommend that further comparative studies designed to understand life history correlates of extinction risk include spawning interval with a comprehensive array of life history characteristics, and we recommend that such studies focus on including atypical species that "break" life history correlations.

When viewed in the context of existing life history classifications, capital breeding adds a new dimension that cuts across recognized strategies. Recognized life history strategies of fishes (Winemiller & Rose, 1992; Villa-Gispert et al., 2002; Villa-Gispert & Moreno-Amich, 2002; King & McFarlane, 2003) include periodic, opportunistic, equilibrium, salmonid, and intermediate strategies. We examined the list of putative capital-breeding species (See Electronic supplementary material—Appendix II) to see whether they tended to occupy similar positions within the existing life history framework. We determined that capital breeders follow every strategy except for the opportunistic strategy. Equilibrium strategists are represented by sharks (Frisk et al., 2001; King & McFarlane, 2003) and the cavefish (Winemiller, 2005). Periodic strategists are represented by long-lived and late-maturing sturgeons and paddlefishes (Jager et al., 2002). The list also includes several intermediate strategists (e.g., cod, halibut, and mackerel) and salmonid strategists (e.g., Atlantic salmon, trout, and Arctic charr).

The observation that extinction risk is higher for capital-breeding species leads to an apparent paradox. If the capital-breeding life history is an adaptive solution that results in higher fitness by minimizing the frequency of costly expenditures on reproduction, how can it also put populations at risk? One possible answer is that the environment has changed since the capital-breeding strategy evolved (multiple times), lowering its fitness. Many capital-breeding species also migrate or have traits characteristic of a 'slow' life history (i.e., small litter, slow growth rate, late maturation, and long gestation period). Empirical studies have confirmed that species that migrate (this study) and those with a 'slow' life history (Johst & Brandl, 1997; Jennings et al., 1998; Dulvy et al., 2003) are at higher risk of extinction than their counterparts are. Changes in the environment might explain why both these groups are now less advantageous than they were in the past. In theory, the evolution of a 'slow' life history is favored when adult mortality is low. When human activities increase adult mortality (e.g., fishing and habitat degradation), these species cannot respond quickly by increasing their spawning frequencies. Likewise, fragmentation and degradation of aquatic habitats has increased the mortality risk associated with migration and the energetic and survival benefits of using multiple habitats. In short, it is likely that the fitness landscape has changed in such a way as to put capital-breeding fishes at a disadvantage.

Acknowledgments Thanks to the organizers of the 2005 Diadfish Conference, Drs. Eric Rochard, Etienne Prevost, Patrick Williott, and others. We owe a great deal of appreciation to Alan Grafen, who helped with fine-tuning his software for phylogenetic regression (Phyreg). Thanks also to several researchers who provided advice on incorporating phylogenetic correlation, including Ann Reed and Dr. Arnold Saxton (University of Tennessee, Knoxville). We appreciate Emmanuel Isang's contributions during his summer internship: running literature searches used to fill in the life history dataset. ORNL is managed by UT-Battelle, LLC, for the DOE under contract DE-AC05-00OR22725.

References

Angermeier, P. L., 1995. Ecological attributes of extinction-prone species: loss of freshwater fishes of Virginia. Conservation Biology 9: 143–158.

Bull, J. J. & R. Shine, 1979. Iteroparous animals that skip reproductive opportunities. The American Naturalist 114: 296–303.

Doroshov, S. I., G. P. Moberg & J. P. V. Eenennaam, 1997. Observations on the reproductive cycle of cultured white sturgeon, *Acipenser transmontanus*. Environmental Biology of Fishes 48: 265–278.

Dulvy, N. K., Y. Sadovy & J. D. Reynolds, 2003. Extinction vulnerability in marine populations. Fish and Fisheries 4: 25–64.

FishBase. 2004. A global information system on fishes. In Froese, R. & D. Pauly (eds), CD-ROM. WorldFish Center—Philippine Office, Los Banos, Philippines.

Frisk, M. G., T. J. Miller & M. J. Fogarty, 2001. Estimation and analysis of biological parameters in elasmobranch fishes: a comparative life history study. Canadian Journal of Fisheries and Aquatic Sciences 58: 969–981.

Gittleman, J. L. & H.-K. Luh, 1992. On comparing comparative methods. Annual Review of Ecology and Systematics 23: 383–404.

Gotelli, N. J. & A. M. Ellison, 2004. A Primer of Ecological Statistics. Sinauer Associates, Inc., Sunderland, Massachussetts, USA: 510 p.

Grafen, A., 1989. The phylogenetic regression. Philosophical Transactions of the Royal Society of London B. 326: 119–157.

Grafen, A., 2006. A User's Guide to the SAS Implementation of the Phylogenetic Regression, Version 0.7. http://www.users.ox.ac.uk/~grafen/phylo/phyman.pdf.

Hilton-Taylor, C., 2000. 2000 IUCN Red List of Threatened Species. IUCN, Gland, Switzerland and Cambridge, UK: 61 p.

Hueter, R. E., M. R. Heupel, E. J. Heist & D. B. Keeney, 2004. Evidence of philopatry in sharks and implications for the management of shark fisheries. e-Journal of Northwest Atlantic Fisheries Science 35: article 7.

Ives, A. R. & J. Zhu, 2006. Statistics for correlated data: phylogenies, space, and time. Ecological Applications 16: 20–32.

Jager, H. I., W. Van Winkle, K. B. Lepla, J. A. Chandler, P. Bates & T. D. Counihan, 2002. Factors controlling white sturgeon recruitment in the Snake River. In AFS Symposium: Biology, Management, and Protection of Sturgeon. American Fisheries Society, Bethesda, MD, USA: 127–150.

Jennings, S., J. D. Reynolds & S. C. Mills, 1998. Life history correlates of responses to fisheries exploitation. Proceedings of the Royal Society of London B 265: 333–339.

Johst, K. & R. Brandl, 1997. Body size and extinction risk in a stochastic environment. Oikos 78: 612–617.

Jonsson, K. I., 1997. Capital and income breeding as alternative tactics of resource use in reproduction. Oikos 78: 57–66.

Jørgensen, C., B. Ernande, O. Fiksen & U. Dieckmann, 2006. The logic of skipped spawning in fish. Canadian Journal of Fisheries and Aquatic Sciences 63: 200–211.

King, J. R. & G. A. McFarlane, 2003. Marine fish life history strategies: applications to fishery management. Fisheries Management and Ecology 10: 249–264.

Littell, R. C., G. A. Milliken, W. P. Stroup & R. D. Wolfinger, 1996. SAS System for Mixed Models. SAS Institute, Cary, NC, 633 pp.

Martins, E. P. & T. F. Hansen, 1997. Phylogenies and the comparative method: a general approach to incorporating phylogenetic information into the analysis of interspecific data. The American Naturalist 149: 646–667.

Murua, H. & F. Saborido-Rey, 2003. Female reproductive strategies of marine fish species of the North Atlantic. Journal of Northwest Atlantic Fisheries Science 33: 23–31.

Musick, J. A., 1999. Ecology and conservation of long-lived marine animals. In J. A. Musick (ed.), Life in the Slow Lane, Ecology and Conservation of Long-lived Marine Animals. American Fisheries Society Symposium 23, American Fisheries Society, Bethesda MD, USA: 1–10.

Pagel, M. D. & P. H. Harvey, 1988. Recent developments in the analysis of comparative data. The Quarterly Review of Biology 63: 413–440.

Paragamian, V. L. & V. D. Wakkinen, 2002. Temporal distribution of Kootenai River white sturgeon spawning events and the effect of flow and temperature. Journal of Applied Ichtyology 18: 542–549.

Parent, S. & L. M. Schriml, 1995. A model for the determination of fish species at risk based upon life-history traits and ecological data. Canadian Journal of Fisheries and Aquatic Sciences 52: 1768–1781.

Reynolds, J. D., T. J. Webb & L. A. Hawkins, 2005. Life history and ecological correlates of extinction risk in European freshwater fishes. Canadian Journal of Fisheries and Aquatic Sciences 62: 854–862.

Rideout R. M, G. A. Rose & M. P. M. Burton, 2005. Skipped spawning in female iteroparous fishes. Fish and Fisheries 6: 50–72.

Smith, C. & R. J. Wootton, 1995. The costs of parental care in teleost fishes. Reviews in Fish Biology and Fisheries 5: 7–22.

Toft, C. A. & P. J. Shea, 1983. Detecting community-wide patterns: estimating power strengthens statistical inference. The American Naturalist 122: 618–625.

Vila-Gispert, A. & R. Moreno-Amich, 2002. Life-history patterns of 25 species from European freshwater fish communities. Environmental Biology of Fishes 65: 387–400.

Vila-Gispert, A., R. Moreno-Amich & E. Garcia-Berthou, 2002. Gradients of life history variation: an intercontinental comparison of fishes. Reviews in Fish Biology and Fisheries 12: 417–427.

Wei, Q., F. Ke, J. Zhang, P. Zhuang, J. Luo, R. Zhou, W. Yang, 1997. Biology, fisheries, and conservation of sturgeons and paddlefish in China. Environmental Biology of Fishes 48: 241–255.

Winemiller, K. O., 2005. Life history strategies, population regulation, and implications for fisheries management. Canadian Journal of Fisheries and Aquatic Sciences 62: 872–885.

Winemiller, K. O. & K. A. Rose, 1992. Patterns of life-history diversification in North American fishes: implications for population regulation. Canadian Journal of Fisheries and Aquatic Sciences 49: 2196–2218.

Hydrobiologia (2008) 602:27–46
DOI 10.1007/s10750-008-9288-1

FISH AND DIADROMY IN EUROPE

Dopaminergic systems in the European eel: characterization, brain distribution, and potential role in migration and reproduction

Marie-Emilie Sébert · Finn-Arne Weltzien · Christine Moisan · Catherine Pasqualini · Sylvie Dufour

Abstract In fish like in mammals, dopamine (DA) is a major catecholaminergic neurotransmitter that contributes to many functions of the nervous system like sensory perception, tuning of sensori-motor cues, and hypothalamic and pituitary functions. In the eel, DA inhibits gonadal development, and juvenile silver eels remain blocked at a prepubertal stage if their reproductive migration does not occur. From data in other teleosts and vertebrates, it is suggested that DA would be involved also in the last steps of eel reproduction (oocyte maturation, ovulation, and spermiation) as well as in eel reproductive migration (locomotion and olfaction). Investigating dopaminergic systems in the eel may help in understanding the mechanisms of its complex life cycle and provide new data for its conservation and reproduction. In this article we review the biosynthesis and catabolism of catecholamines and discuss available methods to investigate brain dopaminergic systems in vertebrates and their application to the eel. Immunocytochemistry, in situ hybridization, and different tracing methods are used to map dopaminergic neurons and projections in the brain and pituitary and infer their potential functions. Moreover, variations in dopaminergic activity may be approached by means of quantitative methods like quantitative real-time RT-PCR and HPLC. These tools are currently used to study dopaminergic systems in the eel brain, their anatomy, regulation, and potential roles with special emphasis on the regulation of reproduction and reproductive migration.

Keywords Teleosts · Catecholamine · Locomotion · Olfaction · Puberty

Guest editors: S. Dufour, E. Prévost, E. Rochard & P. Williot
Fish and diadromy in Europe (ecology, management, conservation)

M.-E. Sébert (✉) · F.-A. Weltzien · S. Dufour
Département des Milieux et Peuplements Aquatiques, Muséum National d'Histoire Naturelle, USM 0401, UMR 5178 CNRS/MNHN/UPMC Biologie des Organismes Marins et Ecosystèmes, 75231 Paris Cedex 05, France
e-mail: sebert@mnhn.fr

C. Moisan
Unité Haute Pression et Métabolisme, UPCI-EA3879, Faculté des Sciences, 6 avenue Le Gorgeu, CS 93837, 29238 Brest Cedex 08, France

F.-A. Weltzien
Department of Molecular Biosciences Department – Programme for Physiology, University of Oslo, P.O. Box 1041, Blindern, 0371 Oslo, Norway

C. Pasqualini
Développement, Evolution et Plasticité du Système Nerveux, UPR CNRS 2197, Institut de Neurobiologie Alfred Fessard, CNRS, 91198 Gif-sur-Yvette Cedex, France

Introduction

First described for their role in the cardiovascular function (Holtz, 1939; McGeer et al., 1978), catecholamines (CAs) were then discovered in the central nervous system (CNS) (Vogt, 1954). Through the various CAs, DA is widely present in the CNS where the differential location of DA neurons between brain areas illustrates an implication in various central functions (for review, Smeets & Gonzalez, 2000). For instance, DA modulates the visual and olfactory sensory perception, tuning of sensory-motor cues, and hypothalamic (food intake, thermoregulation) and pituitary (e.g. prolactin secretion) functions. In fish, various studies report that DA is also involved in such different functions than locomotion (Mok & Munro, 1998b), reproduction (for review, Dufour et al., 2005), and aggressive and dominant behaviors (for review, Winberg & Nilsson, 1993).

In this article we will review methodological approaches and results concerning DA systems in teleosts with special focus on the European eel in which DA plays a critical role in sexual maturation. Indeed DA has been shown to inhibit synthesis and release of gonadotropins, the pituitary hormones which stimulate gonadal development (for review, Dufour et al., 2003). Following this, when silver eels leave continental waters to undertake a 6,000-km long migration to reach their reproductive area they remain blocked at a prepubertal stage. The silver stage is the last stage observed in natural conditions, and migratory and reproductive conditions continue to be a mystery. Moreover DA may also be involved during migration and in the last steps of gonadal development and spawning. Keeping in mind that eel stocks are menaced worldwide (Wirth & Bernatchez, 2003), it is important to elucidate the roles and regulation of DA to develop sustainable conservation plans and commercial aquaculture for this species. In addition, eels belong to the phylogenetically ancient group of Elopomorphs and therefore represent a relevant model to study evolution of dopaminergic control of central functions.

Dopamine metabolism

Biosynthesis

Catecholamines are synthesized from the amino acid tyrosine and therefore have a common structure including a hydroxylated benzene ring and an amine tail. Their synthetic pathway is well conserved among vertebrates as reported in phylogenetical studies (Reiner, 1994). Moreover, the synthetic pathway is similar in various tissues like neurons and non-neural chromaffin cells (Blaschko, 1939; Holtz, 1939). Catecholamine metabolism in fish has been reviewed by Nilsson (1983) and more recently by Randall and Perry (1992).

The synthetic cascade of CAs, also named the "Blaschko pathway" (Blaschko, 1939), begins with the hydroxylation of tyrosine into L-dihydroxyphenylalanine (L-DOPA) by tyrosine hydroxylase (TH), a cytoplasmic enzyme (Fig. 1). This first step is considered as the rate-limiting one (Levitt et al., 1965) in CA synthesis, whatever the end product is. Then, L-DOPA is decarboxylated to DA by the cytoplasmic enzyme aromatic L-amino acid decarboxylase (AADC). Because the available enzymatic equipment determines what end product a particular neuron will produce, dopaminergic neurons would

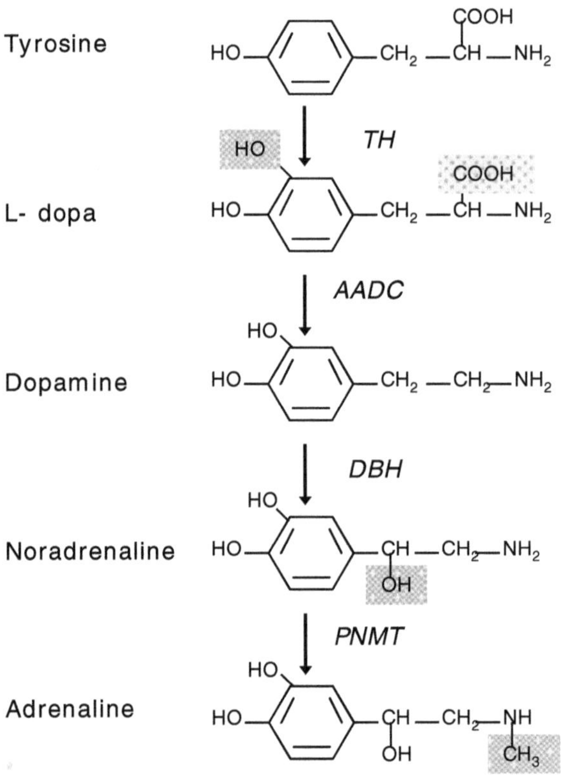

Fig. 1 Biosynthesis of catecholamines. Enzyme abbreviations: TH, tyrosine hydroxylase; AADC: amino acid decarboxylase; DBH, dopamine-β-hydroxylase; PNMT: phenylethanolamine-*N*-methyl transferase

enclose only TH and AADC. In contrast, the neurons using NA or A as neurotransmitters contain additional enzymes and produce DA as an intermediary product. Like TH and AADC, DA is found in the cytosolic space where it is taken up in storage vesicles until its release by dopaminergic neurons. In noradrenergic and adrenergic neurons, the dopamine-β-hydroxylase (DBH) converts DA to NA inside the storage vesicles. The final step in the Blascho pathway consists of the methylation of NA by the cytoplasmic enzyme phenylethanolamine *N*-methyltransferase (PNMT) to produce A. In teleosts, DA and NA are the most frequent CAs in the CNS where they act as neurotransmitters, whereas A is mainly produced in the periphery by the chromaffin tissue.

Many studies have focused on the regulation of TH, the rate-limiting enzyme. Available data report different ways of TH control. First, works in mammals report that synthesized end products like NA, A (Spector et al., 1965, 1967), or L-DOPA (Demarest & Moore, 1980; Reymond & Porter, 1982) may ensure a rapid negative feedback on TH synthesis, thereby regulating CA production. It seems that similar mechanisms exist also in fish as suggested in the trout (*Oncorhynchus mykiss*): Linard and coworkers reported that treatment with an inhibitor of AADC resulted in L-DOPA accumulation in the telencephalon and the hypothalamus of female trout. The L-DOPA content increased in a linear fashion before reaching a plateau, suggesting a negative feedback from L-DOPA on its own production (Linard et al., 1996). Second, the phosphorylation of TH is also a short-term positive regulating way for CA synthesis because it increases TH affinity for its cofactor, oxygen (Joh et al., 1978; Zigmond et al., 1989). Finally, modulation of TH expression, transcription, and translation constitutes a long-term regulation mechanism for CA neurons. Such long-term TH regulation has recently been discussed in the eel in response to steroid feedback (Weltzien et al., 2006). Besides the regulation of the rate-limiting enzyme, the last steps of the "Blaschko pathway" may be submitted to hormonal control in some vertebrate groups. For instance, adrenocortical steroids have been shown to stimulate DBH and PNMT (Nilsson, 1983); in rainbow trout, in chromaffin tissue, cortisol has no effect on the PNMT but significantly increases DBH activity (Jonsson et al., 1983) and thereby NA synthesis.

Degradation

As with their synthesis, CA degradation is a well-conserved process among vertebrates. DA catabolism involves two major enzymes: monoamine oxidase (MAO) and catechol-*O*-methyl transferase (COMT) (Fig. 2). MAO is located in the external membrane of mitochondria. In fish only one form of MAO has been reported (Setini et al., 2005) in contrast to tetrapods in which two isoforms exist (MAO A, MAO B) (McGeer et al., 1978; Cooper et al., 1986; Siegel et al., 1989). COMT is located in the cytoplasm and in the synaptic cleft, but it is also largely involved in the degradation of circulating CAs. During nerve activity DA is released in the synaptic cleft where it may act as a neurotransmitter, binding to postsynaptic G-protein coupled DA receptors (DA receptors 1–5) (for review, Callier et al., 2003). However, about 70% of the released DA will be taken up rapidly by a membrane presynaptic transporter coupled to the co-transport of Na^+ and Cl^- ions (McGeer et al., 1978; Cooper et al., 1986; Siegel et al., 1989; Chen et al., 2004). Once in the presynaptic terminal, free DA may be deaminated and transformed to its corresponding aldehyde by MAO and further converted by aldehyde dehydrogenase into dihydroxyphenylacetic acid (DOPAC, Fig. 2). Alternatively, the remaining fraction of DA released in the synaptic cleft may be inactivated by COMT into O-methylated catabolites. Following this, DOPAC may be O-methylated to the homovanillic acid (HVA) by COMT prior to be released directly or via the cerebrospinal fluid in the circulation.

In the European eel, it seems that MAO is the principal degradative pathway because DOPAC was easily detected, while HVA remained undetectable except in olfactory bulbs (see Section "Estimation via the study of synthesizing enzyme"). The predominance of the MAO or the COMT catabolic pathway follows species-specific variations. DOPAC is found as the primary metabolite of DA catabolism in goldfish (*Carassius auratus*; Dulka et al., 1992) and Atlantic cod (*Gadus morhua*; Ehrenström & Johansson, 1987). In contrast, HVA is preferentially found in the crucian carp (*Carassius carassius*; Nilsson, 1989, 1990), while in a selacian, the dogfish, both O-methylation and deamination appear important (Mazeaud & Mazeaud, 1973). Although these variations are linked to species-specific differences, it

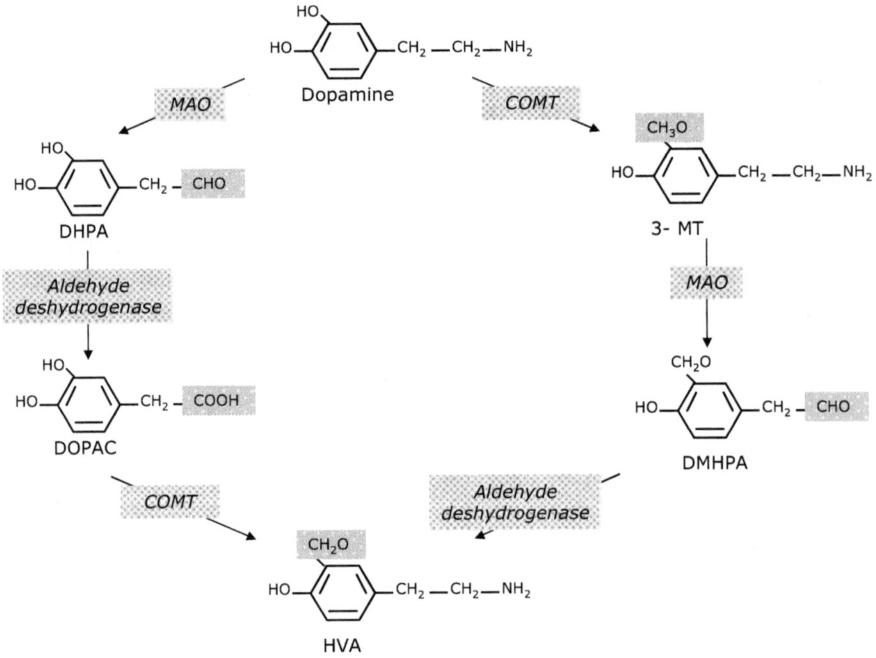

Fig. 2 Catabolism of dopamine. Catabolite abbreviations: DHPA, 3,4-dihydroxyphenyl-acetaldehyde; DMHPA, 3-methoxy-4-hydroxyphenyl-acetaldehyde; DOPAC, 3,4-dihydroxyphenylacetic acid; HVA, homovanillic acid; 3-MT, 3-methoxytyramine. *Enzyme abbreviations*: MAO, monoamine oxidase; COMT, catechol-*O*-methyl transferase

seems that it also depends on the analytical technique used.

Methodological approaches of dopaminergic system investigations

Cartography of dopaminergic systems

Formaldehyde-induced fluorescence method (FIF)

The FIF method was developed following the discovery that sections of formaldehyde-fixed adrenal medulla that produces CAs were fluorescent (Eränkö, 1955). The procedure described by Falck and coworkers consists in the exposure of freeze-dried tissue to formaldehyde vapor (Falck et al., 1962; Erankö, 1967); during this exposure CAs and indoleamines become intensively fluorescent. This technique allowed labeling of catecholaminergic cells and fibers in various tissues including the rat brain (Dahlström & Fuxe, 1964). Later, the treatment of tissue with glyoxylic acid was preferentially used because it was more sensitive compared to the FIF method. However, neither of these methods allow for discrimination between individual CAs or between CAs and indoleamines. With these major limitations, the FIF method is not suitable for studying CA systems in their complexity.

Immunocytochemistry

Immunocytochemistry is the cytological method classically used to map brain cellular bodies and their projections. Because DA is stored in vesicles in the cytosolic space, using antibodies against DA enables the localization of DA neurons and fibers. One may also indirectly label DA neurons and their projections using antibodies against DA-synthesizing enzymes (TH, AADC). However, this indirect labeling necessitates a further distinction between DA, NA, and A neurons because DA is a transitory product in NA and A synthesis.

In the European eel, brain localization of DA and TH immunoreactive (-ir) cells and fibers has been performed (see procedure details in Roberts et al., 1989; Kapsimali et al., 2000; Vidal et al., 2004; Weltzien et al., 2006). In addition to being a useful tool in cartographical studies, immunocytochemistry may be used as a semi-quantitative measure of protein expression. This may in some aspects be biologically more relevant compared to in situ hybridization in which gene expression is visualized.

Like this, it constitutes a nice support for brain CA quantitative assays by HPLC.

In situ hybridization (ISH)

Compared to immunocytochemistry, ISH labels cellular bodies that express the gene of interest, whereas cellular projections are not labeled. Because successful hybridization depends on the exact match between the gene of interest and the synthesized riboprobe, ISH specificity is higher compared to immunocytochemistry especially when using polyclonal antisera. As a result, ISH generally gives higher sensitivity. In addition, ISH can be a precious support for gene expression analyses by quantitative real-time (qrt) RT-PCR. The cloning of a cDNA for European eel TH has opened qualitative and quantitative studies on TH expression in this species (Boularand et al., 1998). Recently, steroid effects on dopaminergic cells in specific brain areas of the European eel have been clearly observed by ISH (see Weltzien et al., 2006).

Retrograde tracing

As a support to the various methods of immunolabeling, the complexity of dopaminergic system organization may be approached through neuronal tract tracing. One of the available tract-tracing methods consists of incorporation of tracers in cell bodies (anterograde tracing) or near-peripheral endings (retrograde tracing). Among available tracers, lipophilic carbocyanine dyes like 1,1′-dioctadecyl-3,3,3′,3′-tetramethylindicarbocyanine perchlorate (DiI) are widely used. In fact, they are not cytotoxic and may be used for labeling live neurons in vivo as well as in vitro (Godement et al., 1987). Such dyes provide intense and long-lasting staining, and their good resolution allows labeling of fine processes. Moreover, it is useful for multi-labeling studies because they are available in different colors. In aldehyde-fixed tissues, the crystals of DiI, inserted into nerve terminals, diffuse along the plasma membrane of cells in a retrograde direction. In the European eel, this method was applied according to previous works in goldfish (Anglade et al., 1993) to trace the origin of hypophysiotropic fibers (Weltzien et al., 2006).

Evaluation of dopaminergic activity

Assay of catecholamine content

Radioenzymatic labeling of catecholamines This technique involved radioactive labeling of amines by the transfer of a tritiated methyl group from *S*-adenosyl methionine by using the enzyme COMT. Labeled CAs were then separated by chromatography, and the radioactivity of the amine derivatives was counted. This method was used in previous studies by Le Bras (1984) to study circadian variations of CAs in various tissues in eel. However, the method was long and constraining, and therefore it is not used anymore.

Quantification by high-performance liquid chromatography (HPLC) DA content in the brain and other tissues can be assayed by high-pressure liquid chromatography (HPLC). This gives a sensitive measurement of CA content. The simultaneous measurement of CA metabolites allows an estimation of the turnover rate. This was recently performed in different brain areas and pituitary of the eel according to the method previously described by Caroff et al. (1986) and is briefly summarized below.

(a) *Sample preparation*. Brains from female silver eel were dissected into five parts (olfactory bulbs, telencephalon including the rostral preoptic area, diencephalic and mesencephalic areas, cerebellum, and medulla oblongata) (Fig. 3) prior to be frozen in liquid nitrogen and stored at −80°C until further processing.

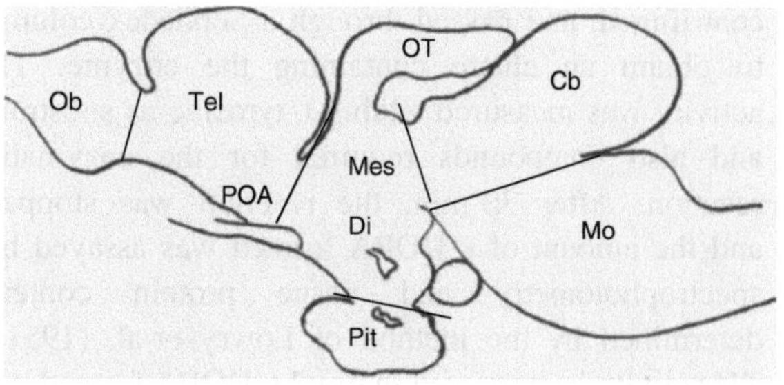

Fig. 3 Dissection of eel brain for sampling. Abbreviations: Ob, olfactory bulbs; Tel, telencephalon; POA, preoptic area; TeO, optic tectum; Di, diencephalon; Mes, mesencephalon; Cb, cerebellum; Mo, medulla oblongata; Pit, pituitary

(b) *Analysis*. Brain parts were weighed, sonicated (4°C) in 225 μl $HClO_4$ 0.1 M (with 25 μl of NAC-5-HT as internal standard), and centrifuged at 12,000 *g* for 10 min (4°C), and the supernatant was filtered (0.45 μm). The analysis was performed with an HPLC system with electrochemical detection. Composition of the mobile phase: citric acid 35 mM, Na_2HPO_4 12.5 mM, sodium octyl sulfate 0.25 mM, EDTA 0.05 mM, and 8% of acetonitrile, pH = 3.35. The injection of 20-μl filtered supernatant in the reversed-phase chromatographic column (Ultrasphere ODS 5 μm) with ion pairing gave a direct measure of the CA content in the brain area concerned by extrapolation to an external standard.

This method is a suitable tool to approach neuronal activity through the measurement of the CA metabolite/CA ratio. Thus, the activity of dopaminergic neurons in the European eel may be estimated by the calculation of DOPAC/DA ratio because DOPAC is the major DA catabolite in this species (see Section "Assay of catecholamine content").

Estimation via the study of synthesizing enzyme

Quantification of enzyme activity Because TH is the rate-limiting enzyme in the biosynthesis of CAs, calculation of TH activity may be used as a measure of CA activity. Chaube and Joy have measured TH enzymatic activity in catfish, *Heteropneustes fossilis*, in different brain parts using the method of Shiman et al. (1971) (Chaube & Joy, 2002a, b, 2003a, b). Briefly, brains were dissected, homogenized, centrifuged, and passed through a Sephadex column to obtain an eluate containing the enzyme. TH activity was measured adding L-tyrosine as substrate and also compounds required for the enzymatic reaction. After 30 min, the reaction was stopped and the amount of L-DOPA formed was assayed by spectrophotometry and tissue protein content determined by the method of Lowry et al. (1951). TH activity is expressed in nmol L-DOPA formed per mg protein per hour. This method allows a direct measure of enzyme activity and has been used to estimate the effect of specific treatments on dopaminergic activity. For instance, Chaube and Joy have studied steroids–CA interactions in female catfish, *H. fossilis* (Chaube & Joy, 2002a, b). They demonstrated that estradiol may modulate TH activity which is of particular interest because both DA and NA may inhibit and stimulate, respectively, GtHs' secretion (see Section "Characterization of dopaminergic systems in the European eel").

Enzyme inhibitors may also be used to indirectly assay the activity of an enzyme. Once the metabolic pathway is blocked, the accumulated product can be assayed. For instance, studies on TH activity may be performed using hydroxybenzyhydrazine, which blocks AADC prior to further assay of the amount of accumulated L-DOPA by HPLC, as reported for rainbow trout (Linard et al., 1996).

Quantification of enzyme gene expression Recently, a qrtRT-PCR assay to measure the gene expression of TH has been developed in the eel. In this qrtRT-PCR assay, sensitive quantification of TH mRNA is given relative to a reference gene, ARP (acidic ribosomal phosphoprotein P0). Because the procedure details have been recently published (Weltzien et al., 2005), we will not discuss this here. As regulation of TH expression is a means to answer an enhanced and sustained neuronal activity, this method is a precious tool to approach DA modulation after special treatments.

Characterization of dopaminergic systems in the European eel

Cartography

Eel is a suitable model to study dopaminergic system organization principally for two reasons. First, eel is a support for phylogenetical studies because, as a member of the Elopomorpha, an ancient teleostean group (Lauder & Liem, 1983), it may provide knowledge on ancestral brain organization in vertebrates. Second, its brain presents well-separated areas which are easier to dissect relative to that of many other fishes.

Several early studies have been performed on the European eel using the FIF method (Lefranc et al., 1969, 1970; L'Hermitte & Lefranc, 1972; Fremberg & van Veen, 1977). Later, these histofluorescence data were elaborated through the distribution of DA- (Roberts et al., 1989) and TH-immunoreactive nuclei (Kapsimali et al., 2000). Moreover, following the cloning of TH in this species, ISH has been performed to characterize the brain distribution of this enzyme (Weltzien et al., 2006; Dufour S. et al., unpublished data). Here, we summarize available literature data on the distribution of dopaminergic cells (Fig. 4) in this species (for details see: Roberts et al., 1989; Kapsimali et al., 2000). The nomenclature used follows the brain atlas of the Japanese eel (Mukuda & Ando, 2003).

Forebrain: eels have well-developed *olfactory bulbs* (Ob), and numerous dopaminergic cells with short processes are located in the external cellular layer, while a few cells with longer processes can be found in the internal glomerular layer. In the *telencephalon* (Tel), a high number of DA cells are found in the ventral area (Va), while fewer cells are found along the anteroposterior axis. These nuclei form a distinctive mediolateral band across each hemisphere from the rostral to the caudal telencephalon until the anterior commissure. This organization is a common feature among vertebrates.

Diencephalon: in the *preoptic area* (POA), DA cells were found in the posterior parvocellular preoptic nucleus (PPp) and in the suprachiasmatic nucleus (SC). Moreover, like in goldfish (Kah et al., 1987), neurons have been observed in the antero-ventral preoptic nucleus (NPOav) and projecting to the proximal pars distalis of the pituitary (Vidal et al., 2004; Weltzien et al., 2006). NPOav is involved in the control of reproduction (see Paragraph 4). In the *thalamus* few TH-ir cells are present in the ventromedial nucleus (Kapsimali et al., 2000) although Roberts et al. (1989) did not report DA-ir cells in this region. In the *pretectum* neither DA- nor TH-ir nuclei have been found. Nevertheless, dopaminergic cells have been reported in the pretectum of other teleosts (Meek, 1994) as well as in reptiles and birds. In the *periventricular posterior tuberculum* (TPp), strong DA- and TH-immunoreactivity was found. In the *hypothalamus*, DA cells formed two groups: one located in the dorsal periventricular hypothalamic nucleus and another associated with the paraventricular organ (PVO). These last neurons line the ventricular walls and extend short, club-like processes to the ventricle. Such cells are DA-ir but not TH labeled, and they are thought to accumulate CAs (presumably DA) from the CSF of the third ventricle. The PVO is considered as a primitive feature (for review: Smeets and Gonzalez, 1990) common to teleost fishes (goldfish: Hornby et al., 1987; zebrafish: Rink & Wulliman, 2001; various species: Bradford & Northcutt, 1983) and other anamniotes (*Rana nigromaculata*, *R. catesbeiana*: Nakai et al., 1977).

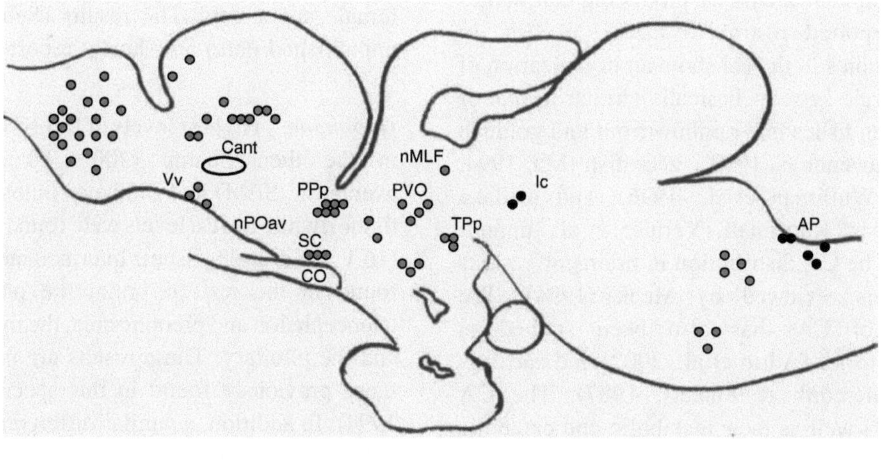

Fig. 4 Schematic organization of dopaminergic systems in the European eel brain. *Abbreviations*: AP, area postrema; Cant, anterior commissure; CO, optic chiasm; lc, locus coeruleus; nMLF, nucleus of medial longitudinal fascicle; Ob, olfactory bulbs; PPp, posterior parvocellular preoptic nucleus; PVO, paraventricular organ; Vv, ventral nucleus of the ventral telencephalic area; SC, suprachiasmatic nucleus; TPp: periventricular posterior tuberculum. Gray points: dopaminergic neurons (DA-ir/TH-ir); dark points: hypothetic noradrenergic neurons

Mesencephalon: in the eel both TH- and DA-ir cells were found in the dorsal tegmentum. This may be associated with the nucleus of the medial longitudinal fascicle (nMLF; Roberts et al., 1989; Kapsimali et al., 2000). In contrast, a dopaminergic group is usually described in the ventral mesencephalon of mammals, but its presence is submitted to great variation among vertebrates and even within the same vertebrate class. In this way, studies have not revealed TH- and DA-ir cells in the ventral mesencephalon either in the eel (Roberts et al., 1989; Kapsimali et al., 2000) or in other teleosts (Smeets & Gonzalez, 2000; Rink & Wullimann, 2001). In contrast, this ventral mesencephalic group was evident in the lungfish *Protopterus* (Sarcopterygian) (Reiner & Northcutt, 1987).

Rhombencephalon: DA-ir cells were found rostrally in the *locus coeruleus* (lc), whereas four other groups were defined caudally. Numerous cells are found in the *area postrema* (AP) at the border between the hindbrain and the spinal cord. Although few immunocytochemical studies have been made for NA in fish (e.g. three-spined stickleback, *Gasterosteus aculeatus*: Ekström et al., 1986), it is generally accepted that the neurons in the lc contain mostly NA, whereas the more caudal groups may also contain A, DA, or L-DOPA as end product. This view is supported by immunocytochemical localization of enzymes involved in NA synthesis (zebrafish, *Danio rerio*: Kaslin & Panula, 2001; *Apteronotus leptorhynchus*: Sas et al., 1990). No NA-producing cell bodies have been reported rostral to the lc in fish. In conclusion, studies in the eel show an organization of the dopaminergic systems basically similar to that of other teleostean fishes like rainbow trout and goldfish (Meek & Nieuwenhuys, 1998.), zebrafish (Ma, 1994, 1997, 2003; Wulliman et al., 1996), and medaka (*Oryzias latipes*; Kapsimali, Vernier et al., unpublished data). The CA distribution in brains of various bony fishes is reviewed by Meek (1994). The organization of CAs has also been studied in chondrostean fishes (Adrio et al., 2002) and cartilaginous fish (Meredith & Smeets, 1987). The CA organization as well as their metabolic and catabolic pathways seems highly conserved in vertebrates. An overview on the catecholaminergic system, including its organization and functional properties among vertebrates, has been recently published (Smeets & Gonzales, 2000).

Dopaminergic activity

The CA contents in the European eel have been investigated by various methods. Seasonal and circadian variations in the CA amount have been first studied using the FIF method (Popek, 1983) or radioimmunoenzymatic labeling (Le Bras, 1984). These authors reported that the CA content was generally higher during the light phase compared to the dark phase (Popek, 1983; Le Bras, 1984). In addition, modifications of acclimatized temperature (and season) induced some variations in CA content, as increases of DA and A and a decrease of NA have been shown when the water temperature increases (15–25°C) (Sébert et al., 1984).

Catecholamine content in eel brain (HPLC assay)

HPLC measurements of DA and DOPAC in the eel indicated variations at different stages of the biological cycle: DA and DOPAC contents in the olfactory bulbs were significantly higher in silver eels (migratory stage) compared to yellow eels (growth stage). However, the DOPAC/DA ratio, reflecting DA turnover, did not change (Giorgi et al., 1994).

We have recently assayed by HPLC various CAs (DA, NA, A) and DA metabolites (DOPAC, HVA) (for details, see Section "Methodological approaches of dopaminergic system investigations") in different brain regions as well as in the pituitary (Fig. 3) of female silver eels. The results (Sebert M. E. et al., unpublished data) are shortly reported below.

Dopamine Highest levels of DA (Fig. 5a) were found in the diencephalon (200 $\pm$ 10 ng/g wet tissue; average $\pm$ SEM) and olfactory bulbs (155 $\pm$ 33 ng/g tissue), while lowest levels were found in the cerebellum (16.1 $\pm$ 3.6 ng/g tissue). Intermediate DA levels were found in the rest of brain: the part including the telencephalon and preoptic area, the medulla oblongata, and the pituitary. These results are in agreement with those previously found in this species (Giorgi et al., 1994). In addition, a similar differential distribution of DA content through the brain is observed in other teleost species. Thus, generally highest DA contents are found in the telencephalon (including olfactory bulbs) and the diencephalon (crucian carp: Nilsson, 1989; goldfish: Dulka et al., 1992).

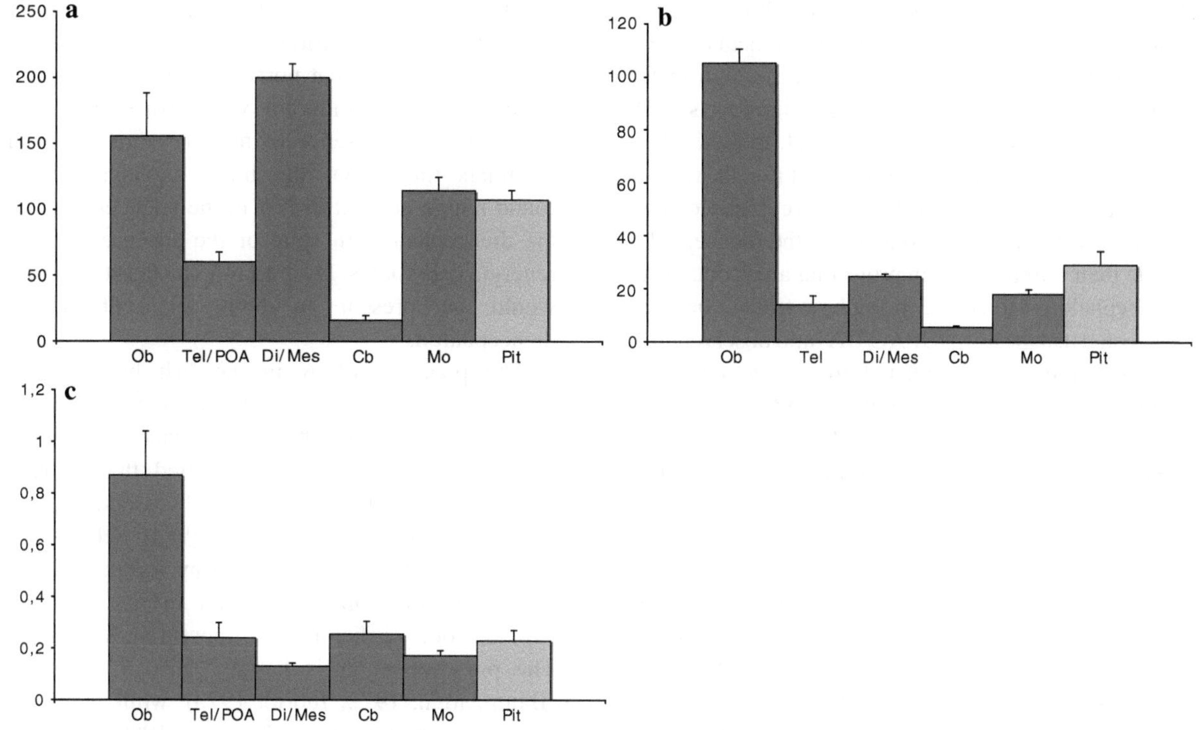

Fig. 5 Dopamine and its metabolite DOPAC contents in the brain and pituitary of the European eel assayed by HPLC. (**a**) Dopamine content; (**b**) DOPAC content; (**c**) DOPAC/dopamine ratio as indicator for dopamine turnover. Brain parts of female silver eels were dissected as shown in Fig. 3. Dopamine and DOPAC concentrations are expressed as ng/g brain tissue. Means are given ± S.E.M. ($n = 8$ eels). *Abbreviations*: Ob, olfactory bulbs; Tel/POA, telencephalon/preoptic area; Di/Mes, diencephalon/mesencephalon; Cb, cerebellum; Mo, medulla oblongata; Pit, pituitary

The high DA content in the olfactory bulbs may be illustrated by immunocytochemical data as shown using antiserum against DA (Roberts et al., 1989). The olfactory bulbs contain numerous cells that are highly TH- and DA-immunoreactive. A majority of these cells are interneurons with DA being synthesized and used locally which is reflected by the high DA contents.

In the diencephalon, elevated DA contents are partly explained by the richness of TH- and DA-ir projections that cross the diencephalon coming from the forebrain and the hindbrain. Moreover, the CSF-contacting cells located in the PVO in the hypothalamus are DA-ir but not TH-ir (Roberts et al., 1989; Kapsimali et al., 2000) indicating that the diencephalon contains cells that take up and concentrate DA from the CSF.

Dopamine metabolites In this study DOPAC and HVA, both DA metabolites, have been assayed, whereas previous works only reported DOPAC content. Our study demonstrates for the first time that DOPAC is the major metabolite of DA in all brain parts of the eel. Large concentrations of DOPAC (Fig. 5b) were found in the olfactory bulbs (105.3 ± 5.4 ng/g tissue). In the other brain areas, DOPAC contents were low with almost undetectable levels in the cerebellum (<10 ng/g tissue). In contrast, HVA was only detectable in the olfactory bulbs (70.5 ± 28.5 ng/g tissue, data not shown).

The DOPAC/DA ratio (Fig. 5c), often used as an estimate of dopaminergic activity (Cooper et al., 1986), differed between brain areas in accordance with previous studies (Giorgi et al., 1994). The DOPAC/DA ratio was four-fold higher in the olfactory bulbs (0.87) compared to other brain areas and the pituitary. This difference would be even higher if the olfactory bulb HVA content was taken into consideration to approximate the DA turnover (HVA + DOPAC/DA). As shown by immunocytochemical data, the olfactory bulbs contain numerous DA cells with short processes, with HVA and DOPAC being produced in an abundant and local manner, in agreement with the high metabolites/DA ratio. In the diencephalon, high levels of DA but low levels of DOPAC were found. In fact, a large part of the diencephalic DA content is due to the richness of dopaminergic fibers in this brain region. However, a majority of these fibers only cross the diencephalon, while their terminals are in projecting areas outside the diencephalon. Moreover, in rainbow trout it has been suggested that the low DOPAC/DA ratio observed in the diencephalon may be related to a low activity of diencephalic neurons themselves. Indeed, treatment with a TH inhibitor did not significantly decrease DA levels in the trout diencephalon (Linard et al., 1996).

Noradrenaline and adrenaline Noradrenaline was found in different brain regions of the eel (Fig. 6), notably in the telencephalon and the diencephalon. In rainbow trout, the highest NA levels were also found in the telencephalon, the POA, and the hypothalamus (Linard et al., 1996). Few NA immunocytochemical studies have been done in fish (three-spined stickleback: Ekström et al., 1986; elephantnose, *Gnathonemus-petersii*: Meek et al., 1993). One can also "indirectly" localize NA-ir neurons by immunocytochemistry using antibodies against DBH. In this way cells that are DBH-ir cells may contain A or NA as end product. If CA assay has revealed low or undetectable A contents these DBH-ir cells may be considered as noradrenergic. This indirect approach has been used in several species (*Apteronotus leptorhynchus*: Sas et al., 1990; goldfish: Hornby et al., 1990; sturgeons, *Acipenser baeri, Huso huso*: Adrio et al., 2002).

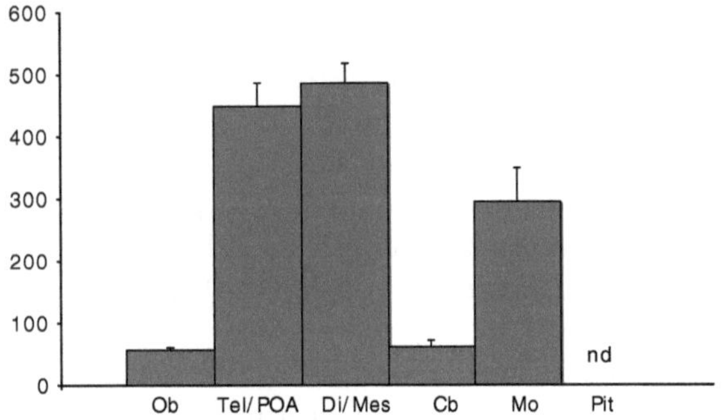

Fig. 6 Noradrenaline content in the brain and pituitary of the European eel brain assayed by HPLC. Brain parts of female silver eels were dissected as shown in Fig. 3. *Abbreviations*: See Fig. 5

In fish it seems that noradrenergic cell bodies are confined to the hindbrain with projections to the forebrain. The presence of noradrenergic fibers and terminals throughout the brain explains why we found a high content of NA in the telencephalon and the diencephalon. In spite of the absence of noradrenergic cell bodies, DA and NA as neurotransmitters would be present in both telencephalon and diencephalon.

The presence of A in the fish brain remains unclear. In our study, A levels were very low or undetectable (<10 ng/g tissue, data not shown). However, previous studies reported that A represented approximately 10% of total CA content in the European eel brain (Le Bras, 1984; Sébert et al., 1984, 1986). The fact that our assay was performed in smaller samples-brain dissected in five parts-may explain our difficulty to measure A contents. The presence of A is largely debated in goldfish (Baumgarten, 1972; Juorio, 1973), while A was not found in crucian carp (Nilsson, 1989, 1990). In addition to these divergent results, mapping of A-ir neurons has not been performed in fish because antibodies against A are not available yet.

Tyrosine hydroxylase expression assayed by quantitative real-time RT-PCR

Recently, the regional TH mRNA expression in the eel brain (Fig. 7) has been described using a newly developed quantitative real-time RT-PCR assay (Weltzien et al., 2005). TH mRNA expression was considerably higher (×4) in the olfactory bulbs relative to the other brain regions. This result is in good agreement with the cartography of TH-positive cell bodies by in situ hybridization and immunocytochemistry (see Section "Cartography of dopaminergic systems"). Specifically, numerous TH-labeled cells are found in the olfactory bulbs (Weltzien et al., 2006). This high expression of TH, the rate-limiting enzyme of DA synthesis, also concords with HPLC measurements of DA and its metabolites (Fig. 5). Indeed, high levels of DA as well as DOPAC in the olfactory bulbs suggest an intensive dopaminergic activity in this region. In the other brain areas, as discussed above, fewer DA cell bodies are present relative to those in the

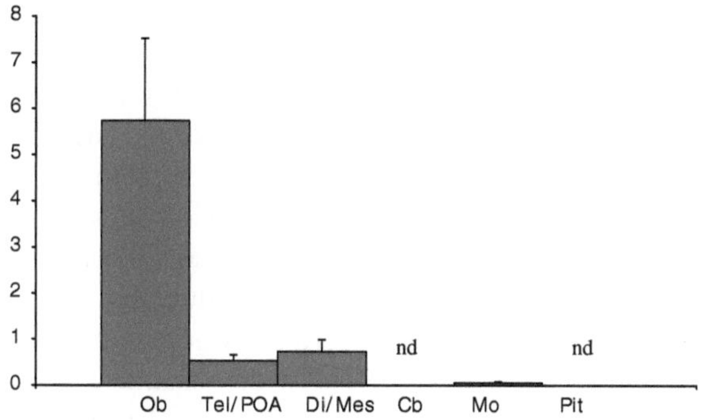

Fig. 7 Distribution of TH messenger RNA in the brain of European eel. TH messenger RNA levels were assayed by qrtRT-PCR and expressed relative to messenger RNA levels of a reference gene: ARP (acidic ribosomal phosphoprotein P0). Brain parts of female silver eels were dissected as shown in Fig. 3. Results are given ± S.E.M. ($n = 7$ eels). *Abbreviations*: See Fig. 5

olfactory bulbs. This explains the observed lower expression of TH in these areas.

Dopamine involvement in reproductive function

Neurohormonal control of reproduction in teleosts

In mammals, the reproductive function is under a complex neurohormonal control ensured by the brain-pituitary-gonad (BPG) axis. Gonadotropin-releasing hormone (GnRH), a hypothalamic decapeptide, stimulates the gonadotropic cells located in the proximal pars distalis (PPD) of the pituitary. The gonadotropes synthesize and release two gonadotropins (GtHs): luteinizing hormone (LH) and follicle-stimulating hormone (FSH) (Weiner et al., 1989). GtHs released in the general circulation act on the gonads and control the different steps of gonadal development (Fig. 8).

In fishes, GnRH plays a similar stimulatory role together with other factors like γ-amino butyric acid (Trudeau et al., 1993), neuropeptide Y (Peng et al., 1993), NA (Chang & Peter, 1983a, b), serotonin (Somoza & Peter, 1991), and glutamate (Trudeau et al., 2000). Unlike tetrapods, fishes are devoid of a hypophysial portal system. Instead, GnRH reaches the endocrine cells of the pituitary directly through hypothalamic nerve fibers (Ball, 1981) where GnRH binds to specific high-affinity receptors (for review: Kah et al., 2004). Moreover, in teleosts, GtHs may be submitted directly to a negative control by DA (Fig. 8; for review: Dufour et al., 2005) as first evidenced in goldfish (Peter et al., 1988). In maturing fishes, DA is involved in the control of spermiation, ovulation, and spawning (for review: Dufour et al., 2005). This role of DA does not seem linked with phylogenetical position within teleosts as it is observed in some species (salmonids, rainbow trout: Saligaut et al., 1999; cyprinids, carp: Lin et al., 1988; perciform, tilapia: Yaron et al., 2003), but not in others (seabream: Zohar et al., 1995). Moreover, when it is observed, the intensity of DA inhibition can be more or less pronounced according to species (for review: Dufour et al., 2005).

DA implicated in the control of pituitary function is secreted by neurons located in the NPOav in the preoptic area (Peter & McKeown, 1974; Vidal et al., 2004). These neurons project directly to the pituitary PPD (Kah et al., 1984, 1987; Anglade et al., 1993) where they release DA that acts on gonadotropes through D2-like receptors (Chang & Peter, 1983a, b; Chang et al., 1984; Omeljaniuk et al., 1987). The DA action on the gonadotropes may be direct or indirect. Indeed, DA inhibits basal and GnRH-induced LH release and acts on the GnRH intracellular signaling pathway (Peter et al., 1986, 1991; De Leeuw et al., 1986; Yu and Peter, 1992, Yaron et al., 2003). Moreover, DA may modulate pituitary sensitivity to GnRH by decreasing the number of GnRH-receptors in goldfish, catfish, and tilapia (De Leeuw et al., 1986, 1988; Omeljaniuk et al., 1989; Levavi-Sivan et al., 2004).

Dopamine involvement in the control of eel reproduction

In contrast to the DA inhibitory control of final maturation in teleosts, relatively few studies have addressed the possible role of DA in the inhibitory control of puberty. The few studies in juveniles showed no evidence for DA involvement in puberty. For example, administration of GnRH agonist and steroids was sufficient to induce precocious sexual maturation in rainbow trout (Crim & Evans, 1983). Treatment with a GnRH agonist alone or in combination with a DA antagonist led to the same conclusion in two percomorphs (striped bass, *Morone saxatilis:* Holland et al., 1998; seabream, *Pagrus major*: Kumakura et al., 2003).

In contrast, preliminary data from Marcano and coworkers indicated that dopaminergic activity (DOPAC/DA ratio) decreases in hypothalamus at puberty in the Atlantic spadefish (*Chaetodipterus faber*; Marcano et al., 1995). Although this work apparently has never been followed up, the results suggest that decreased DA release was a cause of increased GtH release at the onset of puberty in female spadefish.

The involvement of DA in the control of puberty has been studied extensively in our laboratory using the eel as a model. In silver eels, only a combined treatment with GnRH agonist, DA antagonist (pimozide) and also pretreatment with sex steroids induces a significant increase of both LH synthesis and release as well as increased plasma vitellogenin levels (Dufour et al., 1988; Vidal et al., 2004). These results explain that the lack of pubertal development in the non-migratory silver eel has a dual origin: a deficit of GnRH stimulation and a strong DA inhibition (for review: Dufour et al., 2003). When and how the DA inhibition decreases to permit gonadal recrudescence remains enigmatic. In fact, maturing eels are not available in nature because they leave continents at the prepubertal silver stage and have never been caught after this stage. Improved knowledge of the regulation of dopaminergic activity, especially in relation to pubertal development, is of crucial importance, both as a fundamental biological question and also for development of eel aquaculture. This latter aspect has a dual objective: human food production and restocking or reducing the pressure on wild eel populations (Stone, 2003; Wirth & Bernatchez, 2003).

Regulation of dopamine inhibition of eel puberty and perspective works

Internal factors

Among internal factors that are likely to regulate dopaminergic activity, sexual steroids require special interest. Sexual steroids have been shown to exert both negative and positive feedbacks on the BPG axis, e.g. acting on GnRH and GtH synthesis and release (Montero & Dufour, 1996; Schmitz et al., 2005). Recently, their possible effects on the dopaminergic component of the BPG axis have been studied. Using ISH and qrtRT-PCR, an androgen-specific-positive effect on TH expression in the preoptic area in female silver eels was revealed (Weltzien et al., 2006). This suggests that sexual steroids could be involved in modulating the DA inhibitory tone during pubertal development, similar to what has been previously demonstrated during final maturation (female rainbow trout: Linard et al., 1995; Saligaut et al., 1999; Vacher et al., 2002; male goldfish: Dulka et al., 1992). This androgen specificity has not been described in other fish and may be related to species or reproductive stage (Weltzien et al., 2006).

Because other hormones like corticosteroids (Huang et al., 1999) and thyroid hormones (Volkoff, 1999; Pavlidis et al., 2000) are thought to interfere with the reproductive function, it would be interesting to investigate their effects on DA activity in the eel.

Environmental factors

As sexually mature eels have never been observed in natural conditions, it is supposed that silver eels undergo maturation during their reproductive migration and/or at their breeding area. Physiological (Kleckner, 1980; Sébert, 2003) and anatomical (Pankhrust, 1982; Braekevelt, 1988) predispositions as well as tracking studies (Tesch & Rohlf, 2003) indicate that, after leaving coastal waters, silver eels swim at great depth. Among environmental factors linked to a migration at depth, high hydrostatic pressure may be determinant for stimulation of sexual maturation and reproduction. Previous experiments with eels immersed in cages at specific depths showed a significant increase in pituitary LH content and gonadosomatic index (Fontaine et al., 1985). We are currently performing short- and long-term immersion of female and male silver eels, using the hyperbaric chamber with a water-recirculating system set up by Sébert and coworkers at UBO (Brest, France; Sébert et al., 1990). The effects of immersion on dopaminergic activity and other components of the BPG axis will be examined.

The dopamine involvement in the reproductive migration

Dopamine and locomotion

Silver eels are not sexually mature when they leave the continental shelf to reach their breeding area. The European eel swim up to 6,000 km before arriving in the Sargasso Sea, their supposed spawning area (Schmidt, 1923). Even though eels stop feeding at silvering, they demonstrate a striking swimming efficiency that has been recently reported to be 4–6 times higher than for other non-anguilliform fish (Van Ginneken, et al., 2005). The physiological mechanisms, including central regulations ensuring such swimming capacity, remain unknown and require attention.

In mammals, the locomotor function is well studied, and it is established that the neurotransmitters DA, NA, and serotonin are involved in this function. The richness of dopaminergic innervations in the mammalian midbrain was investigated several decades ago (McGeer et al., 1978). The importance of this region for coordination of locomotion was later demonstrated (Björklund & Lindvall, 1984) due to an involvement in various human pathological processes, including Parkinson's disease (Mason, 1984). Central in this respect is the mesostriatal sensorimotor system that consists of DA cells located in the substantia nigra (SN), projecting to the striatum (basal telencephalon). It has also been shown that DA exerts its role in locomotion through D1 and D2 receptors, both located in the striatum.

Recently, Rink and Wulliman have shown in zebrafish that dopaminergic neurons in the periventricular nucleus of the posterior tuberculum (TPp), homologous to the mammalian SN, ascend to the dorsal nucleus of area ventralis in the basal telencephalon (Rink & Wulliman, 2001). These authors identified the area ventralis as the striatum. These data on the similarity in location and distribution of the dopaminergic cells in fishes with those of mammals suggested a basic feature of brain organization for the control of locomotor function.

In eels, immunocytochemical studies have shown TH-ir cells in the TPp (Kapsimali et al., 2000). Furthermore, high expression of D1 receptors has been shown in the dorsal and the ventral nuclei of the area ventralis in the telencephalon (Kapsimali et al., 2000). However, whether neurons in the TPp project effectively to the telencephalon must be confirmed in the eel.

Various experimental studies support the anatomical data to confirm the involvement of DA in locomotor function in fish as in mammals. For example, L-DOPA, a DA precursor (Fig. 1), enhances tail beat frequency and swimming speed in the European eel (Doyle & Roberts, 2004). L-DOPA also induced dominance in the Arctic charr (Salmonid), while high doses led to dyskinesia (Winberg and Nilsson, 1992). Administration of dopaminergic drugs like apomorphine, a DA agonist, stimulates swimming in *Oreochromis niloticus* and *O. Mossambicus* (Cichlids), and this effect is abolished by D1-antagonist (Mok & Munro, 1998). Apomorphine has been reported to modulate swimming and behavior in lampreys (Kemnitz et al., 1995).

Besides this direct implication of DA in locomotion, experiments show that DA may mediate the effect of various endocrine factors. In juvenile rainbow trout, the administration of GH increased swimming activity (Jönsson et al., 2003; Johansson et al., 2004) and dopaminergic activity (DOPAC/DA ratio) in both telencephalon and hypothalamus (Johansson et al., 2004). Moreover, the GH-stimulatory effects on swimming were abolished when fishes were treated with GH and D1 dopamine antagonist (Johansson et al., 2005), leading to the conclusion that DA is implicated in GH-induced swimming in trout.

Thyroid hormones are proposed to play a role in migratory behavior in salmonids (for review: Iwata, 1995, Specker et al., 2000) and cod (Comeau et al., 2000), and also in birds (Pathak & Chandola, 1984; Pant & Chandola-Saklani, 1993). Moreover, enhanced locomotor activity has been correlated with elevated thyroid hormone levels in American eels, *A. rostrata* (Castonguay et al., 1990), and most recently in *A. anguilla* glass eels (Edeline et al., 2005). Chaube and Joy have demonstrated in catfish *H. fossilis* that thyroid hormones modulate TH activity (Chaube & Joy, 2003b), which may suggest that DA could mediate thyroid hormone effects on migration as it may be the case for GH-induced swimming (see the above paragraph).

Further studies should aim at investigating the possible role of DA in the signaling pathways of endogenous factors (GH, thyroid hormones, and other factors such as sexual steroids, cortisol, etc.), leading to the striking swimming capacity observed in silver eels. The role of dopaminergic activity in eel swimming could be investigated through swim trials performed using the swim tunnel system developed by Van den Thillart and coworkers (University of Leiden).

Dopamine and olfaction

Regulation of DA in the olfactory bulbs needs also attention because olfaction plays a major role in fishes for feeding (for review: Hara, 1994), reproduction, and migration (for review: Doving and Stabell, 2003). Like salmonids, eels return to their birth area for breeding, but the mechanisms of this homing behavior remain to be discovered. Various experiments suggest that adult salmons use olfaction in finding their way back to their original streams (for review: Doving and Stabell, 2003). In eels, olfaction has been shown to play a critical role in orientation during their spawning migration. Westin (1990) reported that silver eels, originating from glass eels imported from France for stocking in Sweden, showed a lower swimming speed as compared to indigenous silver eels, and the imported eels also failed to find their way out of the Baltic Sea. Barbin and coworkers (1998) tracked anosmic and control silver American eels during their estuarine migration. Compared to controls, the anosmic fishes showed longer migration times, and many of them did not leave the estuary. Because other relevant environmental cues were limited (no gradients of salinity and temperature, low electric-field, etc.), these eels seemed to use mainly the olfaction for orientation (Fig. 8).

In mammals, studies have shown that DA release in the olfactory bulbs is enhanced during learning, and DA antagonists inhibit olfactory memory formation (Coopersmith et al., 1991; Hsia et al., 1999; Davila et al., 2003; Pavlis et al., 2006). Available experimental data indicate that DA may also modulate the discrimination of olfactory input (Nowycky et al., 1983; Wilson et al., 1995; Hsia et al., 1999; Ennis et al., 2001).

As for locomotion, DA may take part in an intricate neurohormonal network that controls olfaction in fish. Thyroid hormones would be involved in homing in salmon (for review: Dittman & Quinn, 1996) and affect DA turnover in the olfactory bulbs (Morin et al., 1997). Besides the involvement of olfaction in homing this sensory function is required at the last step of reproduction. The detection of pheromones in water is crucial during the breeding phase to promote spawning success (goldfish: Demski & Dulka, 1984). Miranda and coworkers have recently demonstrated that male tilapia are able to discriminate between females of different reproductive stages (Miranda et al., 2005). Moreover, data from Bhatt and coworkers (2002) in a cyprinid, *Barilius bendelisis*, support this hypothesis that during the reproductive phase, pheromones as olfactive stimuli dominate over feeding stimuli to promote reproduction.

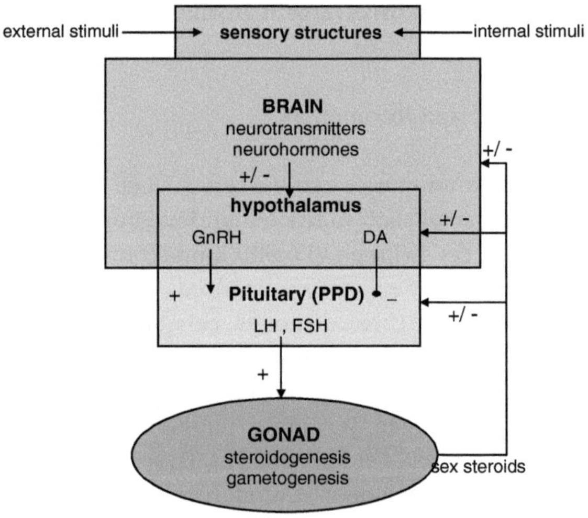

Fig. 8 Neurohormonal control of reproduction by the brain-pituitary-gonad (BPG) axis. Brain GnRH (gonadotropin-releasing hormone) stimulates the synthesis and release of pituitary gonadotropins (LH, luteinizing hormone; and FSH, follicle stimulating hormone), which activate gonadal activity (gametogenesis and steroidogenesis). An additional brain control is exerted in some teleosts by DA, which may inhibit the synthesis and release of gonadotropins. Sex steroids exert positive and negative feed at different levels of the BPG axis. Environmental and internal stimuli as sex steroids are integrated by the central nervous system and lead to modulation of the activity of GnRH and DA neurons

As silver eels are fasting, the olfactory dopaminergic system could be involved in two crucial final steps of the eel life cycle, namely, to reach the spawning area and during the actual spawning act. Our recent qrtRT-PCR studies showed a strong positive effect of androgens on TH expression in the silver eel olfactory bulbs, indicating that sex steroids may modulate DA mediation of olfactory processes in the eel (Weltzien et al., 2006). Further experimental studies, on artificially matured female and male eels, are necessary to investigate the role of DA in the mediation of pheromone effects, spawning behavior, and gamete release.

Conclusion

Dopamine has been shown to inhibit eel pubertal development and may also control the last steps of reproduction as in some other adult teleosts. In addition, DA is likely involved in locomotion and olfaction in fish as in mammals, functions implicated in reproductive migration and spawning. The recent development of various methodological tools allows investigation of dopaminergic systems in the eel. Anatomical data obtained by ISH and immunocytochemistry show that the organization of dopaminergic systems in the European eel is relatively similar to those of other fishes. Moreover, this organization is well conserved among vertebrates. Quantitative methods (HPLC, qrtRT-PCR) allow evaluation of variations in DA activity, and current studies investigate the effects of endogenous and environmental factors on these systems. A better understanding of the role and regulation of DA systems will contribute to gain basic and applied knowledge necessary to reproduce the eel life cycle in captivity and allow stock renewal.

Acknowledgments We are grateful to Dr. P. Sébert and Dr. P. Vernier for their hospitality and for their advice, and to Dr. M. Wulliman and Dr. M. Kapsimali for helpful discussions on brain neuroanatomy. This work was partly supported by the European project EELREP (Project No. Q5RS-2001-01836). M.-E. Sébert is a recipient of an MNRT Ph.D. fellowship and Dr. F.-A. Weltzien of a postdoctoral European Marie Curie Individual Fellowship (No. MCFI-2002-01609).

References

Adrio, F., R. Anadon & I. Rodriguez-Moldes, 2002. Distribution of tyrosine hydroxylase (TH) and dopamine beta-hydroxylase (DBH) immunoreactivity in the central nervous system of two chondrostean fishes (*Acispenser baeri* and *Huso huso*). The Journal of Comparative Neurology 448: 280–297.

Anglade, I., T. Zandbergen & O. Kah, 1993. Origin of the pituitary innervation in the goldfish. Cell and Tissue Research 273: 345–355.

Ball, J. N., 1981. Hypothalamic control of the pars distalis in fishes, amphibians and reptiles. General and Comparative Endocrinology 44: 135–170.

Barbin, G. P., S. J. Parker & J. D. McCleave, 1998. Olfactory clues play a critical role in the estuarine migration of silver-phase American eels. Environmental Biology of Fishes 53: 283–291.

Baumgarten, H. G., 1972. Biogenic monoamines in the cyclostome and lower vertebrate brain. Progress in Histochemistry and Cytochemistry 4: 1–90.

Bhatt, J. P., J. S. Kandwall & R. Nautiyal, 2002. Water temperature and pH influence olfactory sensitivity to preovulatory and post-ovulatory ovarian pheromones in male *Barilius bendelisis*. Journal of Biosciences 27: 273–281.

Björklund, A. & O. Lindvall, 1984. Dopamine-containing systems in the CNS. In Björklund, A. & T. Hökfelt (eds), Handbook of Chemical Neuroanatomy, Vol. 2: Classical Transmitters in the CNS, Part 1. Elsevier, Amsterdam: 55–122.

Blaschko, H., 1939. The specific action of L-DOPA decarboxylase. Journal of Physiology (London) 96: 50–51.

Boularand, S., N. F. Biguet, B. Vidal, M. Veron, J. Mallet, J. D. Vincent, S. Dufour & P. Vernier, 1998. Tyrosine hydroxylase in the European eel (*Anguilla anguilla*): cDNA cloning, brain distribution, and phylogenetic analysis. Journal of Neurochemistry 71: 460–470.

Bradford, M. R. Jr. & R. G. Northcutt, 1983. Organization of the diencephalon and pretectum of the ray-finned fishes. In Davis, R. E. & R. G. Northcutt (eds), Fish Neurobiology, Vol. 2: Higher Brain Areas and Functions. The University of Michigan Press, Ann Arbor: 17–163.

Braekevelt, C. R., 1988. Retinal fine structure in the European eel, *Anguilla anguilla*. VI. Photoreceptors of the sexually immature silver eel stage. Anatomischer Anzeiger 166: 23–31.

Callier, S., M. Snapyan, S. Le Crom, D. Prou, J.-D. Vincent & P. Vernier, 2003. Evolution and cell biology of dopamine receptors in vertebrates. Biology of the Cell 95: 489–502.

Caroff, J., L. Barthélémy & P. Sébert, 1986. Brain and plasma biogenic amines analysis by the EC-HPLC technique: application to fish. Comparative Biochemistry and Physiology, Part C 84: 151–153.

Castonguay, M., J. D. Dutil, C. Audet & R. Miller, 1990. Locomotor activity and concentration of thyroid hormones in migratory and sedentary juvenile American eels. Transactions of the American Fisheries Society 119: 946–956.

Chang, J. P. & R. E. Peter, 1983a. Effects of dopamine on gonadotropin release in female goldfish, *Carassius auratus*. Neuroendocrinology 36: 351–35.

Chang J. P. & R. E. Peter, 1983b. Influences of norepinephrine and alpha-adrenergic mechanisms on gonadotropin secretion in female goldfish, *Carassius auratus*. General and Comparative Endocrinology 55: 89–95.

Chang, J. P., R. E. Peter, C. S. Nahorniak & M. Sokolowska, 1984. Effects of catecholaminergic agonists and antagonists on serum gonadotropin concentrations and ovulation in goldfish: evidence for specificity of dopamine inhibition of gonadotropin secretion. General and Comparative Endocrinology 55: 351–360.

Chaube, R. & K. P. Joy, 2002a. Effects of altered photoperiod and temperature, serotonin-affecting drugs, and melatonin on brain tyrosine hydroxylase activity in female catfish, *Heteropneustes fossilis*: a study correlating ovarian activity changes. Journal of Experimental Zoology 293: 585–593.

Chaube, R. & K. P. Joy, 2002b. Effects of ovariectomy and oestradiol-17β replacement on brain tyrosine hydroxylase in the catfish *Heteropneustes fossilis*: changes in in vivo activity and kinetic parameters. Journal of Endocrinology 175: 329–342.

Chaube, R. & K. P. Joy, 2003a. Brain tyrosine hydroxylase in the catfish *Heteropneustes fossilis*: annual and circadian variations, and sex and regional differences in enzyme activity and some kinetic properties. General and Comparative Endocrinology 130: 29–40.

Chaube, R. & K. P. Joy, 2003b. Thyroid hormone modulation of brain in vivo tyrosine hydroxylase activity and kinetics in the female catfish *Heteropneustes fossilis*. Journal of Endocrinology 179: 205–215.

Chen, N. H., M. E. Reith & M. W. Quick, 2004. Synaptic uptake and beyond: the sodium and chloride-dependent neurotransmitter transporter family SLC6. Pflugers Archives 447: 519–531.

Comeau, L. A., S. E. Campana, J. M. Hanson & G. A. Chouinard, 2000. Seasonal changes of thyroid hormones in field-collected Atlantic cod in relation to condition indices, water temperature and photoperiod. Journal of Fish Biology 57: 571–588.

Cooper, J. R., F. E. Bloom & R. H. Roth, 1986. The Biochemical Basis of Neuropharmacology. Oxford University Press, New York.

Coopersmith, R., F. B. Weihmuller, C. L. Kirstein, J. F. Marshall & M. Leon, 1991. Extracellular dopamine increases in the neonatal olfactory bulb during odor preference training. Brain Research 564: 149–153.

Crim, L. W. & D. M. Evans, 1983. Influence of testosterone and/or luteinizing-hormone releasing hormone analogue on precocious sexual development in the juvenile rainbow trout. Biology of Reproduction 29: 137–142.

Dahlström, A. & K. Fuxe, 1964. Evidence for the existence of monoamine containing neurons in the central nervous system. Acta Physiologica Scandinavica 62 Supplement 232: 1–55.

Davila, N. G., L. J. Blakemore & P. Q. Trombley, 2003. Dopamine modulates synaptic transmission between rat olfactory bulb neurons in culture. Journal of Neurophysiology 90: 395–404.

De Leeuw, R., H. J. Th. Goos & P. G. W. J. Van Oordt, 1986. The dopaminergic inhibition of the gonadotropin-releasing hormone-induced gonadotropin release: an in vitro study with fragments and cell suspensions from pituitaries of the African catfish, *Clarias gariepinus* (Burchell). General and Comparative Endocrinology 63: 171–177.

De Leeuw, R., C. Van'tVeer, H. J. Th. Goos & P. G. W. J. Van Oordt, 1988. The dopaminergic regulation of gonadotropin-releasing hormone receptor binding in the pituitary of the African catfish, *Clarias gariepinus*. General and Comparative Endocrinology 72: 408–415.

Demarest, K. T. & K. E. Moore, 1980. Accumulation of L-DOPA in the median eminence: an index of tuberoinfundibular dopaminergic nerve activity. Endocrinology 106: 463–468.

Demski, L. S. & J. G. Dulka, 1984. Functional-Anatomical studies on sperm release evoked by electrical stimulation of the olfactory tract in goldfish. Brain Research 291: 241–247.

Dittman, A. & T. Quinn, 1996. Homing in Pacific salmon: mechanisms and ecological basis. The Journal of Experimental Biology 199: 83–91.

Døving, K. B. & O. B. Stabell, 2003. Trails in open waters: sensory cues in salmon migration. In Collin, S. P. & N. J. Marshall (eds), Sensory Processing in Aquatic environments. Springer-Verlag, New york: 39–52.

Doyle, L. M. F. & B. L. Roberts, 2004. Functional recovery and axonal growth following spinal cord transection is accelerated by sustained L-DOPA administration. European Journal of Neuroscience 20: 2008–2014.

Dufour, S., E. Lopez, F. Le Menn, N. Le Belle, S. Baloche & Y. A. Fontaine, 1988. Stimulation of gonadotropin release and of ovarian development, by the administration of a gonadoliberin agonist and of dopamine antagonists, in female silver eel pretreated with estradiol. General and Comparative Endocrinology 70: 20–30.

Dufour S., E. Burzawa-Gérard, N. Le Belle, M. Sbaihi & B. Vidal, 2003. Reproductive endocrinology of the European eel, *Anguilla anguilla*. In Aida, K., K. Tsukamoto & K. Yamauchi (eds), Eel Biology. Springer Verlag, Tokyo, 373–383.

Dufour, S., F.-A. Weltzien, M.-E. Sébert, N. Le Belle, B. Vidal, P. Vernier & C. Pasqualini, 2005. Dopaminergic inhibition of reproduction in teleost fishes: ecophysiological and evolutionary implications. Annals of New York Academy of Sciences 1040: 9–21.

Dulka, J. G., B. D. Sloley, N. E. Stacey & R. E. Peter, 1992. A reduction in pituitary dopamine turnover is associated with sex pheromone-induced gonadotropin secretion in male goldfish. General and Comparative Endocrinology 86: 496–505.

Edeline, E., A. Bardonnet, V. Bolliet, S. Dufour & P. Elie, 2005. Endocrine control of *Anguilla anguilla* glass eel dispersal: effect of thyroid hormones on locomotor activity and rheotactic behavior. Hormones and Behavior 48: 53–63.

Ehrenström, F. & P. Johansson, 1987. Circadian rhythms and contents of catechols in different brain structures, peripheral organs and plasma of the Atlantic cod, *Gadus morhua*. Comparative Biochemistry and Physiology, Part C 56: 85–88.

Ekström, P., M. Reschke, H. Steinbusch & T. van Veen, 1986. Distribution of noradrenaline in the brain of the teleost *Gasterosteus aculeatus* L.: an immunohistochemical analysis. The Journal of Comparative Neurology 254: 297–313.

Ennis, M., F. M. Zhou, K. J. Ciombor, V. Aroniadou-Anderjaska, A. Hayar, E. Borrelli, L. A. Zimmer, F. Margolis & M. T. Shipley, 2001. Dopamine D2 receptor-mediated presynaptic inhibition of olfactory nerve terminals. Journal of Neurophysiology 86: 2986–2997.

Eränkö, O., 1955. Histochemistry of the adrenal medulla in rats and mice. Endocrinology 57: 363–368.

Eränkö, O., 1967. The practical histochemical demonstration of catecholamines by formaldehyde-induced fluorescence. Journal of the Royal Microscopical Society 87: 259–276.

Falck, B., N. A. Hillarp, G. Thieme & A. Thorp, 1962. Fluorescence of catecholamines and related compounds with formaldehyde. Journal of Histochemistry and Cytochemistry 10: 348–354.

Fontaine, Y. A., S. Dufour, J. Alinat & M. fontaine, 1985. A long immersion in deep sea stimulates the pituitary gonadotropic function of the female European eel (*Anguilla anguilla* L.). Comptes Rendus de l'Academie des Sciences, Paris 300: 83–87.

Fremberg, M. & Th. van Veen, 1977. Formaldehyde-induced fluorescence in the telencephalon and diencephalon of the eel (*Anguilla anguilla* L.). Cell and Tissue Research 176: 1–22.

Giorgi, O., A. M. Deiana, S. Salvadori, D. Lecca & M. G. Corda, 1994. Developmental changes in the content of dopamine in the olfactory bulb of the European eel (*Anguilla anguilla*). Neuroscience Letters 172: 35–38.

Godement, P., J. Vanselow, S. Thanos & F. Bonhoeffer, 1987. A study in developing visual systems with a new method of staining neurones and their processes in fixed tissue. Development 101: 697–713.

Hara, T. J., 1994. Olfaction and gestation in fish: an overview. Acta Physiologia Scandinavica 152: 207–217.

Hsia, A. Y, J. D. Vincent & P. M. Lledo, 1999. Dopamine depresses synaptic inputs into the olfactory bulb. Journal of Neurophysiology 82: 1082–1085.

Holland, M. C., S. Hassin & Y. Zohar, 1998. Effects of long-term testosterone, gonadotropin-releasing hormone agonist, and pimozide treatments on gonadotropin II levels and ovarian development in juvenile female stripped bass (*Morone saxatilis*). Biology of Reproduction 59: 1153–1162.

Holtz, P., 1939. Dopadecarboxylase. Naturwissenschaften 27: 72–75.

Hornby, P. J. & D. T. Piekut, 1990. Distribution of catecholamine-synthesizing enzymes in goldfish brains: presumptive dopamine and norepinephrine neuronal organization. Brain, Behavior and Evolution 35: 49–64.

Hornby, P. J., D. T. Piekut & L. S. Demski, 1987. Localization of immunoreactive tyrosine hydroxylase in the goldfish brain. The Journal of Comparative Neurology 261: 1–14.

Huang, Y. S., K. Rousseau, M. Sbaihi, N. Le Belle, M. Schmitz & S. Dufour, 1999. Cortisol selectively stimulates pituitary gonadotropin beta-subunit in a primitive teleost, *Anguilla anguilla*. Endocrinology 140: 1228–1235.

Iwata, M., 1995. Downstream migratory behaviour of salmonids and its relationship with cortisol and thyroid hormones: a review. Aquaculture 135: 131–139.

Joh, T. H., D. H. Park & D. J. Reis, 1978. Direct phosphorylation of brain tyrosine hydroxylase by cyclic ATP-dependent protein kinase: mechanism of enzyme activation. Proceedings of the National Academy of Science of the U.S.A. 75: 4744–4748.

Johansson, V., S. Winberg, E. Jönsson, D. Hall & B. J. Björnsson, 2004. Peripherally administered growth hormone increases brain dopaminergic activity and swimming in rainbow trout. Hormones and Behavior 46: 436–443.

Johansson, V., S. Winberg & B. T. Björnsson, 2005. Growth hormone-induced stimulation of swimming and feeding behaviour of rainbow trout is abolished by the D1 dopamine antagonist SCH23390. General and Comparative Endocrinology 141: 58–65.

Jönsson, E., V. Johansson, B. J. Björnsson & S. Winberg, 2003. Central nervous system actions of growth hormone on brain monoamine levels and behavior of juvenile rainbow trout. Hormones and Behavior 43: 367–374.

Jonsson, A. C., I. Wahlqvist & T. Hansson, 1983. Effects of hypophysectomy and cortisol on the catecholamine biosynthesis and catecholamine content in chromaffin tissue from rainbow trout, *Salmo gairdneri*. General and Comparative Endocrinology 51: 278–285.

Juorio, A. V., 1973. The distribution of catecholamines in the hypothalamus and other brain areas of some lower vertebrates. Journal of Neurochemistry 20: 641–645.

Kah, O., P. Chambolle, J. Thibault & M. Geffard, 1984. Existence of dopaminergic neurons in the preoptic region of the goldfish. Neuroscience Letters 48: 293–298.

Kah, O., J. G. Dulka, P. Dubourg, J. Thibault & R. E. Peter, 1987. Neuroanatomical substrate for the inhibition of gonadotropin secretion in goldfish: Existence of a dopamine preoptic-hypophyseal pathway. Neuroendocrinology 45: 451–458.

Kah, O., C. Lethimonier & J. J. Lareyre, 2004. Gonadotrophin-releasing hormone (GnRH) in the animal kingdom. Journal de la Société de Biologie 198: 53–60.

Kapsimali, M., B. Vidal, A. Gonzalez, S. Dufour & P. Vernier, 2000. Distribution of the mRNA encoding the four dopamine D_1 receptor subtypes in the brain of the European eel (*Anguilla anguilla*): Comparative approach to the function of D_1 receptors in vertebrates. The Journal of Comparative Neurology 419: 320–343.

Kaslin, J. & P. Panula, 2001. Comparative anatomy of the histaminergic and other aminergic systems in zebrafish (*Danio rerio*). The Journal of Comparative Neurology 440: 342–377.

Kemnitz, C. P., T. R. Strauss, D. M. Hosford & J. T. Buchanan, 1995. Modulation of swimming in the lamprey, *Petromyzon marinus*, by serotonergic and dopaminergic drugs. Neuroscience Letters 201: 115–118.

Kleckner, R. C., 1980. Swim bladder volume maintenance elated to initial oceanic migratory depth in silver-phase *Anguilla rostrata*. Science 208: 1481–1482.

Kumakura, N., K. Okuzawa, K. Gen & H. Kagawa, 2003. Effects of gonadotropin-releasing hormone agonist and dopamine antagonist on hypothalamus-pituitary-gonadal axis of prepubertal female red seabream (*Pagrus major*). General and Comparative Endocrinology 131: 264–273.

Lauder, G. V. & K. F. Liem, 1983. Patterns of diversity and evolution in ray-finned fishes. In Northcutt, R. G. & R. E. Davis (eds), Fish Neurobiology, Vol. I. Brain Stem and Sense Organs. University of Michigan Press, Ann Arbor, 2–24.

Le Bras, Y. M., 1984. Circadian variations of catecholamine levels in brain, heart, and plasma in the eel, *Anguilla anguilla* L., at three different times of year. General and Comparative Endocrinology 55: 472–479.

Lefranc, G., A. L'Hermite & J. Tusques, 1969. Mise en évidence de neurones monoaminergiques par la technique de fluorescence dans l'encéphale d'Anguille. Comptes Rendu de la Société de Biologie (Paris) 163: 1193–1196.

Lefranc, G., A. L'Hermite & J. Tusques, 1970. Etude topographique et cytologique des différents noyaux monoaminergiques de l'encéphale d'*Anguilla vulgaris*.

Comptes Rendu de la Société de Biologie (Paris) 164: 1629–1632.
L'Hermite, A. & G. Lefranc, 1972. Recherches sur les voies monoaminergiques de l'encéphale *d'Anguilla vulgaris*. Archives d'Anatomie Microscopique 61: 139–152.
Levavi-Sivan, B., H. Safarian, H. Rosenfeld, A. Elizur & A. Avitan, 2004. Regulation of gonadotropin-releasing hormone (GnRH)-receptor gene expression in tilapia: effect of GnRH and dopamine. Biology of Reproduction 70: 1545–1551.
Levitt, M., S. Spector, A. Sjoerdsma & S. Udenfriend, 1965. Elucidation of the rate-limiting step in norepinephrine biosynthesis in the perfused Guinea-pig heart. The Journal of Pharmacology and Experimental Therapeutics 148: 1–8.
Lin, H. R., G. Van der Kraak, X. J. Zhou, J. Y. Liang, R. E. Peter, J. E. Rivier & W. W. Vale, 1988. Effects of [D-Arg6, Trp 7, Leu8, Pro9Net]-luteinizing hormone releasing hormone (sGnRH-a) and [D-Ala6, Pro9Net]-luteinizing hormone-releasing hormone (LHRH-a), in combination with pimozide or domperidone, on gonadotropin release and ovulation in the Chinese loach and common carp. General and Comparative Endocrinology 69: 31–40.
Linard, B., S. Bennani, P. Jego & C. Saligaut, 1996. Tyrosine hydroxylase activity and dopamine turnover of rainbow trout (*Oncorhynchus mykiss*) brain: the special status of the hypothalamus. Fish Physiology and Biochemistry 15: 41–48.
Linard, B., S. Bennani & C. Saligaut, 1995. Involvement of estradiol in a catecholamine inhibitory tone of gonadotropin release in rainbow trout (*Oncorhynchus mykiss*). General and Comparative Endocrinology 99: 192–196.
Lowry, O. H., N. J. Rosenbrough, A. L. Farr & R. J. Randall, 1951. Protein measurement with Folin-phenol reagent. Journal of Biological Chemistry 193: 265–275.
Ma, P. M., 1994. Catecholaminergic systems in the zebrafish. II. Projection pathways and pattern of termination of the locus coeruleus. The Journal of Comparative Neurology 344: 256–269.
Ma, P. M., 1997. Catecholaminergic systems in the zebrafish. III. Organization and projection pattern of medullary dopaminergic and noradrenergic neurons. The Journal of Comparative Neurology 381: 411–427.
Ma, P. M., 2003. Catecholaminergic systems in the zebrafish. IV. Organization and projection pattern of dopaminergic neurons in the diencephalon. The Journal of Comparative Neurology 460: 13–37.
Mason, S. T., 1984. Catecholamines and Behaviour. Cambridge University Press, Cambridge.
Marcano, D., H. Y. Guerrero, N. Cago, E. Cardillo, M. Requena & L. Ruiz, 1995. Monoamine metabolism in the hypothalamus of the juvenile teleost fish, *Chaetodipterus faber*. In Goetz, F. W. & P. Thomas (eds), Proceedings of the Fifth International Symposium on the Reproductive Physiology of Fish. Fish Symposium, Austin: 64–66.
Mazeaud, M. M. & F. Mazeaud, 1973. Excretion and catabolism of catecholamines in fish. II. Catabolites. Comparative and General Pharmacology 4: 209–217.
McGeer, P., J. C. Eccles & E. G. McGeer, 1978. Molecular Neurobiology of the Mammalian Brain. Plenum Press, New York.
Meredith, G. E. & W. J. Smeets, 1987. Immunocytochemical analysis of the dopamine system in the forebrain and midbrain of *Raja radiata*: evidence for a substantia nigra and ventral tegmental area in cartilaginous fish. The Journal of Comparative Neurology 265: 530–548.
Miranda, A., O. G. Almeida, P. C. Hubbard, E. N. Barata & A. V. M. Canario, 2005. Olfactory discrimination of female reproductive status by male tilapia (*Oreochromis mossambicus*). The Journal of Experimental Biology 208: 2037–2043.
Meek, J., 1994. Catecholamines in the brains of Osteichthyes (bony fishes). In Smeets W. J. A. J. & A. Reiner (eds), Phylogeny and Development of Catecholamine Systems in the CNS of Vertebrates. Cambridge University Press, Cambridge: 49–76.
Meek, J., H. W. J. Joosten & T. G. M. Hafmans, 1993. Distribution of noradrenaline-immunoreactivity in the brain of the mormyrid teleost *Gnathonemus-petersii*. Journal of Comparative Neurology 328: 145–160.
Meek, J., R. Nieuwenhuys, 1998. Holosteans and teleosts. In Nieuwenhuys, R., H. J. Ten Donkelaar & C. Nicholson (eds), The Central Nervous System of Vertebrates, Vol. 2. Springer-Verlag, Berlin: 759–937.
Mok, E. Y.-M. & A. D. Munro, 1998. Effects of dopaminergic drugs on locomotor activity in teleost fish of the genus *Oreochromis* (Ciclidae): involvement of the telencephalon. Physiology and Behavior 64: 227–234.
Montero, M. & S. Dufour, 1996. Gonadotropin-releasing hormones (GnRH) in fishes: evolutionary data on their structure, localization, regulation, and function. Zoological Studies 35: 149–160.
Morin, P.-P., S. Winberg, G. E. Nilsson, T. J. Hara & J. G. Eales, 1997. Effect of 1-thyroxine on brain monoamines during parr-smolt transformation of Atlantic salmon (*Salmo salar* L.). Neuroscience Letters 224: 216–218.
Mukuda, T. & M. Ando, 2003. Brain atlas of the Japanese eel: comparison to other fishes. Memories Faculty of Integrated Arts and Sciences, Hiroshima University, Series IV 29: 1–25.
Nakai, Y., H. Ochiai & S. Shioda, 1977. Cytological evidence for different types of cerebrospinal fluid-contacting subependymal cells in the preoptic and infundibular recesses of the frog. Cell Tissue Research 176: 317–334.
Nilsson, S., 1983. Autonomic Nerve Function in the Vertebrates. Zoophysiology, Vol. 13. Springer-Verlag, Berlin: 49–57.
Nilsson, G. E., 1989. Regional distribution of monoamines and monoamine metabolites in the brain of crucian carp, *Carassius carassius* L. Comparative Biochemistry and Physiology, Part C 94: 223–228.
Nilsson, G. E., 1990. Long term anoxia in crussian carp: changes in the levels of amino acid and monoamine neurotransmitters in the brain, catecholamines in chromaffin tissue, and liver glycogen. Journal of Experimental Biology 150: 295–320.
Nowycky, M. C., N. Halasz & G. M. Shepherd, 1983. Evoked field potential analysis of dopaminergic mechanisms in the isolated turtle olfactory bulb. Neuroscience 8: 717–722.
Omeljaniuk, R. J., H. R. Habibi & R. E. Peter, 1989. Alterations in pituitary GnRH and dopamine receptors associated with the seasonal variation and regulation of

gonadotropin release in the goldfish (*Carassius auratus*). General and Comparative Endocrinology 74: 392–399.

Omeljaniuk, R. J., S. H. Shih & R. E. Peter, 1987. In-vivo evaluation of dopamine receptor-mediated inhibition of gonadotrophin secretion from the pituitary gland of the goldfish, *Carassius auratus*. Journal of Endocrinology 114: 449–458.

Pavlidis, M., L. Greenwood, B. Mourot, C. Kokkari, F. Le Menn, P. Divanach & A. P. Scott, 2000. Seasonal variations and maturity stages in relation to differences in serum levels of gonadal steroids, vitellogenin, and thyroid hormones in the common dentex (*Dentex dentex*). General and Comparative Endocrinology 118: 14–25.

Pavlis, M., C. Feretti, A. Levy, N. Gupta & C. Linster, 2006. L-DOPA improves odor discrimination learning in rats. Physiology and Behavior 87: 109–113.

Pankhrust, N. W., 1982. Relation of visual changes to the onset of sexual maturation in the European eel *Anguilla anguilla* (L.). Journal of Fish Biology 21: 127–140.

Pant, K. & A. Chandola-Saklani, 1993. A role for thyroid hormones in the development of premigratory disposition in redheaded bunting, *Emberiza bruniceps*. Journal of Comparative Physiology, Part B 163: 389–394.

Pathak, V. K. & A. Chandola, 1984. Variations in circulating thyroxine and triiodothyronine concentration in relation to spring migration in rosy pastor, *Sturnus roseus*. Hormones and Behavior 18: 111–116.

Peng, C., J. P. Chang, K. L. Yu, A. O. Wong, F. Van Goor, R. E. Peter & J. E. Rivier, 1993. Neuropeptide-Y stimulates growth hormone and gonadotropin-II secretion in the goldfish pituitary: involvement of both presynaptic and pituitary cell actions. Endocrinology 132: 1820–1829.

Peter, R. E., J. P. Chang, C. S. Nahorniak, R. J. Omeljaniuk, M. Sokolowska, S. H. Shih & R. Billard, 1986. Interactions of catecholamines and GnRH in regulation of gonadotropin secretion in teleost fish. Recent Progress in Hormone Research 42: 513–548.

Peter, R. E., H. E. Lin & G. Van der Kraak, 1988. Induced ovulation and spawning of cultured freshwaterfish in China: advances in application of GnRH analogues and dopamine antagonists. Aquaculture 74: 1–10.

Peter, R. E. & B. A. McKeown, 1974. Effects of hypothalamic and thalamic lesions on prolactin secretion in goldfish (*Carassius auratus*). General and Comparative Endocrinology 23: 438–452.

Peter, R. E., V. Trudeau, B. D. Sloley, C. Peng & C. S. Nahorniak, 1991. Actions of catecholamines, peptides and sex steroids in regulation of gonadotropin-II in the goldfish. In Scott, A. P., J. P. Sumpter, D. E. Kime & M. S. Rolfe (eds), Reproductive Physiology of Fish. Fish Symposium, Sheffield, 30–34.

Popek, W., 1983. Seasonal variations in circadian rhythm of hypothalamic catecholamine content in the eel (*Anguilla anguilla* L.). Comparative Biochemistry and Physiology, Part C 75: 193–198.

Randall, D. J. & S. F. Perry, 1992. Catecholamines. In Hoar, W. S., D. J. Randall & A. P. Farell (eds), Fish Physiology, Vol. XII, Part B. The Cardiovascular System. Academic Press, New York: 255–275.

Reiner, A., 1994. The study of catecholaminergic perikarya and fibers in the nervous system: methodological considerations, technical limitations. In Smeets, W. & A. Reiner (eds), Phylogeny and Development of Catecholamine Systems in the CNS of Vertebrates. Cambridge University Press, London.

Reiner, A. & R. G. Northcutt, 1987. An immunohistochemical study of the telencephalon of the African lungfish *Protopterus annectens*. Journal of Comparative Neurology 256: 463–481.

Rink, E. & M. F. Wullimann, 2001. The teleostean (zebrafish) dopaminergic system ascending to the subpallium is located in the basal diencephalons (posterior tuberculum). Brain Research 889: 316–330.

Reymond, M. J. & J. C. porter, 1982. Hypothalamic secretion of dopamine after inhibition of aromatic L-amino acid decarboxylase activity. Endocrinology 111: 1051–1056.

Roberts, B. L., G. E. Meredith & S. Maslam, 1989. Immunocytochemical analysis of the dopamine system in the brain and spinal cord of the European eel, *Anguilla anguilla*. Anatomy and Embryology 180: 401–412.

Saligaut, C., B. Linard, B. Breton, I. Anglade, T. Bailhache, O. Kah & P. Jego, 1999. Brain aminergic systems in salmonids and other teleosts in relation to steroid feedback and gonadotropin release. Aquaculture 177: 13–20.

Sas, E., L. Maler & B. Tinner, 1990. Catecholaminergic systems in the brain of a gymnotiform teleost fish: an immunohistochemical study. The Journal of Comparative Neurology 292: 127–162.

Schmidt, J., 1923. Breeding places and migration of the eel. Nature 111: 51–54.

Schmitz, M., S. Aroua, B. Vidal, N. Le Belle, P. Elie & S. Dufour, 2005. Differential feedback by sexual steroids is involved in the opposite regulation of luteinizing hormone and follicle-stimulating hormone expression during ovarian development in the European eel. Neuroendocrinology 81: 107–119.

Sébert, P., 2003. fish adaptations to pressure. In Val, A. L. & B. G. Kapoor (eds), Fish Adaptations. Science Publisher, Enfield: 73–95.

Sébert, P., L. Barthélémy & J. Caroff, 1986. Catecholamine content (as measured by the HPLC method) in brain and blood plasma of the eel: effects of 101 ATA hydrostatic pressure. Comparative Biochemistry and Physiology, Part C 84: 155–157.

Sébert, P., L. Barthélémy & B. Simon, 1990. Laboratory system enabling long-term exposure to hydrostatic pressure of fishes or other animals breathing water. Marine Biology 104: 165–168.

Sébert, P., Y. Le Bras, L. Barthélémy & C. Peyraud, 1984. Effects of high hydrostatic pressure on catecholamine contents in tissues of the eel acclimatized at two temperatures. Aviation, Space, and Environmental Medicine 55: 931–934.

Setini, A., F. Pierucci, O. Senatori & A. Nicotra, 2005. Molecular characterization of monoamine oxidase in zebrafish (*Danio rerio*). Comparative Biochemistry and Physiology, Part B 140: 153–161.

Shiman, R., M. Akino & S. Kaufman, 1971. Solubilization and partial purification of tyrosine hydroxylase from bovine adrenal medulla. The Journal of Biological Chemistry 246: 1330–1340.

Siegel, G., B. Agranoff, R. W. Albers & P. Molinoff, 1989. Basic Neurochemistry. Raven Press, New York.

Smeets, W. J. A. J. & A. Gonzalez, 1990. Are putative dopamine-accumulating cell bodies in the hypothalamic periventricular organ a primitive brain character of non-mammalian vertebrates? Neuroscience Letter 114: 248–252.

Smeets, W. J. A. J. & A. Gonzalez, 2000. Catecholamine systems in the brain of vertebrates: new perspectives through a comparative approach. Brain Research Reviews 33: 308–379.

Somoza, G. M. & R. E. Peter, 1991. Effects of serotonin on gonadotropin and growth hormone release from in vitro perifused goldfish pituitary fragments. General and Comparative Endocrinology 82: 103–110.

Specker, J. L., J. G. Eales, M. Tagawa & W. A. Tyler III, 2000. Parr-smolt transformation in Atlantic salmon: thyroid deiodination in liver and brain and endocrine correlates of change in rheotactic behavior. Canadian Journal of Zoology 78: 696–705.

Spector, S., R. Gordon, A. Sjoerdsma & S. Undenfriend, 1967. Endproduct inhibition of tyrosine hydroxylase as a possible mechanism for regulation of norepinephrine synthesis. Molecular Pharmacology 3: 549–555.

Spector, S., A. Sjoerdsma & S. Undenfriend, 1965. Blockade of endogenous norepinephrine synthesis with α-methyltyrosine, an inhibitor of tyrosine hydroxylase. The Journal of Pharmacology and Experimental Therapeutics 147: 86–95.

Stone, R., 2003. Freshwater eels are slip-sliding away. Science 302: 221–222.

Tesch, F. W. & N. Rohlf, 2003. Migration from continental waters to the spawning grounds. In Aida, K., K. Tsukamoto & K. Yamauchi (eds), Eel Biology. Springer-Verlag, Tokyo: 223–234.

Trudeau, V. L., D. Spanswick, E. J. Fraser, K. Lariviere, D. Crump, S. Chiu, M. MacMillan & M. Schulz, 2000. The role of amino acid neurotransmitters in the regulation of pituitary gonadotropin release in fish. Biochemistry and Cell Biology 78: 241–259.

Trudeau, V. L., B. D. Sloley & R. E. Peter, 1993. GABA stimulation of gonadotropin-II release in goldfish: involvement of GABA receptors, dopamine, and sex steroids. The American Journal of Physiology 265: 348–355.

Vacher, C., F. Ferriere, M. H. Marmignon, E. Pellegrini & C. Saligaut, 2002. Dopamine D2 receptors and secretion of FSH and LH: role of sexual steroids on the pituitary of the female rainbow trout. General and Comparative Endocrinology 127: 198–206.

Van Ginneken, V., E. Antonissen, U. K. Müller, R. Booms, E. Eding, J. Verreth & G. van den Thillart, 2005. Eel migration to the Sargasso: remarkably high swimming efficiency and low energy costs. The Journal of Experimental Biology 208: 1329–1335.

Vidal, B., C. Pasqualini, N. Le Belle, M. C. H. Holland, M. Sbaihi, P. Vernier, Y. Zohar & S. Dufour, 2004. Dopamine inhibits luteinizing hormone synthesis and release in the juvenile European eel: a neuroendocrine lock for the onset of puberty. Biological Reproduction 71: 1491–1500.

Vogt, M., 1954. The concentration of sympathin in different parts of the central nervous system under normal conditions and after the administration of drugs. Journal of Physiology 123: 451–481.

Volkoff, H., J. P. Wourms, E. Amesbury & F. F. Snelson, 1999. Structure of the thyroid gland, serum thyroid hormones, and the reproductive cycle of the Atlantic stingray, *Dasyatis sabina*. The Journal of Experimental Zoology 284: 505–516.

Weiner, R. I., P. R. Findell & C. Kordon, 1989. Role of classic and peptide neuromediators in the neuroendocrine regulation of LH, Prolactin. In Knobil, E. & J. Neill (eds), The Physiology of Reproduction. Raven Press, New York: 1235–1281.

Weltzien, F.-A., C. Pasqualini, P. Vernier, S. Dufour, 2005. A quantitative real-time RT-PCR assay for European eel tyrosine hydroxylase. General and Comparative Endocrinology 142: 134–142.

Weltzien, F.-A., C. Pasqualini, M.-E. Sébert, B. Vidal, N. Le Belle, O. Kah, P. Vernier & S. Dufour, 2006. Androgen-dependent stimulation of brain dopaminergic systems in the female European eel (*Anguilla anguilla*). Endocrinology 147: 2964–2973.

Westin, L., 1990. Orientation mechanisms in migrating European silver eel (*Anguilla anguilla*): temperature and olfaction. Marine Biology 106: 175–179.

Wilson, D. A. & R. M. Sullivan, 1995. The D2 antagonist spiperone mimics the effect of olfactory deprivation on mitral/tufted cell odor response patterns. Journal of Neuroscience 15: 5574–5581.

Winberg, S. & G. E. Nilsson, 1992. Induction of social dominance by L-dopa treatment in Arctic charr. Neuroreport 3: 243–246.

Winberg, S. & G. E. Nilsson, 1993. Roles of brain monoamine neurotransmitters in agonistic behaviour and stress reactions, with particular reference to fish. Comparative Biochemistry and Physiology, Part C 106: 597–614.

Wirth, T. & L. Bernatchez, 2003. Decline of North Atlantic eels: a fatal synergy? Proceedings of the Royal Society of London 270: 681–688.

Wulliman, M. F., B. Rupp & H. Reichert, 1996. Neuroanatomy of the Zebrafish Brain: A Topological Atlas. Birkauser Verlag, Switzerland.

Yaron, Z., G. Gur, P. Melamed, H. Rosenfeld, A. Elizur & B. Levavi-Sivan, 2003. Regulation of fish gonadotropins. International Review of Cytology 225: 131–185.

Yu, K. L. & R. E. Peter, 1992. Adrenergic and dopaminergic regulation of gonadotropin-releasing hormone release from goldfish preoptic-anterior hypothalamus and pituitary in vitro. General and Comparative Endocrinology 85: 138–146.

Zigmond, E. R., M. A. Schwarzschild & A. R. Rittenhouse, 1989. Acute regulation of tyrosine hydroxylase by nerve activity and by neurotransmitters via phosphorylation. Annual Review of Neuroscience 12: 415–461.

Zohar, Y., M. Harel, S. Hassin & A. Tandler, 1995. Broodstock management and manipulation of spawning in the gilthead seabream, *Sparus aurata*. In Bromage, N. & R. J. Roberts (eds), Broodstock Management and Egg and Larval Quality. Blackwell Science Press, London: 94–117.

Hydrobiologia (2008) 602:47–59
DOI 10.1007/s10750-008-9285-4

Morphological analysis of geographic variation of sea lamprey ammocoetes in Portuguese river basins

P. R. Almeida · G. Tomaz · N. O. Andrade · B. R. Quintella

Abstract The main objective of this study was to compare the morphological variability of sea lamprey (*Petromyzon marinus* L.) larvae from the main Portuguese river basins. Samples were collected in rivers Minho, Lima, Cávado, Vouga, Mondego, Tejo and Guadiana. Specimens were analysed in terms of morphometric (linear body measures) and meristic (number of myomeres) characters to investigate the hypothesis of population fragmentation between river basins caused by some degree of homing behaviour. The discriminant analysis showed a morphological segregation of the studied populations based on the characters head, tail and branchial length. The discriminatory power of the meristic characters was comparatively weaker, with the number of trunk myomeres, and to some extent the head myomeres, being responsible for the reduced separation between groups. Both analyses were consistent in identifying the cephalic region as the most important morphological feature to discriminate populations of sea lamprey larvae in the Portuguese territory. The largest cephalic region of the ammocoetes sampled in the northern river basins may be responsible for a better feeding efficiency and, consequently, higher values of condition factor.

Keywords *Petromyzon marinus* · Morphometric · Meristic · Condition factor · Transformers

Guest editors: S. Dufour, E. Prévost, E. Rochard & P. Williot
Fish and diadromy in Europe (ecology, management, conservation)

P. R. Almeida (✉)
Department of Biology, University of Evora,
7002-554 Evora, Portugal
e-mail: pmraposo@mail.telepac.pt

P. R. Almeida · G. Tomaz · N. O. Andrade ·
B. R. Quintella
Institute of Oceanography, Faculty of Sciences,
University of Lisbon, Campo Grande, 1749-016 Lisbon,
Portugal

Introduction

The sea lamprey (*Petromyzon marinus* L.) is the only species of the genus *Petromyzon*, and the largest anadromous parasitic cyclostome, reaching 1.2 m in length and 2.3 kg of weight (Hardisty, 1986). It is widely distributed along the temperate zone of the North Atlantic (Hardisty, 1986). In Portugal sea lampreys can be found in all major hydrographical basins, being more abundant in the central and northern regions of the country (Almeida et al., 2002; Quintella et al., 2003; Rogado et al., 2005).

Lampreys' success as a species is due to a particular life cycle, a prolonged larval period (4–5 years) (Beamish & Medland, 1988; Quintella et al., 2003), during which ammocoetes live buried in the soft sediments of fresh water systems feeding on microscopic detritus and microorganisms in suspension

(Hardisty & Potter, 1971a; Manion & McLain, 1971; Almeida & Quintella, 2002), followed by an adult parasitic phase of ca. 2 years in the sea feeding on blood and soft tissues of other fishes (Hardisty & Potter, 1971b). Larval phase duration is, in part, the result of the time needed to attain a critical size and gather the necessary energetic reserves to initiate metamorphose (Youson, 1980). In Portugal, metamorphose occurs when ammocoetes attain a minimum size around 130 mm (average 168 mm) commencing in late summer until middle winter, with a peak in the months of October–November (Quintella et al., 2003).

The study of morphological characters, meristic and morphometric, with the objective of defining and characterizing populations, has a long tradition in ichthyology. The physiology and anatomy of an individual is sensitive both to genetic and environmental control, which are responsible for phenotypical variation. The effect of abiotic factors during ontogeny may result in significant differences between individuals of the same population, in meristic terms (Tåning, 1952; Barlow, 1961). Morphometric characters are exposed to the same abiotic factors, for an even longer period of time, which may increase the susceptibility of having more differences (Melvin et al., 1992). If those differences are significant and constant, during time, they may allow the identification of individuals of different populations (Ihssen et al., 1981).

It is not yet fully understood if sea lampreys tend to migrate to their natal streams (homing behaviour) to spawn like salmonids. The genetic similarity between sea lamprey specimens collected in two Iberian rivers (River Minho and River Sella), whose mouths are separated by almost 500 km, suggests that homing does not occur in the anadromous *P. marinus* (Rodríguez-Muñoz et al., 2004). Bergstedt & Seelye (1995) found evidences for lack of homing in the landlocked form of this species in the North American Great Lakes Region, and suggested instead that sea lampreys seemed to select spawning streams through innate attraction to other sensory cues. The existence of unique bile acids released by ammocoetes that function as migratory pheromones is widely recognized (Li et al., 1995; Bjerselius et al., 2000; Polkinghorne et al. 2001; among others). Recently Li et al. (2002) also found that reproductively mature male sea lampreys release a bile acid that acts as a potent sex pheromone, inducing preference and searching behaviour in ovulated female lampreys. Conspecifics population olfactory cues combined with innately recognized environmental cues of natal rivers may act as stimuli to some type of homing behaviour, which may facilitate morphological differentiations among river populations of sea lamprey ammocoetes.

Sea lamprey is on the list of the endangered species in Portuguese rivers, along with the rest of the indigenous migratory fishes (Rogado et al., 2005). Geographic biometric variation may identify distinct populations suggesting that river populations should be managed as distinct stocks. On the contrary, if we are dealing with the same population for the Portuguese territory it is possible to direct some mitigation measures to particular river basins that due to their biological and morphological characteristics (e.g., viable habitats for the establishment of ammocoete beds and for adults to reproduce, high river flows, low anthropogenic intrusions such as weirs and dams, dredging, gravel extraction, channelization and pollution) might serve as a support to the rest of the population spread out through the remaining watersheds.

The main objective of this study was to examine the geographic variation in morphological and meristic characters of *P. marinus* ammocoetes from the main Portuguese river basins, identifying at the same time the best set of characters to establish the separation of the eventual groups.

Materials and methods

Field work

Sampling occurred during the months of September and October of 2003 in seven hydrographical river basins (Minho, Lima, Cávado, Vouga, Mondego, Tejo and Guadiana) selected due to the recognized annual spawning migration of adults (Fig. 1). The selection of the sampling sites was facilitated by the study of Almeida & Quintella (2002) which identified a group of physical and environmental characteristics strongly associated with ammocoete beds. Ammocoetes and metamorphosing sea lampreys were collected by electro-shocker (Hans Grassl EL62, 600V-DC). Metamorphosing individuals (i.e.,

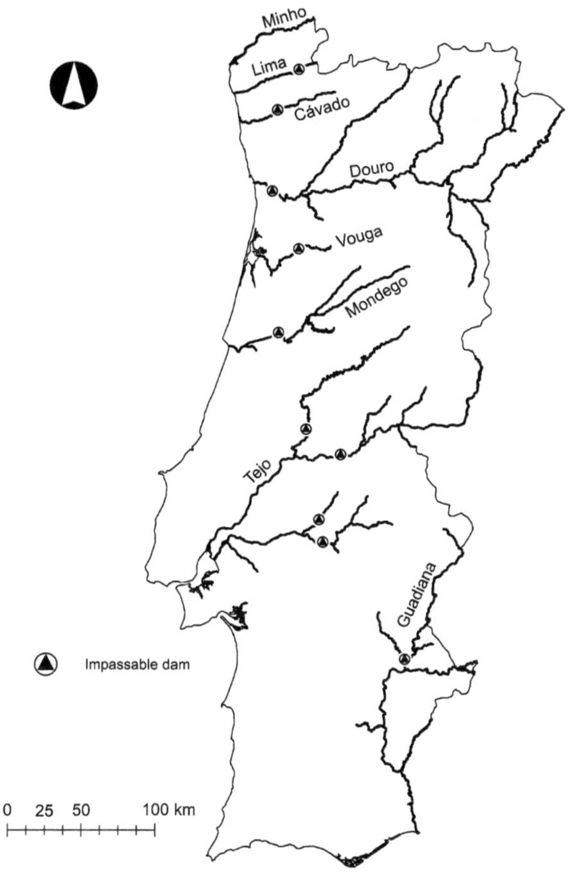

Fig. 1 Map of Portugal with the location of the sampling sites and the representation of the river basins where sea lampreys are known to occur. The first impassable obstacle defines the habitat available to this species in each river basin

transformers) were identified according to the criteria defined by Youson & Potter (1979), i.e., structure of oral aperture and appearance of teeth, the eruption of the eyes, enlargement of the fins and changes in pigmentation.

A total of 1,402 sea lampreys were sampled (#255 Minho, #321 Lima, #4 Cávado, #446 Vouga, #226 Mondego, #154 Tejo and #6 Guadiana). In rivers Cávado and Guadiana an insufficient number of specimens were captured and consequently these two sampling sites were removed from further analysis. The length of all captured individuals were measured in the field with the help of an acrylic transparent graduated tube, permitting an accuracy to the nearest millimetre without anaesthesia. The weight (±0.01 g) of each captured sea lamprey was also measured with a Kern 440-36 precision balance, and a sub-sample of ca. 90 ammocoetes equitably distributed in four length-classes: <60 mm, [60–100[mm, [100–140[mm and >160 mm was selected afterwards. They were killed and immediately frozen for further morphometric and meristic analyses in laboratory. The remaining captured specimens were returned to the wild.

Laboratorial work

The morphological characters of each individual were measured and registered according to Holčik (1986), with the help of a Leica MZ6 stereomicroscope and callipers (±0.05 mm). A total of 11 characteristics were used: eight morphometric and three meristic. The morphometric characters measured were (Fig. 2):

- Total length (TL)—distance between the anteriormost tip of the oral hood and the posteriormost tip of the caudal fin;
- Head length (HL)—distance between the anteriormost tip of the oral hood and the posterior edge of the last branchial opening;
- Trunk length (TrL)—distance from the posterior edge of the last branchial opening to the anterior edge of the cloacal slit;
- Tail length (TaL)—distance between the anterior edge of the cloacal slit and the posteriormost tip of the caudal fin;
- Prenostril length (PnL)—distance between the anteriormost tip of the oral hood and the nasohypophysial opening;
- Branchial length (BL)—distance between the anterior edge of the first and the posterior edge of the last branchial opening;

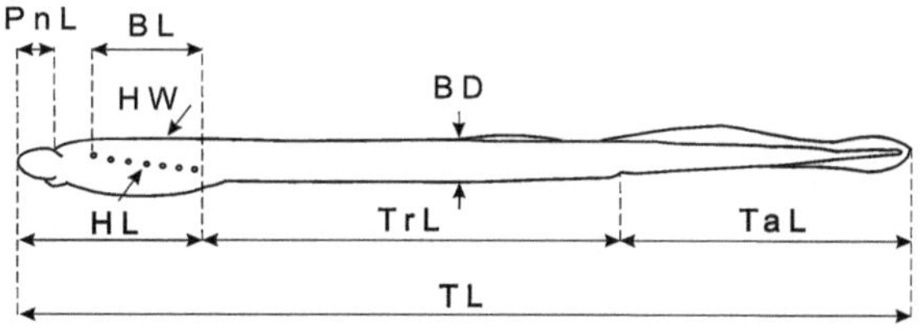

Fig. 2 Morphometric characters measured and acronyms used for analysis of geographic variation of sea lamprey ammocoetes in Portugal. TL, total length; HL, head length; TL, trunk length; TaL, tail length; PnL, prenostril length; BL, branchial length; HW, head width; BD, body depth

- Head width (HW)—distance between the left and right side of the head measured at the level of the fourth branchial opening;
- Body depth (BD)—distance between the dorsal and ventral faces of the trunk measured at the level of the anterior insertion of the first dorsal fin.

The meristic characters analysed were the number of myomeres (body muscle segments) of the head (myHe), trunk (myTr) and tail (myTa). The myomeres of the tail were counted between the beginning of the caudal peduncle and the anterior edge of the cloacal slit, the trunk myomeres were counted between the anterior edge of the cloacal slit and the posterior edge of the last branchial opening and the myomeres of the head were counted from posterior edge of the last branchial opening forward.

Statistical analysis

The statistical analysis was applied following Costa et al. (2003). The morphological variation was analysed considering each ammocoete as one multivariate observation, because a univariate approach would ignore the joint effect of variates (Thorpe, 1976). Analyses were performed separately for morphometric and meristic characters, since these two variables are biologically and statistically different. Morphometric characters are represented by continuous data and are more affected by the surrounding environment. Conversely, meristic characters are discrete and predetermined early during the embryonic development (Ihssen et al., 1981).

Morphometric data were statistically adjusted to eliminate the influence of allometric growth so that their variation is independent of the specimen's size. The morphometric character and the total length of the individual were converted to a logarithmic to produce a linear relationship between the two variates (Thorpe, 1976; Hair et al., 1998). Outliers were detected by regression analysis of morphometric characters against total length and by scatter plots of residual versus predicted values (Schaefer, 1991). Outliers were eliminated with SPSS casewise diagnostics, any value greater than three standard deviations from the mean was considered an outlier. When an outlier was found, all morphometric data (but not the meristic counts) of that ammocoete were withdrawn from further consideration (Schaefer, 1991). This procedure resulted in the elimination of morphometric data for 39 lampreys (#8 Minho, #8 Lima, #11 Vouga, #3 Mondego and #9 Tejo).

Each of the seven morphometric characters showed a linear relationship with total length (significant at $P < 0.001$). Analysis of covariance (ANCOVA) was employed to test for differences in allometric relationships among samples and to estimate the common within-group regression slopes (Schaeffer, 1991). According to Claytor & MacCrimmon (1987) it is advisable to determine the common within-group regression slope and not the slope of the data pooled irrespective of locality. If there is both geographic variation in the allometric character and the specimen sizes differ between samples, the regression slope computed from all the data, pooled irrespective of geographic origin, will usually differ from the within-group regression slopes and consequently will not be the most suitable for eliminating the allometric influence of size (Thorpe, 1976). Within-group regression slopes were significantly different ($\mathrm{df} = 4,\ 374$; $P < 0.05$) for four of the morphometric characters (i.e., TaL, PnL, BL and BD), and thus size adjustment was based on the common within-group slopes. Coefficients from the common within-group regression are used to allometrically adjust variates when between-group heterogeneity exists (Thorpe, 1976; Reist, 1986). The morphometric character is thus adjusted applying a modification of the formula given by Claytor & MacCrimmon (1987):

$$\mathrm{AC}_{ij} = \mathrm{LnOC}_{ij} - [\beta \times (\mathrm{LnTL}_j - \mathrm{Ln}\overline{\mathrm{TL}})];$$

where AC_{ij} is the adjusted logarithmic character measurement i of the j specimen, OC_{ij} is the unadjusted character measurement i of the j specimen, β is the common within-group regression coefficient of that character i against total length after the logarithmic transformation of both variables, TL_j is the total length of the j specimen and $\overline{\mathrm{TL}}$ is the overall mean total length.

Meristic characters, which are discrete variables, are not normally distributed even after variable transformations. However, the analyses applied in this investigation are robust to deviations from normality (Sokal & Rohlf, 1995). Hence, we followed

Claytor et al. (1991) and proceed as if raw variables were normally distributed but provide this information as a cautionary note (Costa et al., 2003). No significant relationship (at level $P < 0.05$) was found between the meristic characters and total length. Outliers were detected by the SPSS Boxplot procedure, all cases that had values more than three hspreads (one hspread corresponds to the interquartile range) below the lower hinge or above the upper hinge were eliminated causing the removal of that lamprey from further consideration in the meristic analysis (but not for morphometric studies). The application of this procedure resulted in the elimination of 33 specimens (#2 Minho, #9 Lima, #11 Vouga, #9 Mondego and #2 Tejo) from this analysis.

Morphometric and meristic data from the five river basins were compared separately by means of multiple discriminant analysis (MDA). The computational method utilized to derive the discriminant function was the stepwise method which involves entering the independent variables into the discriminant function one at a time on the basis of their discriminating power. The selection rule in this procedure is to maximize Mahalanobis D^2 between groups (Hair et al., 1998). Discriminant *Z* scores, score defined by the discriminant function for each object in the analysis, were used to examine the effectiveness of the size adjustment procedures. Thus, *Z* scores from discriminant functions 1 and 2 were regressed against total length, and size was considered to be effectively removed if regressions were not significant at $P < 0.001$ (Claytor & MacCrimmon, 1987). Discriminant *Z* scores and group centroids were plotted for visual inspection of differences between groups on each function. The resultant discriminant functions were used to classify individual fish into groups. For assessing group membership prediction accuracy, the expected actual error rates of the classification functions were estimated using cross-validation by the leaving-one-out procedure in which the discriminant function is fitted to repeatedly drawn samples of the original sample. The predictive accuracy relative to chance, which would have a correct classification rate of 50%, was measured with Press's *Q* statistic. Discriminant loadings and potency index were used to assess the relative importance of each independent variable for discriminating between groups (Hair et al., 1998). Discriminant loadings measure the simple linear correlation between each independent variable and the discriminant *Z* score for each discriminant function. The potency index is a relative measure used for comparing the overall discrimination provided by each independent variable across all significant discriminant functions.

We used a Mantel test to estimate the degree of association between morphometric or meristic distances (D^2) and geographic separation (in kilometres) of the river basin mouths. A total of 249 permutations were performed for each Mantel test (Sokal & Rohlf, 1995).

As a complement to the multivariate analysis, significant differences in adjusted morphometric characters and meristic characters were determined using the ANOVA analysis. A posteriori comparisons were used to test significant differences among all possible pairs, SS-STP test was used when equal variance was assumed and Tamhane's T2 when moderate heterogeneity of variances was detected (Sokal & Rohlf, 1995).

All mathematical procedures were applied using SPSS 11.0 statistical package with the exception of ANCOVA analysis and the Mantel test performed with BIOMstat programs (Rohlf & Slice, 1995).

In addition, relative condition factor (K') was calculated for ammocoetes and transformers, allowing the comparison of the animals' well-being between the sampling sites. The formula used was an extension of Fulton's condition factor developed by Ricker (1975):

$$K' = \frac{\mathrm{TW}}{\mathrm{TL}^b} \times 100,$$

where TW and TL are the observed total weight (in g) and total length (in cm) of a lamprey and *b* the slope of the ammocoetes weight–length relationship for each river basin. Kruskal–Wallis tests were used to compare the K' of the ammocoetes and transformers captured in the five watersheds, and to compare the K' of the ammocoetes belonging to the different length-classes for each of the sampled river basins (SPSS 11.0).

Results

A total of 1,402 lampreys were captured, 1,291 ammocoetes and 111 transformers (Fig. 3). The total length (TL) of the sampled ammocoetes ranged from

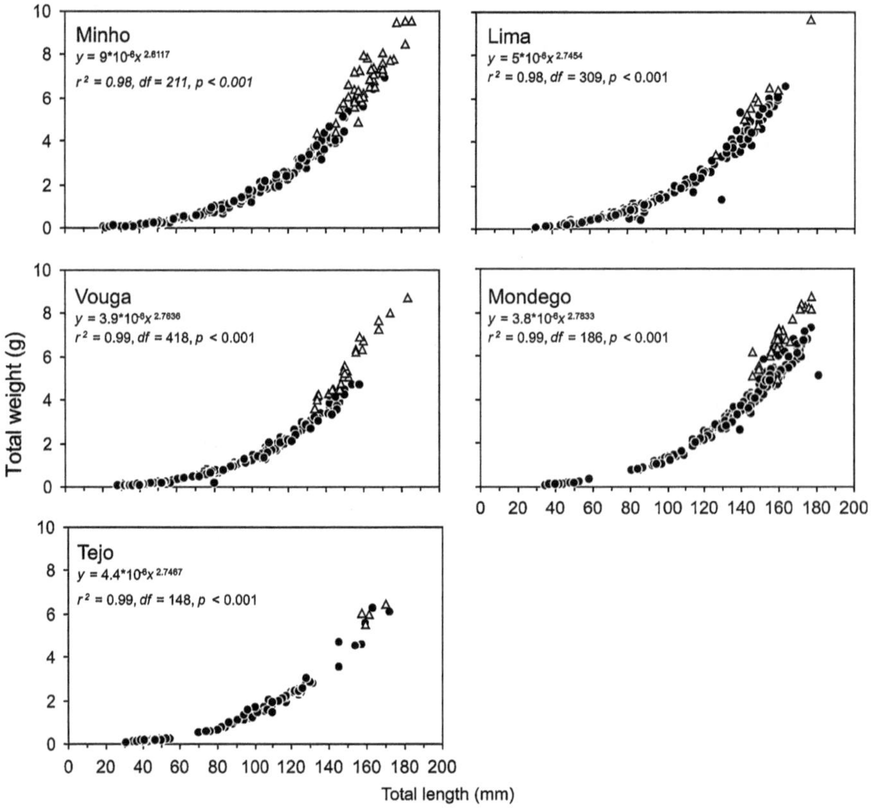

Fig. 3 Plot of length–weight relationship for ammocoetes (●) and transformers (△) captured in the five sampling sites. Also represented the curve estimation parameters of the power regressions for the ammocoete length–weight relationship

21 to 181 mm, while the total weight (TW) varied between 0.02 and 7.26 g (Table 1). Only 8% of the captured lampreys had already initiated the metamorphosis process, with the transformers ranging from 127 to 185 mm ($\overline{\mathrm{TL}} = 158$ mm) and from 3.44 to 9.66 g ($\overline{\mathrm{TW}} = 6.38$ g) (Table 1).

The relative condition factor (K') was applied to compare differences in the fatness of the sea lampreys sampled in the five studied river basins (Table 2). The K' of the ammocoetes captured in the five watersheds was compared with a Kruskal–Wallis test ($H = 882$, df = 4, $P < 0.001$) which revealed statistical significant differences between sampling sites. A similar test was performed with the transformers ($H = 84$, df = 4, $P < 0.001$), and once more significant differences were found between the K' of the metamorphosing individuals. The K' of the transformers was always higher when compared with the ammocoetes (Table 2). Generally, both ammocoetes and transformers K' increased with latitude with the weightier animals captured in the Minho river basin (Table 2). The exception were the ammocoetes captured in the southern river basin, River Tejo, which had higher K' than the larvae sampled in the central region of Portugal (i.e., rivers Vouga and Mondego). The ammocoetes' condition factor increased in general with growth (Table 2). Kruskal–Wallis tests were also used to compare the K' of the ammocoetes belonging to the different length-classes (i.e., groups) for each of the sampled river basins. Significant differences ($P < 0.01$) were found for the Minho ($H = 55$, df = 13), Lima ($H = 29$, df = 11) and Mondego river basins ($H = 63$, df = 12) on which larger individuals appear to be proportionally heavier, whereas for Vouga ($H = 19$, df = 13, $P = 0.110$) and Tejo ($H = 13$, df = 10, $P = 0.176$) the condition of the ammocoetes appears to statistically not differ with growth.

Morphometric analysis

The regressions for Z scores from discriminant functions 1 and 2 of discriminant analysis against total length were not significant ($r^2 = 0.009$, df = 373,

Table 1 Summary statistics for total length and total weight for ammocoetes and transformers captured in the sampled river basins

	N	Total length (mm)				Total weight (g)			
		Mean	SD	Min.	Max.	Mean	SD	Min.	Max.
Ammocoetes									
Minho	213	87	37	21	171	1.51	1.51	0.02	6.92
Lima	311	94	28	32	164	1.61	1.39	0.07	6.58
Vouga	420	67	33	28	158	0.73	0.96	0.04	4.71
Mondego	188	123	42	35	181	3.16	2.07	0.08	7.26
Tejo	150	76	36	31	172	1.03	1.24	0.05	6.30
Transformers									
Minho	42	161	10.75	135	185	6.77	1.27	4.35	9.55
Lima	10	150	13.01	127	177	5.86	1.61	3.44	9.66
Vouga	26	152	12.14	134	183	5.52	1.37	3.61	8.73
Mondego	28	161	8.73	146	177	6.85	1.00	4.93	8.74
Tejo	4	162	5.74	157	170	6.00	0.39	5.53	6.48

N, number of individuals; SD, standard deviation; Min., minimum; Max., maximum

$P = 0.070$; and $r^2 = 0.005$, df = 373, $P = 0.163$), indicating that size effects had been removed from the adjusted morphometric variables. Discriminant functions are statistically significant (Table 3), and all pairs of groups showed statistically significant differences (df = 7, 363; $P < 0.001$). The stepwise analysis revealed that all adjusted morphometric characters contributed significantly to the MDA. The *Z* scores and centroids from discriminant functions 1 and 2 were plotted against each other to develop a graphic representation of the relationship among groups (Fig. 4a). The two discriminant functions account for 49.6 and 32.2% of total variation, and the total classification rate estimated from cross-

Table 2 Mean relative condition factor ($\overline{K}'$) of ammocoetes (Amo.) and transformers (Tran.) captured in the five studied river basins and distributed in length–frequency classes of 10 mm

Length–frequency class (mm)	$\overline{K}'$-Minho		$\overline{K}'$-Lima		$\overline{K}'$-Vouga		$\overline{K}'$-Mondego		$\overline{K}'$-Tejo	
	Amo.	Tran.	Amo.	Tran.	Amo.	Tran.	Amo.	Tran.	Amo.	Tran.
]20–30]	0.35	–	–	–	0.24	–	–	–	–	–
]30–40]	0.33	–	0.30	–	0.23	–	0.25	–	0.25	–
]40–50]	0.35	–	0.30	–	0.23	–	0.23	–	0.25	–
]50–60]	0.33	–	0.28	–	0.22	–	0.23	–	0.22	–
]60–70]	0.35	–	0.28	–	0.22	–	–	–	0.26	–
]70–80]	0.34	–	0.28	–	0.22	–	–	–	0.23	–
]80–90]	0.35	–	0.28	–	0.22	–	0.21	–	0.24	–
]90–100]	0.36	–	0.28	–	0.22	–	0.21	–	0.26	–
]100–110]	0.38	–	0.28	–	0.22	–	0.21	–	0.24	–
]110–120]	0.37	–	0.28	–	0.23	–	0.23	–	0.25	–
]120–130]	0.38	–	0.28	–	0.23	–	0.23	–	0.25	–
]130–140]	0.40	0.49	0.29	0.32	0.23	0.30	0.23	–	0.24	–
]140–150]	0.40	0.47	0.29	0.35	0.24	0.29	0.23	0.31	0.27	–
]150–160]	0.43	0.47	0.30	0.33	0.24	0.31	0.24	0.30	0.25	0.30
]160–170]	0.42	0.47	0.30	–	–	0.31	0.24	0.29	0.30	0.28
]170–180]	0.42	0.46	–	0.36	–	0.30	0.24	0.29	0.25	–
]180–190]	–	0.46	–	–	–	0.28	–	–	–	–
Mean	0.36	0.47	0.28	0.34	0.23	0.30	0.23	0.30	0.25	0.29

Table 3 Results of Wilk's lambda (Λ) tests to verify the hypothesis that the means (centroids) of all functions are equal in the five groups when their morphometric and meristic characters were separately compared by stepwise MDA. The other successive tests were useful for identifying whether or not additional functions reflect populations differences or only random variation

Test of function(s)	Λ	χ^2	df
Morphometric analysis			
1–4	0.31	428.32***	28
2–4	0.54	228.97***	18
3–4	0.79	87.91***	10
4	0.97	11.07*	4
Meristic analysis			
1–3	0.48	291.21***	12
2–3	0.76	117.89***	6
3	0.92	34.22***	2

$^*P < 0.05$; $^{***}P < 0.001$

validation procedure was 54.8% (Table 4). Some degree of overlap between individuals from the five groups was detected, with correct classifications rate ranging from 44.7 to 67.3% (Table 4). Press's Q test revealed that the classification accuracy is statistically significant better than chance (Press's $Q = 283.289$, df = 1, $P < 0.001$).

The interpretation of the plot in Fig. 4a indicates that the first discriminant function (1) is the primary source of separation between groups formed by rivers Vouga and Mondego vs. the group formed by the northern river basins (Minho and Lima) and the southern (Tejo). The second discriminant function (2) provides the distinction between the group formed by the Tejo and Vouga river basins from the remaining sampling sites. Discriminant loadings and potency index were used to assess the contributions of the seven morphometric variables (Table 5). High correlations between the discriminant loadings of the variables HL and BL with the first function, and TaL and TrL with the second function identified the variables with the best discriminatory power for each axis (Table 5). HL, TaL and BL were the variables with the highest potency index values and can be considered the most important morphometric characters to distinguish sea lamprey ammocoete populations in the main Portuguese river basins (Table 5).

The Mantel test showed that the morphometric distances (D^2) between ammocoete populations were not associated with their geographic distances ($r = -0.146$, $t = -0.418$, $P = 0.361$).

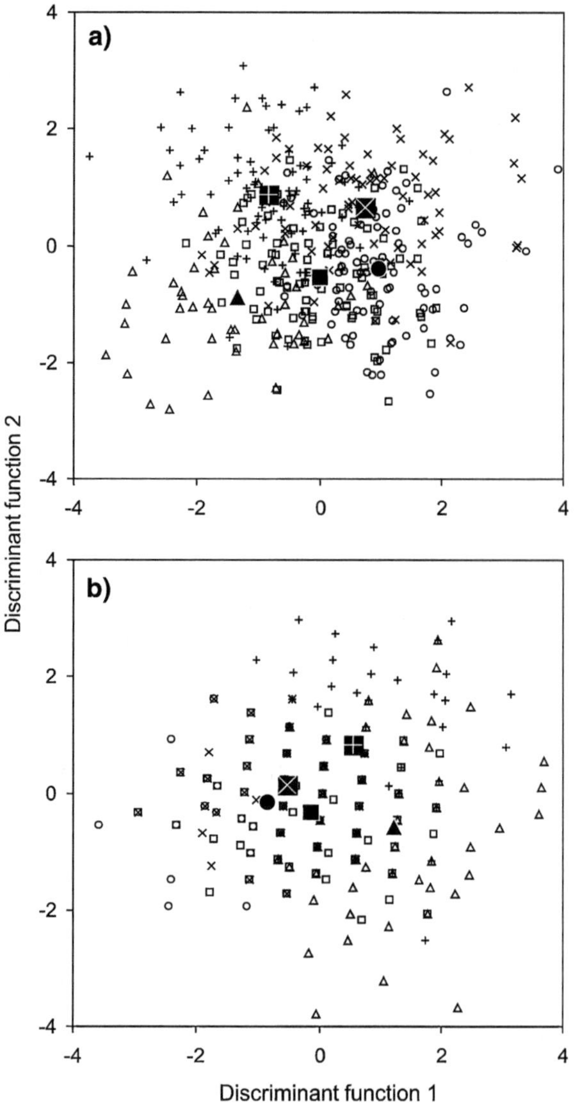

Fig. 4 Plot of the discriminant Z scores and group centroids on discriminant functions 1 and 2 for the five groups of sea lamprey ammocoetes based on the morphometric (**a**) and meristic (**b**) characters. Symbols for groups: ○, Minho; □, Lima; +, Vouga; △, Mondego; ×, Tejo

Univariate comparisons of the adjusted morphometric characters revealed significant statistical differences for the five groups (df = 4,369; $P < 0.001$). A posteriori multiple comparisons revealed that ammocoetes captured in the northern (Minho and Lima) and southern (Tejo) river basins have significantly longer cephalic and branchial regions when compared with the animals

Table 4 Classification results obtained with the stepwise discriminant analysis cross-validation for morphometric and meristic characters to determine the predictive accuracy level of the discriminant functions

Groups	*N*	Percent correct	Number of individuals classified into group				
			Minho	Lima	Vouga	Mondego	Tejo
Morphometric analysis							
Minho	82	47.6	39	14	5	4	20
Lima	85	44.7	20	38	9	11	7
Vouga	83	65.1	1	11	54	5	12
Mondego	49	67.3	5	5	5	33	1
Tejo	75	54.7	10	5	12	7	41
Total	374	54.8	–	–	–	–	–
Meristic analysis							
Minho	88	54.5	48	8	5	4	23
Lima	84	26.2	24	22	11	19	8
Vouga	83	57.8	7	0	48	16	12
Mondego	69	58.0	6	3	16	40	4
Tejo	82	32.9	27	5	16	7	27
Total	406	45.6	–	–	–	–	–

N, number of individuals

Table 5 Summary of discriminant loadings and potency index for morphometric and meristic variables

Variables	Discriminant loadings		Potency index
	Function 1	Function 2	
Morphometric analysis			
HL	0.67[a]	0.14	0.27
TaL	0.15	0.75[a]	0.23
BL	0.59[a]	−0.03	0.18
BD	0.38	−0.31	0.15
HW	−0.04	0.09	0.13
PnL	0.29	−0.46	0.09
TrL	−0.04	−0.52[a]	0.06
Meristic analysis			
myTr	0.89[a]	−0.24	0.53
myHe	−0.47	−0.33	0.24
myTa	0.118	0.86[a]	0.23

Acronyms of variables as defined in "Materials and methods" section

[a] Largest absolute correlation between each variable and any discriminant function

sampled in the central region of Portugal (Vouga and Mondego). The TaL of the animals captured in the Rivers Tejo and Vouga is significantly longer when compared with the ammocoetes of Minho, Lima and Mondego (Table 6).

Meristic analysis

The number of myomeres from the head, trunk and tail region varied significantly along the five groups according to univariate comparisons of meristic characters (df = 4, 401; $P < 0.001$) (Table 6). Ammocoetes sampled in Minho, Lima and Tejo river basins have a significantly larger number of head myomeres than the remaining two watersheds, while ammocoetes from River Mondego appear to have additional trunk myomeres. The number of trunk myomeres is more reduced in the Minho and Tejo river basins. The number of tail myomeres does not differ significantly between river basins, except for the River Mondego which appears to have a relatively reduced number of muscle segments.

Wilk's lambda test indicated differences among the five groups when their meristic characters were compared by means of discriminant analysis (Table 3). Stepwise MDA revealed that the three meristic characters contributed significantly to the multivariate discrimination of the five sampled groups (Table 3). The plot of the two discriminant functions, which account for 62.7 and 26.9% of the total variation, shows a noticeable overlap of individuals from the five groups, in particular ammocoetes captured in Minho, Lima and Tejo basins (Fig. 4b). The correct

Table 6 Univariate comparisons of adjusted morphometric and unadjusted meristic characters for the five groups

Variables	Mean value for each group					ANOVA + Post Hoc tests
	Minho	Lima	Vouga	Mondego	Tejo	
Morphometric						
HL	3.00	2.99	2.97	2.92	2.99	35.48*** (1) – 1 2 5 3 4
TrL	4.06	4.07	4.05	4.07	4.04	12.49*** (1) – 1 2 4 3 5
TaL	3.32	3.28	3.34	3.29	3.36	30.92*** (1) – 1 4 2 3 5
PnL	0.95	0.86	0.78	0.90	0.85	17.38*** (1) – 1 4 2 5 3
BL	2.54	2.51	2.47	2.46	2.54	23.86*** (1) – 1 5 2 3 4
HW	1.11	1.08	1.12	1.20	1.20	7.30*** (1) – 1 2 3 4 5
BD	1.65	1.60	1.52	1.62	1.65	19.56*** (1) – 1 4 5 2 3
Meristic						
myHe	14.16	14.30	13.87	13.83	14.09	20.49*** (2) 2 1 5 3 4
myTr	72.88	74.25	74.48	75.64	73.20	45.88*** (2) 1 5 4 2 3
myTa	43.44	44.07	45.77	43.20	44.10	20.16*** (2) 1 2 4 5 3

Acronyms of variables as defined in "Materials and methods" section

***$P < 0.001$; (1) df = 4, 369;.(2) df = 4, 401; rules join non-significant subsets

classification rate estimated from the cross-validation ranged from 26.2 to 58.0%, with an overall rate of 45.6% which is a poor discrimination rate, although significantly better than would have occurred only by chance (Press's $Q = 165.86$, df $= 1, P < 0.001$). This low rate is partially due to the low classification rate of individuals from the Lima and Tejo basins which are mainly classified in other groups (Table 4). Nevertheless, the complete separation of the centroid values for each group was indicated by all pairwise group comparison (df = 3, 399; $P < 0.05$). According to the potency index, the number of trunk myomeres can be considered the most important meristic characters for discriminating groups (Table 5). The myTr also shows the highest correlation with the first discriminant function (1), which allows the separation between Minho/Lima/Tejo basins with Mondego and Vouga.

As for morphometrics, the Mantel test showed that the meristic distances (D^2) between ammocoetes captured in the five river basins were not associated with their geographic distances ($r = -0.376$; $t = -1.059$; $P = 0.189$).

Discussion

From the initial seven hydrographical basins selected for sampling, it was only possible to capture a workable number of lampreys in five sites, rivers Minho, Lima, Vouga, Mondego and Tejo. Extreme levels of pollution might be responsible for the reduced numbers of ammocoetes captured in the Cávado river basin. However, during the spawning season a considerable amount of adults are captured near the mouth of this watershed in sufficient numbers to constitute the base for the professional fishery in this river estuary (unpublished data). The River Guadiana is near the southern natural distribution of this species (Hardisty, 1986), and the reduced numbers of adults entering every year to spawn are probably insufficient to maintain ammocoete beds with a high density of individuals (unpublished data).

The mean length of the transformers captured (158 mm) is similar to the one found by Quintella et al. (2003) for the River Mondego (168 mm), but considerably higher than most literature on which the mean length of metamorphosing lampreys is generally around 140 mm (Manion & Stauffer, 1970; Beamish, 1980; Purvis, 1980; Hardisty, 1986; Morman, 1987; Bird et al., 1993). Lampreys are poikilotherm organism and, as might be expected, the growth rate of the ammocoete is correlated with water temperatures which can vary considerably between geographic regions with different climatic regimes (Potter, 1980). More favourable climatic conditions, similar to the ones observed in Portuguese river basins, are likely to induce higher growth rates.

Sea lamprey ammocoetes' and transformers' condition factor in the Portuguese river basins increases with latitude. Differences between the condition factor of the lampreys sampled in the five river basins are probably caused by abiotic factors. Young et al. (1990) found that out of 14 physicochemical variables, water temperature was one of only two variables that explained the variation in larval length among 17 survey streams, with longer larvae associated with warmer waters. Warmer water temperatures are known for increasing larval lampreys' growth (Manion & McLain, 1971) through increasing food consumption and conversion efficiency. Recently, Griffiths et al. (2001) found that the growth of sea lamprey larvae was highest in streams with high conductivity (>300 μS) and annual mean water temperature around 8°C, and was lower at both lower and higher temperatures. This author finding suggests an optimal temperature for ammocoete growth which in Portugal corresponds to the colder river basins in the north. Conductivity is highest in River Minho (89 μS/cm), followed by River Mondego (82 μS/cm), River Vouga (56 μS/cm), River Tejo (41 μS/cm) and finally River Lima (31 μS/cm) (SNIRH, 2004). This parameter was highest in the most northern river basin, River Minho, where the heavier sea lampreys were found. Curiously, the conductivity measured in River Lima was the lowest, but the individuals collected in this basin had the second highest condition factor.

Initiation of metamorphosis in Portugal typically occurs in mid-summer (August/September) and extends to winter period (January/February) (Quintella et al., 2003). Since metamorphosing larvae do not feed, ammocoetes must accumulate during the last months of the larval phase sufficient lipid reserves, their primary source of energy (Youson et al., 1979), to meet their nutritional requirements throughout metamorphosis. Thus, larvae increase body fat from 4 to 14% of body weight in the spring and summer prior to metamorphosis, when food is most abundant (Potter, 1980), which is the most probable explanation for the generally observed increase of ammocoetes' condition factor with growth in this study. The possibility of an arrested or retarded growth phase in the year preceding metamorphosis may also explain this phenomenon, during which the metabolism of the premetamorphic ammocoete is switched to lipid accumulation (Hardisty & Potter, 1971a). In our study we found results similar to Youson et al. (1993), metamorphosing animals had higher condition factors than larvae that did not metamorphose but which were of similar lengths. The increased weight is a consequence of the immense lipid reserves stored in the tissues prior to entering metamorphosis (Youson et al., 1979). The extra reserve of lipid materials will clearly be necessary to enable the juvenile to survive until the onset of the trophic phase (Hardisty & Potter, 1971a).

The low discriminatory power of the morphometric and meristic analyses suggests small morphological differences between the ammocoetes captured in the five studied river basins. Morphometric characters are apparently more suitable than meristic for a good discrimination between *P. marinus* ammocoetes, as was found before for other fish species (e.g., Sharp et al., 1978; Costa et al., 2003). Both analyses were statistically significant and congruent in the formation of a group constituted by the lampreys caught in the Minho, Lima and Tejo river basins. The ammocoetes from River Vouga and Mondego are the ones more discriminate by the morphometric and meristic analyses. The best discriminant variables for the morphometric analyses were the head, tail and branchial length, while the trunk myomeres and, to some extent, the head myomeres were the meristic characters more important to separate the groups. According to Holčik (1986), the number of trunk myomeres is a very important determination criterion employed in the taxonomy of lampreys. Therefore, it is not surprising that it can be as well an important meristic character to distinguish populations. There seems to be some consistency in identifying the cephalic region as the most important morphological feature to discriminate populations of sea lamprey larvae in the Portuguese territory. Lampreys captured in Minho, Lima and Tejo rivers have bigger heads with longer branchial regions and larger number of myomeres. The microphagous mode of feeding is probably enhanced with these characteristics by increasing the rate of food intake and thus the body size that the ammocoete has been able to attain. This hypothesis is supported by the higher K' values found in the larval lampreys sampled in rivers Minho, Lima and Tejo.

It is not clear however why less morphological similarities were found in groups which are geographically closer. Several hypotheses could be raised related with environmental factors or genetics. The conditions for growth and the establishment of ammocoete beds may vary considerably between river basins according to their abiotical and morphological characteristics. However, river basins might share particular characteristics, even if they are more distant geographically, that will condition the morphology of sea lamprey populations. Within a group of river systems, the distribution of lampreys is performed by the adults when entering the rivers to spawn. Recently, biochemical, physiological and behavioural studies have suggested that migratory sea lamprey use conspecific cues comprising bile acids to locate spawning streams (Li et al., 1995; Bjerselius et al., 2000; Polkinghorne et al., 2001). Homing behaviour within lampreys might be related with the bile acids released by the ammocoetes and not by other sensory cues related to specific physical characteristics of the natal watersheds as for salmonids.

Morphometric variables were found to be more suitable for a morphological analysis of geographic variation of sea lamprey ammocoetes. Variation in shape is usually more related to genetic causes, while meristic characters are more influenced by abiotic factors (Barlow, 1961). To complement this study results it would be important to perform genetic studies with ammocoetes and adults, since previous genetic studies on sea lampreys have not shown significant differences among populations or between basins (Jacobson et al., 1984; Rodríguez-Muñoz et al., 2004). The accumulated information from morphometric characters and genetic analysis could be used to assess evidences for spawners' homing behaviour. It is essential to understand in full extent the life cycle of threatened species to be possible to efficiently manage populations. The existence or not of homing behaviour might be of particular importance for the drawing of conservation measures of this species as well as a more profound knowledge of the parasitic adult phase in the sea.

Acknowledgements The authors especially thank J. L. Costa for advice and assistance with the statistical analysis. This research was financially supported by two Portuguese institutions, the Water Institute (INAG) of the Ministry of Environment and by the Foundation for Science and Technology (FCT) with Project funding (PNAT/1999/BIA/15019/P) and a Ph.D. grant to B. R. Quintella (SFRH/BD/9125/2002).

References

Almeida, P. R. & B. R. Quintella, 2002. Larval habitat of the sea lamprey (*Petromyzon marinus*) in the river Mondego (Portugal). In Collares-Pereira, M. J., M. M. Coelho & I. G. Cowx (eds), Freshwater Fish Conservation: Options for the Future. Fishing News Books, Blackwell Science, Oxford: 121–130.

Almeida, P. R., H. T. Silva & B. R. Quintella, 2002. The spawning migration of the sea lamprey (*Petromyzon marinus* L.), in the River Mondego (Portugal). In Pardal, M. A., J. C. Marques & M. A. Graça (eds), Aquatic Ecology of the Mondego River Basin. Global Importance of Local Experience. Imprensa da Universidade de Coimbra, Coimbra: 381–386.

Barlow, G. W., 1961. Causes and significance of morphological variation in fishes. Systematic Zoology 10: 105–117.

Beamish, F. W. H., 1980. Biology of the North American anadromous sea lamprey, *Petromyzon marinus*. Canadian Journal of Fisheries and Aquatic Sciences 37: 1924–1943.

Beamish, F. W. H. & T. E. Medland, 1988. Age determination for lampreys. Transactions of the American Fisheries Society 117: 63–71.

Bergstedt, R. A. & J. G. Seelye, 1995. Evidence for lack of homing by sea lampreys. Transactions of the American Fisheries Society 124: 235–239.

Bird, D. J., I. C. Potter, M. W. Hardisty & B. I. Baker, 1993. Morphology, body size and behaviour of recently-metamorphosed sea lampreys, *Petromyzon marinus*, from the lower River Severn, and their relevance to the onset of parasitic feeding. Journal of Fish Biology 44: 67–74.

Bjerselius, R., W. Li, J. H. Teeter, J. G. Seelye, P. B. Johnsen, P. J. Maniak, G. C. Grant, C. N. Polkinghorne & P. W. Sorensen, 2000. Direct behavioral evidence that unique bile acids released by larval sea lamprey (*Petromyzon marinus*) function as a migratory pheromone. Canadian Journal of Fisheries and Aquatic Sciences 57: 557–569.

Claytor, R. R. & H. R. MacCrimmon, 1987. Partitioning size from morphometric data: a comparison of five statistical procedures used in fisheries stock identification research. Canadian Technical Report of Fisheries and Aquatic Services 1531: 1–23.

Claytor, R. R., H. R. MacCrimmon & B. Gots, 1991. Continental and ecological variance components of European and North American Atlantic salmon (*Salmo salar*) phenotypes. Biological Journal of the Linnean Society 44: 203–229.

Costa, J. L., P. R. Almeida & M. J. Costa, 2003. A morphometric and meristic investigation of Lusitanian toadfish *Halobatrachus didactylus* (Bloch & Schneider, 1801): evidence of population fragmentation on Portuguese coast. Scientia Marina 67: 219–231.

Griffiths, R. W., F. W. H. Beamish, B. J. Morrison & L. A. Barker, 2001. Factors affecting larval sea lamprey growth and length at metamorphosis in lampricide-treated streams. Transactions of the American Fisheries Society 130: 289–306.

Hair, J. F., R. E. Anderson, R. L. Tatham & W. C. Black, 1998. Multivariate Data Analysis, 5th ed. Prentice Hall, Upper Saddle River, USA.

Hardisty, M. W., 1986. *Petromyzon marinus* Linnaeus, 1758. In Holčik, J. (ed.), The Freshwater Fishes of Europe –

Petromyzontiformes, Vol. 1. Part 1. Aula-Verlag, Wiesbaden, 94–116.

Hardisty, M. W. & I. C. Potter, 1971a. The behaviour, ecology and growth of larval lampreys. In Hardisty, M. W. & I. C. Potter (eds), The Biology of Lampreys, Vol. 1. Academic Press, London: 85–126.

Hardisty, M. W. & I. C. Potter, 1971b. The general biology of adult lampreys. In Hardisty, M. W. & I. C. Potter (eds), The Biology of Lampreys, Vol. 1. Academic Press, London: 127–206.

Holčik, J., 1986. Determination criteria. In Holčik, J. (ed.), The Freshwater Fishes of Europe – Petromyzontiformes, Vol. 1. Part 1. Aulag-Verlag, Wiesbaden: 24–28.

Ihssen, P. E., H. E. Booke, J. M. Casselman, J. M. McGlade, N. R. Payne & F. M. Utter, 1981. Stock identification: materials and methods. Canadian Journal of Fisheries and Aquatic Sciences 38: 1838–1855.

Jacobson, L. D., R. L. Torblaa & R. H. Morman, 1984. Problems with samples of sea lamprey (*Petromyzon marinus*) ammocoetes used in stock identification studies. Canadian Journal of Fisheries and Aquatic Sciences 41: 1421–1427.

Li, W., A. P. Scott, M. J. Siefkes, H. Yan, Q. Liu, S. Yun & D. Gage, 2002. Bile acid secreted by male sea lamprey that acts as a sex pheromone. Science 296: 138–141.

Li, W., P. W. Sorensen & D. D. Gallaher, 1995. The olfactory system of migratory adult sea lamprey (*Petromyzon marinus*) is specifically and acutely sensitive to unique bile acids released by conspecific larvae. Journal of General Physiology 105: 569–587.

Manion, P. J. & A. L. McLain, 1971. Biology of larval sea lampreys (*Petromyzon marinus*) of the 1960 year class, isolated in the Big Garlic River, Michigan, 1960–65. Great Lakes Fisheries Commission Technical Report 16: 1–35.

Manion, P. J. & T. M. Stauffer, 1970. Metamorphosis of the landlocked sea lamprey, *Petromyzon marinus*. Journal of the Fisheries Research Board of Canada 27: 1735–1746.

Melvin, G. D., M. J. Dadswell & A. McKenzie, 1992. Usefulness of meristic and morphometric characters in discriminating populations of American shad (*Alosa sapidissima*) (Ostreichthyes:Cluoeidae) inhabiting a marine environment. Canadian Journal of Fisheries and Aquatic Sciences 49: 266–280.

Morman, R. H., 1987. Relationship of density to growth and metamorphosis of caged larval sea lampreys, *Petromyzon marinus* Linnaeus, in Michigan streams. Journal of Fish Biology 30: 173–181.

Polkinghorne, C. N., J. M. Olson, D. G. Gallaher & P. W. Sorensen, 2001. Larval sea lamprey release two unique bile acids to the water at a rate sufficient to produce detectable riverine pheromone plumes. Fish Physiology and Biochemistry 24: 15–30.

Potter, I. C., 1980. The Petromyzoniformes with particular reference to paired species. Canadian Journal of Fisheries and Aquatic Sciences 37: 1595–1615.

Purvis, H. A., 1980. Effects of temperature on metamorphosis and the age and length at metamorphosis in sea lamprey (*Petromyzon marinus*) in the great lakes. Canadian Journal of Fisheries and Aquatic Sciences 37: 1827–1834.

Quintella, B. R., N. O. Andrade & P. R. Almeida, 2003. Distribution, larval stage duration and growth of the sea lamprey ammocoetes, *Petromyzon marinus* L., in a highly modified river basin. Ecology of Freshwater Fish 12: 286–293.

Reist, J. D., 1986. An empirical evaluation of coefficients used in residual and allometric adjustment of size covariation. Canadian Journal of Zoology 64: 1363–1368.

Ricker, W. E., 1975. Computation and interpretation of biological statistics of fish populations. Bulletin of the Fisheries Research Board of Canada 191: 1–382.

Rodríguez-Muñoz, R., J. R. Waldman, C. Grunwald, N. K. Roy & I. Wirgin, 2004. Mitochondrial DNA variation in sea lamprey between North American and Spanish rivers. Journal of Fish Biology 64: 783–787.

Rogado, L. (coord.), P. Alexandrino, P. R. Almeida, J. Alves, J. Bochechas, R. Cortes, I. Domingos, F. Filipe, J. Madeira & F. Magalhães, 2005. *Petromyzon marinus* Lampreia-marinha. In M. J. Cabral et al., Livro vermelho dos Vertebrados de Portugal Instituto de Conservação da Natureza, Lisboa, 65–66.

Rohlf, F. J. & D. E. Slice, 1995. Biomstat for Windows. Statistical Software for Biologists. Version 3.0. Exeter Software, New York.

Schaefer, K. M., 1991. Geographic variation in morphometric characters and gill-raker counts of Yellowfin Tuna, *Thunnus albacares*, from the Pacific Ocean. Fishery Bulletin 89: 289–297.

Sharp, J. C., K. A. Able, W. C. Leggett & J. E. Carscadden, 1978. Utility of meristic and morphometric characters for identification of capelin (*Mallotus villosus*) stocks in Canadian Atlantic waters. Journal of the Fisheries Research Board of Canada 35: 124–130.

SNIRH, 2004. Sistema Nacional de Informação de Recursos Hídricos. INAG, MAOTDR. Available: http://snirh.inag.pt (January 2006).

Sokal, R. R., & F. J. Rohlf, 1995. Biometry: The Principles and Practice of Statistics in Biological Research, 2nd ed. W.H. Freeman and Company, New York.

Tåning, A. V., 1952. Experimental study of meristic characters in fishes. Biological Reviews of the Cambridge Philosophical Society 27: 169–193.

Thorpe, R. S., 1976. Biometric analysis of geographic variation and racial affinities. Biological Reviews of the Cambridge Philosophical Society 51: 407–452.

Young, R. J., J. R. M. Kelso & J. G. Weise, 1990. Occurrence, relative abundance, and size of landlocked sea lamprey (*Petromyzon marinus*) ammocoetes in relation to stream characteristics in the Great Lakes. Canadian Journal of Fisheries and Aquatic Sciences 47: 1773–1778.

Youson, J. H., 1980. Morphology and physiology of lamprey metamorphosis. Canadian Journal of Fisheries and Aquatic Sciences 37: 1687–1710.

Youson, J. H., J. A. Holmes, J. A. Guchardi, J. G. Seelye, R. E. Beaver, J. E. Gersmehl, S. A. Sower & F. W. H. Beamish, 1993. The importance of condition factor and the influence of water temperature and photoperiod on metamorphosis of sea lampreys, *Petromyzon marinus*. Canadian Journal of Fisheries and Aquatic Sciences 50: 2448–2456.

Youson, J. H., J. Lee & I. C. Potter, 1979. The distribution of fat in larval, metamorphosing, and young adult anadromous sea lampreys, *Petromyzon marinus*. Canadian Journal of Zoology 57:237–246.

Youson, J. H. & I. C. Potter, 1979. A description of the stages of metamorphosis in the anadromous sea lamprey, *Petromyzon marinus* L. Canadian Journal of Zoology 57: 1808–1817.

Hydrobiologia (2008) 602:61–78
DOI 10.1007/s10750-008-9286-3

FISH AND DIADROMY IN EUROPE

Life history changes in Atlantic salmon from the River Dee, Wales

Miran W. Aprahamian · Ian C. Davidson · Richard J. Cove

Abstract The change in life history of Atlantic salmon (*Salmo salar* L.) on the River Dee over the last 60 years is described. Over the last 60 years, salmon have shown a change in run timing, the majority currently entering the river between August and October compared with prior to June. This has coincided with a change in the sea age composition, which was dominated by multi-sea winter salmon prior to the 1980s after which the proportion of 1sea-winter fish increased until they now dominate the mature population. Growth rates of salmon in fresh water remained relatively stable until the mid-1980s and then increased. By the end of the 1990s juvenile salmon were, by the end of their first and second year, respectively, ~60 and ~19%, on average, larger than they were between the late 1930s and mid-1980s. This has been reflected in a change in the age composition of smolts where the mean smolt age has declined from ~2 years prior to the 1980s to ~1.6 years in the late 1990s. There was no observed trend in post-smolt (marine) growth for salmon. Size at return for 1SW salmon appeared stable while there is some evidence of an increase in mean length of 2SW salmon at the end of the 1990s. A steady state life history model was developed which suggests an increase in the instantaneous rate of mortality by 2.9% from 1.495 $year^{-1}$ in 1937/1938 to 1.538 $year^{-1}$ in 1967/1969 and by 21.6% to 1.870 $year^{-1}$ in 1997/1999. This is considered to explain the shift in mean age at maturity from 5.2 to 4.8 to 3.9 years for the three periods examined. There is close agreement between the observed mean age at maturity and that predicted by the model suggesting optimal lifetime reproductive success.

Keywords *Salmo salar* · Life history · Sea age · Smolt age · Growth · Run timing

Guest editors: S. Dufour, E. Prévost, E. Rochard & P. Williot
Fish and diadromy in Europe (ecology, management, conservation)

M. W. Aprahamian (✉)
Environment Agency, Richard Fairclough House, Knutsford Road, Warrington WA4 1HG, UK
e-mail: miran.aprahamian@environment-agency.gov.uk

I. C. Davidson · R. J. Cove
Environment Agency, Chester Road, Buckley CH7 3AJ, UK

Introduction

Change in life histories of salmon (*Salmo salar* L.), specifically the sea age composition has been reported for numerous populations throughout their geographic range (Martin & Mitchell, 1985; Prouzet, 1990; Churchward & Hickley, 1991; Gough et al., 1992; Turrell & Shelton, 1993; Anon., 1994; Gudjonsson et al., 1995; Moore et al., 1995; Summers, 1995; Youngson, 1995; Webb & Campbell, 2000; García de Leániz et al., 2001; Youngson et al., 2002; Baglinière et al., 2004; Heddell-Cowie, 2005; Quinn et al., 2006). A number of possible causes have been postulated. These include changes in the freshwater environment affecting the growth of the juveniles and the subsequent age at maturity of the adults (Gardner, 1976; Anon., 1994), selective exploitation of certain stock components (Gee & Milner, 1980; Moore et al., 1995; Consuegra et al., 2005) and climatic changes affecting the ocean environment (Friedland & Hass, 1996; Bisbal & McConnaha, 1998; Friedland, 1998; Friedland et al., 2003).

Age at maturity in salmon has shown considerable changes over the last 100 years or more, with periods where the stock was dominated by multi-sea winter (MSW) salmon and others by grilse (1sea winter salmon) (George, 1984, 1991; Martin & Mitchell, 1985; Summers, 1995; Heddell-Cowie, 2005). The fact that the changes occurred at approximately the same time amongst rivers, suggests that change in the marine environment was the main cause for change in the age at maturity. The general consensus, as advocated by ICES (2004), is that the trend results from an increase in marine mortality. In relation to freshwater processes, growth and mortality, the influence of growth has been considered and discounted (Anon., 1994) and fresh water mortality only rarely considered (Riddell, 1986).

The change in age at maturity can be brought about through phenotypic, as suggested by Rago & Goodyear (1987) or by genetic adaptation as proposed by Youngson (1995) and by Consuegra et al. (2005) to a change in environmental conditions. Life-history theory suggests that a trait will change in relation to changes in age-specific mortality, growth and fecundity to ensure fitness is maximised (Roff, 1992; Stearns, 1992). The aims of this study were to describe any changes in life history characteristics over a 60-year period and reconstruct the total mortality and maturation schedule from the final age composition assuming the population is in equilibrium. The latter was estimated for three periods 1937/1938 and 1967/1969 when detailed studies on the biology of salmon were undertaken (Jones, 1939; Lees, 1972) and for 1997/1999 as this gave a 30-year period between investigations. These findings are then discussed in relation to changes in the fresh water and marine environments. The study also aims to predict the age at maturity that maximised fitness using observed growth rates and estimated total mortality rates and discuss the implications this might have for salmon management, specifically in relation to biological reference points.

Background

Study area

The Dee (Fig. 1) drains a watershed of 2,137 km^2, rising in the Cambrian Mountains and flowing eastward before heading north to enter the Irish Sea in Liverpool Bay, a distance of 160 km. The Dee is the most heavily regulated river in England and Wales. The main period of regulation started with the completion in 1956 of the new sluice gates (replacing those constructed at the start of the 19th century) downstream of Llyn Tegid (400 ha). At the same time, the Tryweryn was diverted so that it discharged upstream of the new sluice gates. In 1964, Llyn Celyn (325 ha) was constructed in the headwaters of the Tryweryn and in 1976 Llyn Brenig (370 ha), in the upper Alwen catchment, was completed.

The effect of the scheme was to reduce the amount of winter flooding and thus the magnitude of peak flows (Mayall, 2003) as well as the frequency of spring, summer and autumn freshets (APEM, 1998). The scheme was designed to support abstraction in the lower reaches and maintain a flow at Chester of 4.2 $m^3 s^{-1}$. Under certain conditions this flow is greater than would have occurred naturally and thus extreme low flows during the summer are prevented. The elevated flow has improved the estuarine environment and has virtually eliminated the kills of salmon that used to occur during the summer months as a result of poor water quality (Hodgson et al., 1980). The elevated flow has also resulted in a reduction in water temperature in the main Dee when compared to the naturalised situation (Hodgson, 1993). However, over the period 1965–1999, there has been a significant ($P < 0.05$) increase in mean annual river temperature of 0.02°C $year^{-1}$ (ANOVA) for details see Davidson et al. (2006). This could lead to enhanced growth of the juveniles (Brett, 1979) and salmon migrating to sea at a younger age (Metcalf et al., 1989; Økland et al., 1993).

The construction of the two reservoirs Celyn and Brenig had the effect of reducing the amount of nursery habitat available to salmon by a total of

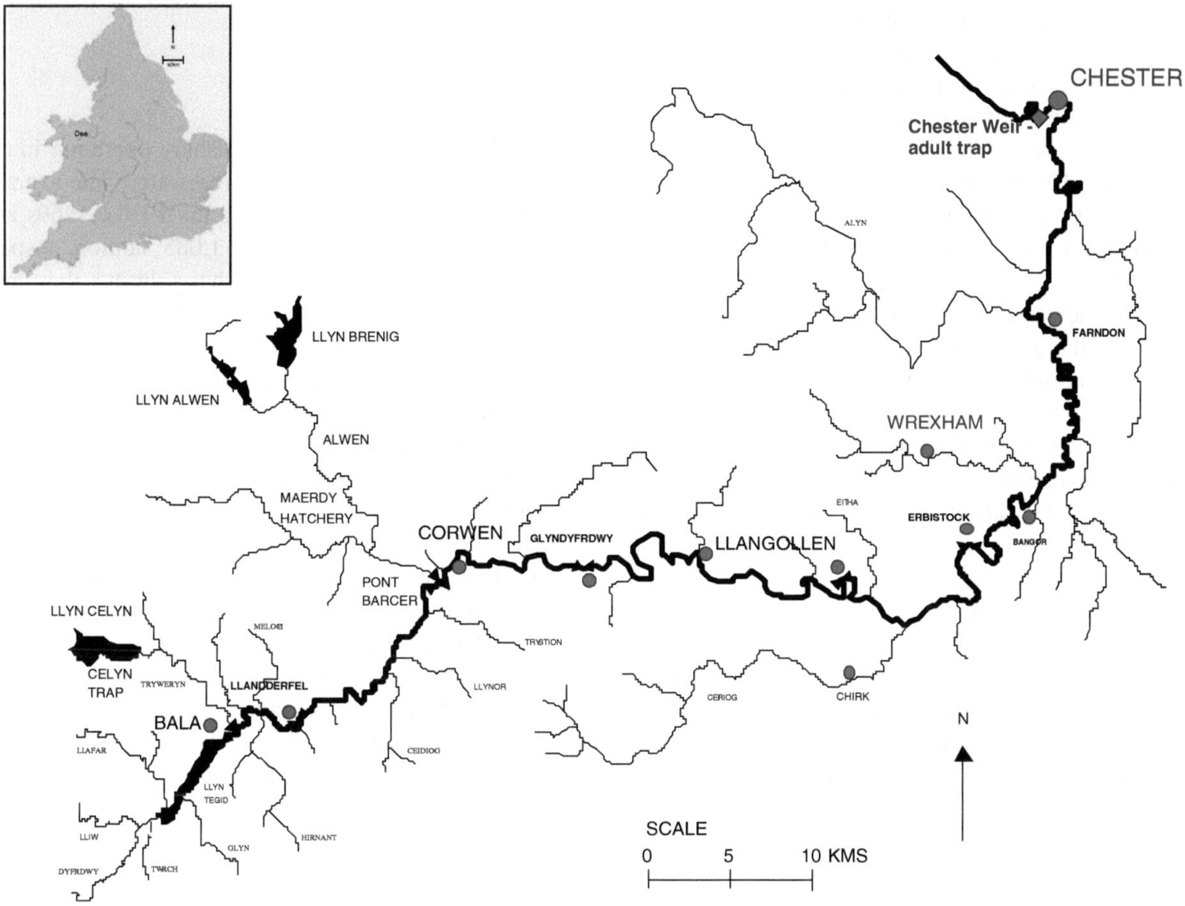

Fig. 1 The Dee catchment

25 ha, in equal proportion. There is, presently, 617 ha of habitat accessible to salmon. This may have impacted more on the MSW component as there is a tendency for the older sea age classes to spawn in the upper reaches of river catchments (Hawkins & Smith, 1986; Webb, 1989; Laughton & Smith, 1992).

The catchment is mainly rural supporting a mixture of beef and sheep farming on high ground and intensive dairy farming in the lowland plains (APEM, 1998). The main industrial areas are centred on the estuary and the towns of Chester, Wrexham and Ruabon. Landuse upstream of Llyn Tegid consists of marginal upland with the balance between intensive farming and rough grazing being dependent on market forces and agriculture subsidies (Thomson & Fuller, 2003). Agricultural statistics indicate that there has been a 2.5-fold increase in sheep numbers across North Wales between 1950 and 1995. Sheep numbers remained relatively stable between 1950 and 1970 at 1.5–2.0 million before increasing steadily during the 1970s and 1980s to a plateau of 3.5 million in the early 1990s. This would suggest an increase in grazed pasture and/or an increase use of fertiliser to enhance herbage production. Over the last 20–30 years management practices have changed, there has been an increase in the use of herbicides and fertilisers at the expense of ploughing and reseeding (Thomson & Fuller, 2003). There is evidence from the diatom record of nutrient enrichment in that part of the watershed that drains into Llyn Tegid, the level of diatom-inferred total phosphorous was relatively stable until the 1980s and then increased three fold (Bennion et al., 2003).

Forestry, specifically the development of coniferous forest, has increased over the last 50 years, especially during the 1980s. In 1995, 11.6% of the upper Dee catchment was estimated to be covered by forest an increase of 4.2% compared to the situation in 1985 (Thomson & Fuller, 2003). The upper catchment and the upper reaches of the rivers Alwen and Ceidiog experience problems with acidity (APEM, 1998) and elevated levels of aluminium (Heller, 1992). This may be related to the increased coverage of coniferous forest (Blackie & Newson, 1986).

The increased use of herbicides and insecticides as well as the elevated levels of aluminium may well have reduced the survival of juvenile salmon in fresh water and also at sea (Staurnes et al., 1996; Moore & Waring, 2001). While nutrient enrichment may have lead to increased growth rate and a reduction in mean smolt age (Metcalf et al., 1989; Økland et al., 1993).

Stock enhancement

There has been some stocking of hatchery-reared salmon, however, low return rates of tags from reared fish indicate that the vast majority of returning salmon are of wild origin (Cragg-Hine et al., 2006).

Fisheries

Home water fisheries

The Dee supports both a net fishery operating in the estuary and a rod fishery in freshwater. The average catch in the net fishery during the 1930s, 1960s and 1990s was 3,580, 3,278 and 1,085 salmon, respectively (Fig. 2) in the rod fishery during the same decades the average catch was 870, 1,208 and 462 salmon, respectively. There are two types of net fishery operating in the Dee, draft (consists of a wall of netting with a weighted foot rope and floated head rope, one end is held on the shore while the rest is paid out from a boat to enclose an area of water between two points on the shore, the net is then retrieved and any fish enclosed drawn up to the shore) and trammel (consists of 3 sheets of netting which hang from a floated head rope to a weighted foot rope the outer nets have a larger mesh size than the inner net and the net is designed to drift with the current or tide) net fisheries. The number of trammel nets has remained constant at four while the draft nets have declined from 35 in the 1930s to 31–33 nets during

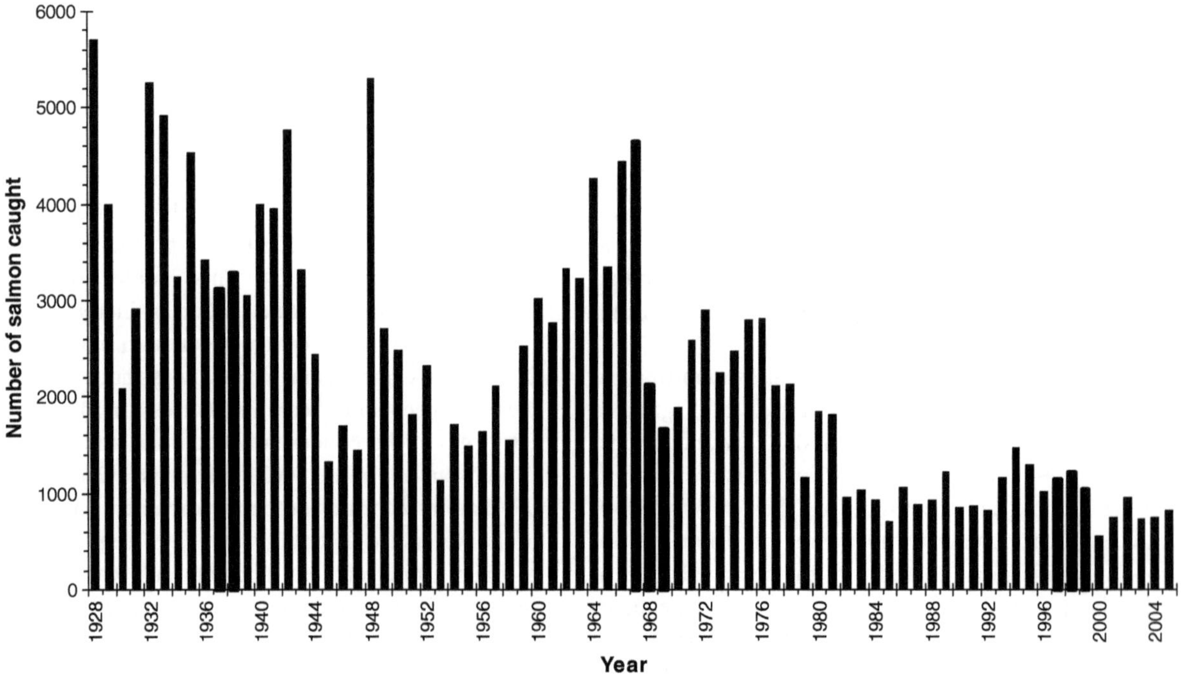

Fig. 2 The declared net catch of salmon from the River Dee (1928–2005), columns in bold represent reference years; 1937–1938, 1967–1969 and 1997–1999

the 1960s to 12–22 nets during the 1990s. From the 1930s to 1989, the netting season ran from March 15th to August 31st. In 1990 the start of the season was delayed until April 1st, from 1997 it moved to May 1st and from 2000 to present the season now operates from June 1st to August 31st. This will have reduced the level of exploitation on the early running (MSW) component of the salmon population. In the 1990s net exploitation is estimated to be around 19.5%. The mean rod exploitation in the early 1990s was estimated to be between 16.9 and 20.1% (Davidson et al., 1996). Exploitation was highest on the older sea age groups being ~31% for 3SW, ~22% for 2SW and between 15.2 and 18.5% for 1SW salmon.

High seas fisheries

Over the period, there has been an expansion of the high seas fisheries off West Greenland, the Faroes and the West coast of Ireland. The West Greenland fishery expanded during the 1960s, reaching a peak in the early 1970s of between 2,000 and 2,600 tonnes before declining to its current low level of ~15 tonnes. The fishery exploits salmon that are destined to spawn in United Kingdom waters as MSW salmon and the selective loss of this component of the stock may have lead to salmon maturing earlier (Gee & Milner, 1980; Consuegra et al., 2005). The exploitation level of the West Greenland fishery on English and Welsh salmon has been estimated at between 10 and 20% during the late 1980s/early 1990s (Russell & Potter, 1996), currently it is estimated to be below 1% (Anon., 2006).

The Faroes fishery commenced slightly later at the end of the 1960s, reaching a peak in the 1980s of between 500 and 1,300 tones since then it has declined to currently where no fishery operates. Exploitation of salmon in English and Welsh waters during the 1980s was estimated at ~1% (Russell & Potter, 1996), and is on both 1SW and MSW with a high preponderance of females (Shearer, 1992). The Irish fishery commenced around the same time as the West Greenland fishery and exploits English and Welsh salmon that are destined to spawn as 1SW. Exploitation of salmon from the Dee during the period 1992–1997 was 16.8 ± 5.9% after 1997 the rate dropped to 2.9 ± 1.3% (Anon., 2006). The fishery was at its peak during the 1960s until the end of the 1980s.

Materials and methods

Data sources

Catch and sea age composition were available for the period 1928–1973 from the Dee and Clwyd River Board, from 1974 to 1989 from Welsh Water Authority and more recently from the Environment Agency and its predecessor the National Rivers Authority (Fig. 2). Jones (1939, 1949, 1950) provides information on freshwater and sea age composition and back-calculated length-at-age, from adults caught in the net fishery. Lees (1972) provides similar data for the period 1967–1969 as well as data on smolt age composition. Additional information on freshwater residency was also available from Woolland (1972) and Laird (1976), for the early 1970s, from smolts sampled on their outmigration.

Since 1991 the Dee has supported an intensive programme of study (DSAP) into salmonid population dynamics, for more details see Davidson et al. (1996). Data were available on freshwater and sea age composition, back-calculated length at age, smolt size and age composition and adult run size. All the data with the exception of smolt size were derived from adult samples.

The data are based on single year samples and thus represent contributions from various cohorts. In this study, it has been assumed that the population is stable and that the data does represent cohort-specific data. It was not possible to calculate cohort specific life tables for years prior to 1991 as no abundance data is available.

Estimation of mortality and maturation schedule

For salmon, data from three time periods (1937–1938 (Jones, 1939), 1967–1969 (Lees, 1972) and 1997–1999 (Davidson & Cove, unpublished data)) were used to estimate the instantaneous total mortality from egg to adult and the maturation schedule pertaining at that time. As total abundance is unknown, it was assumed that the population was in equilibrium in each of these periods in order to obtain a unique solution for total mortality and the maturation probability. Thus, the number of eggs spawned in a particular year equals the number of eggs deposited by the cohort over its lifetime. A steady state life history model was used to determine the expected number of spawners ($n_{t,x}$) as follows:

$$n_{t,x} = \text{eggs}_{\text{spawn}}\, e^{(-Z(t+x+1))} \beta_x\, \alpha_t \prod_{i=0}^{x-1} (1 - \beta_i) \prod_{j=0}^{t-1} (1 - \alpha_j) \qquad (1)$$

where: t = sea age at spawning; x = age at smoltification; β_x = probability of a fish of freshwater age x becoming a smolt ($0 \leq \beta_x \leq 1$; $\beta_x \geq \beta_{x-1}$; β oldest age $x = 1$; $\beta_0 = 0$); α_t = probability of a fish of sea age t maturing ($0 \leq \alpha_t \leq 1$; $\alpha_t \geq \alpha_{t-1}$; α_t for oldest age $x = 1$; $\alpha_0 = 0$); Z = instantaneous total mortality per year.

The model was fitted subject to the following four conditions. First, that the population is semelparous the evidence from Jones (1939), Lees (1972) and Davidson and Cove (unpublished data) is that previous spawners contribute <5% of the spawning stock (Table 1). Second, that the population is stable such that eggs recruited by a cohort equals the number of eggs spawned which produced that cohort:

$$\text{eggs}_{\text{spawn}} = \sum_t \sum_x \text{eggs}_{t,x} \qquad (2)$$

$$\text{eggs}_{t,x} = n_{t,x}\, \delta_{t,x}\, \gamma_t \qquad (3)$$

where: δ = proportion of females (in the absence of data on sex ratio, the proportion female has been taken as 0.5); γ_t = age-specific fecundity as follows:

$$\gamma_t = 0.0012 L_s^{2.3345} \qquad (4)$$

(Pope et al. 1961).

L_s = length at return at sea age s (mm). The values of L_s for 1937–1938 were taken from Jones (1939) for the period 1967–1969 from Lees (1972) and for 1997–1999 from Davidson & Cove (unpublished data).

Thirdly, that the probability of a fish of freshwater age x becoming a smolt (β_x) equates to that reported during the three periods (Table 1). Fourthly, that the values of α_t and Z ensured that the number of fish in the various sea-age categories predicted from the model most closely resembled the proportion observed during the three time periods examined (Table 1).

Table 1 Age at migration in relation to age at return to the river for salmon from the River Dee in 1937–1938, 1967–1969 and between 1997 and 1999

Smolt age	Sea age				
	1SW	2SW	3SW	4SW	PS
(a) 1937–1938 (Jones, 1939)					
1	1.3%	3.8%	2.1%	0.3%	
2	12.4%	47.1%	23.9%	1.2%	4.7%
3	0.4%	2.2%	0.6%	0.1%	
N	154	578	292	21	51
(b) 1967–1969 (Lees, 1972)					
1	1.1%	2.3%	0.4%	>0.1%	
2	27.6%	49.1%	12.4%	0.5%	1.1%
3	2.2%	3.0%	0.2%		
N	774	1358	328	13	28
(c) 1997–1999 (Davidson & Cove, Unpublished)					
1	30.0%	11.3%	0.1%		0.3%
2	42.3%	12.9%	0.3%		1.6%
3	0.9%	0.2%			
N	2111	705	14	0	53

SW, Sea winter; PS, previous spawner; *N*, sample size

The model was fitted using solver in Microsoft excel.

Growth

Data were available on mean length at age, back-calculated from adult scales (Jones, 1939; Lees 1972; Davidson & Cove, unpublished data).

Fitness

Charlesworth (1980), Roff (1992) and Stearns (1992) have regarded lifetime fecundity, defined as the average number of eggs per female surviving to spawn, as the most suitable measure of fitness for stable populations. The optimal age at maturity was calculated using Eqs. 1–3 for each of the three periods, using the same levels of mortality (as determined above), growth history, smolt age (Table 1), fecundity and sex ratio but with no constraint on the final sea age structure of the spawning population such that the relationship:

$$\sum_t \sum_x \mathrm{eggs}_{t,x}/n_{t,x} \quad (5)$$

was maximised. The optimal age at maturity is defined as that age (total) which maximises the average number of eggs per female.

Results

The net fishery provided the main source of data used for the 1930 and 1960 samples, however, it was not possible to assess bias. The 1997–1999 samples were taken from the trap operated at Chester (Davidson et al., 1996), as the net fishery was effectively only operating over a 3-month period (June–August). A two sample *t*-test comparison of the mean weight of fish caught in the trap and net fishery, for each month, suggested no significant difference ($P < 0.05$), and that for these 3 months the net fishery accurately reflected the size composition of the run.

Run timing

The change in run timing (Fig. 3) was inferred from the net catch taken each month between 1934 and 1998. Up until the mid 1950s the pattern, though varying between years, remained relatively stable with around 12% of the salmon caught in March, approximately 20% taken in each month between April and July and only 7% in August. After the mid-1950s there is a gradual progression towards later running, so that by the 1990s above 85% of the catch was taken in July and August. The delay in the start of the fishing season until May 1st between 1995 and 1998 does not appear to have affected this trend.

Sea age composition

There has been a decline in the proportion of multi sea-winter (MSW) and a corresponding increase in 1SW salmon over the last 60 years (Fig. 4). Table 1 shows that in the late 1930s MSW salmon dominated the run with 2SW salmon being pre-eminent. By the late 1960s 2SW salmon were still dominant but there has been a decrease in the percentage of 3SW fish and an increase in 1SW salmon. At the end of the 1990s 1SW salmon were the most abundant sea age group constituting ca. 70% of the adult run, with 3SW and older fish being virtually absent.

Smolt age

The trend in smolt age was determined from adult samples (Fig. 5). It is possible that they may be biased reflecting age (size) selective mortality. However, the age composition of smolts and those derived from adults caught during the same period were similar (Table 2), suggesting that the life stage on which the trend is based (adults) has not biased the results. From the late 1930s until the late 1970s approximately 90% of the smolts were migrating at age 2 (Jones, 1939, 1950; Lees, 1972) in contrast at

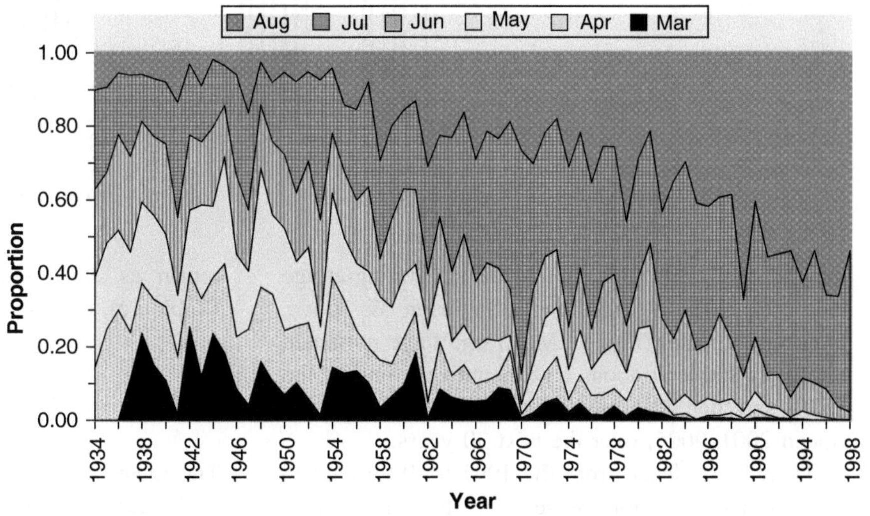

Fig. 3 Proportion of total salmon caught per month by the net fishery between 1934 and 1998

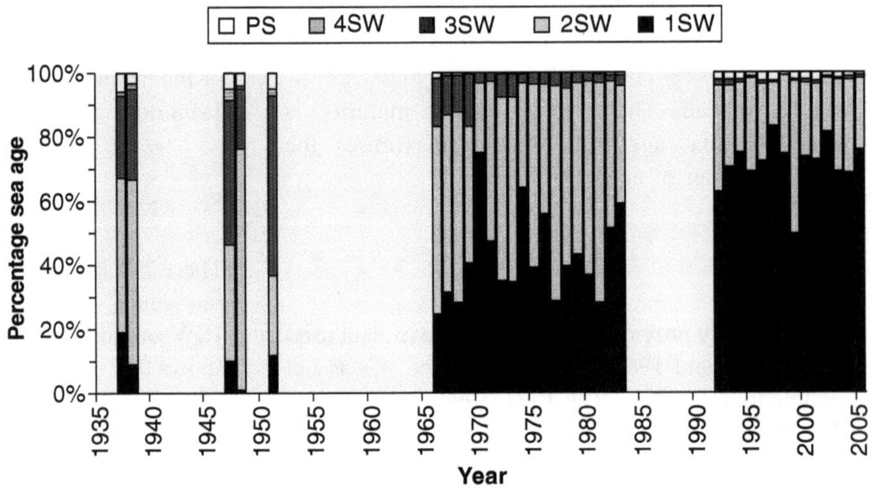

Fig. 4 Sea age composition of salmon between 1937 and 2005 (PS = previous spawner; SW = sea winter)

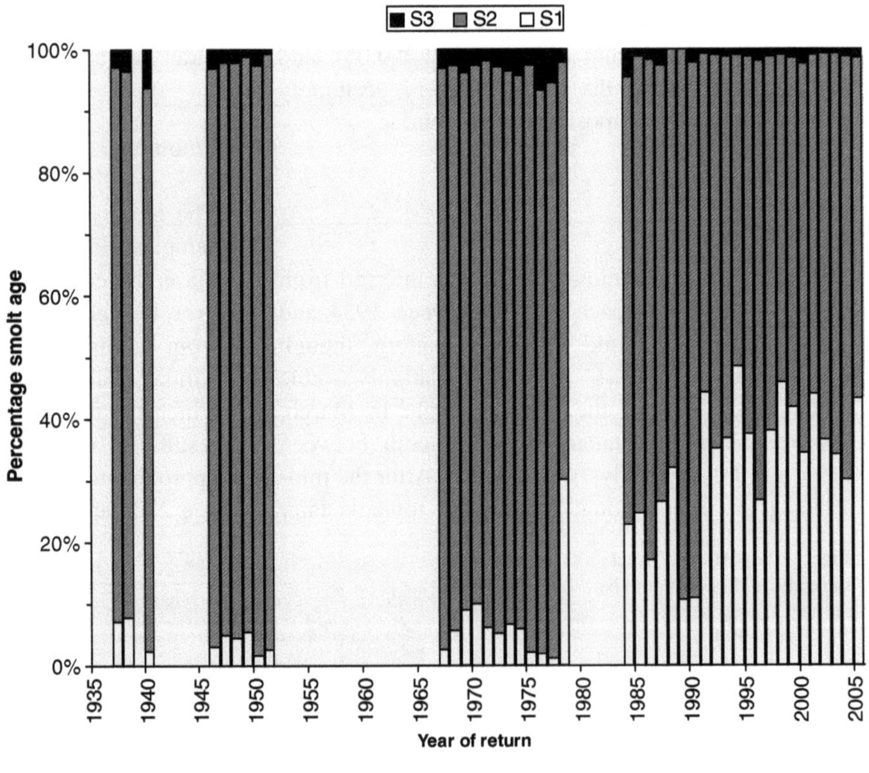

Fig. 5 Smolt age composition, based on year of return, between 1937 and 2005 (S refers to smolt age at year 1, 2 or 3)

the end of the 1990s ca. 60% were emigrating at age two and 40% at age one (Davidson & Cove, unpublished data) (Fig. 5). Mean smolt age remained relatively stable at about 2 years until the end of the 1970s and then decreased by ca. 18.5% to 1.63 years (mean 2001–2005) over the next 30 years.

There is evidence from the 1997 to 1999 data to suggest that S1 smolts have a greater tendency to return as 2SW salmon as opposed to S2 smolts (Table 1c), (Chi-square $P < 0.05$).

Growth

The mean size of parr at age 1 and 2 (Fig. 6) was back-calculated from adult scales. The mean length at

Table 2 Percentage smolt-age class categories for salmon from the River Dee

Smolt age	1967–1969	1969	1969–1970	1971–1972
Reference	Lees (1972)	Lees (1972)	Woolland (1972)	Laird (1976)
1	3.9	1.7	1.3	3.6
2	90.7	80.7	93.8	87.1
3	5.4	17.6	4.8	9.3
4			0.1	
N	2,501	1,247	1,339	588
No. of years	3	1	2	2
Life stage	Adult	Smolt	Smolt	Smolt

N, Number of fish aged

age 1 was around 50 mm during the late 1930s and 1960s, ranged between 55 and 70 mm during the 1980s and increased to between 80 and 90 mm during the 1990s and in the early part of 2000. A similar upward trend in size over the time period is evident for fish at the end of the second winter (which emigrated as 2-year-old smolts). For these fish, the mean size remained relatively stable at ca. 132 mm until the end of the 1980s after which it increased to ca. 156 mm. No trend was evident in post-smolt growth (growth in length between the last river annulus and the first sea water annulus (Friedland & Hass, 1996)), the values evident during the 1930s and 1960s being in the same range as that observed since 1982 (Fig. 7a). The length at return of 1SW and 2SW salmon has remained relatively stable for 1SW fish though for 2SW salmon there is an indication that size at return has increased, with the mean length at the end of the 1990s being in excess of 800 mm (Fig. 7b).

Mortality and maturation

The egg to adult mortality and the maturation schedule for the three periods was estimated using the steady state life history model. The approach is dependent on the population being at equilibrium. Although there are no data on the size of the population in the 1930s and 1960s, the indication from Fig. 8 is that currently the population is relatively stable with a mean (±95% CI) of 5,668 ± 836 salmon in the spawning run.

The instantaneous mortality rate was estimated to have increased from 1.495 year^{-1} in 1937–1938 to 1.538 year^{-1} in 1967–1969 to 1.870 year^{-1} in 1997–1999. The model was tested against the 1992–2003 data. The mean number of eggs deposited during the period was 15 million. (This assumes a spawning stock of 5,668 salmon, that fecundity was estimated according to Pope et al. (1961), that the sex ratio was 1:1 and that there is a 9% mortality rate between

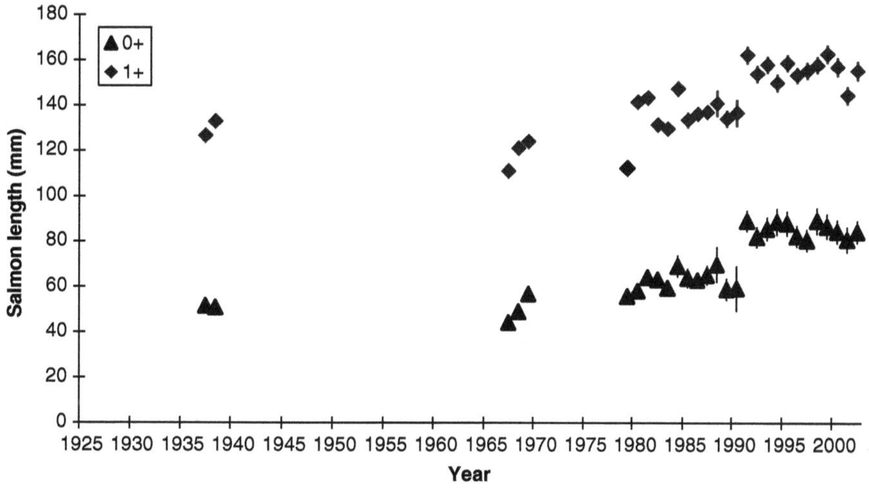

Fig. 6 Mean length of 0+ and 1+ salmon parr at the end of their first and second winter, respectively, between 1937 and 2002, 95% confidence limits presented for data post 1983

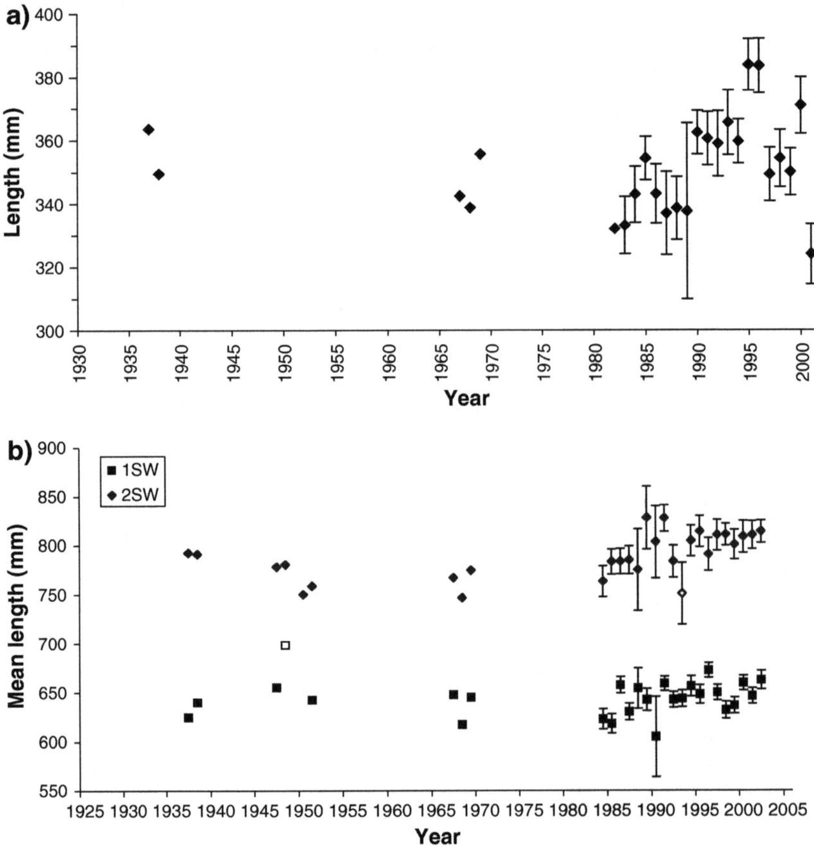

Fig. 7 Mean post-smolt growth of salmon between 1937 and 2002, 95% confidence limits presented for data post 1983 (**a**); mean length of 1SW and 2SW returning adults, 95% confidence limits presented for data post 1983 (**b**). Open symbol equals small sample size <10 individuals

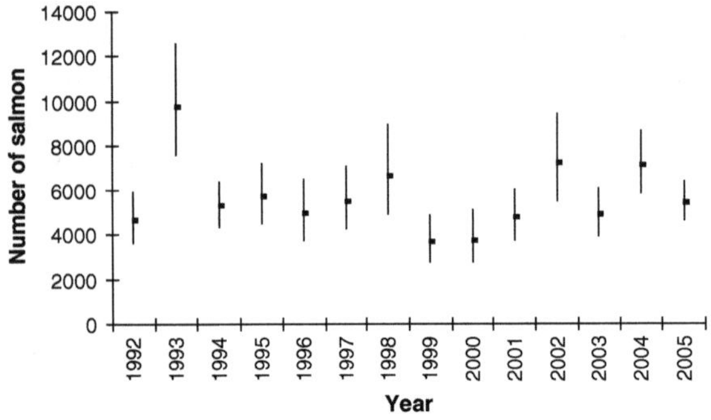

Fig. 8 Mean (±95%CI) run estimate of salmon between 1992 and 2005 (Davidson & Cove, unpublished)

when fish return to fresh water and spawn. For details of methodology see Milner et al. (2000).) The model predicted that 15 million eggs would produce a spawning stock of 5,562 salmon, this assumed a S1:S2 smolt ratio in the adult population of 2:3 (Davidson & Cove, unpublished data). The instantaneous mortality rate after the 0+ stage for the S1s was 2.402 year^{-1}, higher than for S2s of 1.860 year^{-1}, the overall estimate being 1.870 year^{-1}.

The mean age at maturity was estimated to have decreased during the time period, this excludes those males which will have matured precociously. The mean age declined from 5.2 years in 1937–1938 to 4.8 years in 1967–1969 to 3.9 years in 1997–1999. The maturation schedule as indicated from the model is shown in Table 3. There is a major shift in the

Table 3 Maturation schedule for salmon as estimated from the steady state life history model

Age (years)	1937–1938	1967–1969	1997–1999
3	0.000	0.000	0.020
4	0.015	0.048	0.286
5	0.262	0.406	0.953
6	0.794	0.781	1.000
7	1.000	1.000	

Age is in whole years taken from the time the eggs were laid to when the surviving adults spawn reflecting the complete life cycle

proportion of the population maturing at age 4 from <5% in the 1930s and 1960s to 28.6% in the 1990s, similarly at age 5 there has been a steady increase from 26.2 to 40.6 to 95.3% over the time period. The consequence of this is to reduce the number of eggs per female by 28.6% over the whole period; from 7,603 eggs in the late 1930s to 6,157 eggs in the late 1960s to 5,431 eggs in 1997–1999.

Fitness

The optimal age at maturity was predicted from the model under the same conditions of growth, mortality and fecundity as that considered to be pertaining to the three periods. The close agreement between the observed mean age at maturity and that predicted by the model (Fig. 9), suggests that the observed decrease in age at maturity from the 1930s to the 1990s has ensured that lifetime reproductive success, defined as the number of eggs produced per female, remains maximised.

Discussion

Over the period, there has been a shift in run timing from spring-summer to summer-autumn, a decline in the proportion of MSW salmon, an increase in growth rate of juvenile salmon and a concomitant decrease in the age of smolts. There is also an indication that the size of returning 2SW salmon has increased from the end of the 1990s, no trend was apparent for the 1SW salmon. It was estimated that total mortality over the life cycle of the salmon had increased by 25% between the late 1930s and the late 1990s.

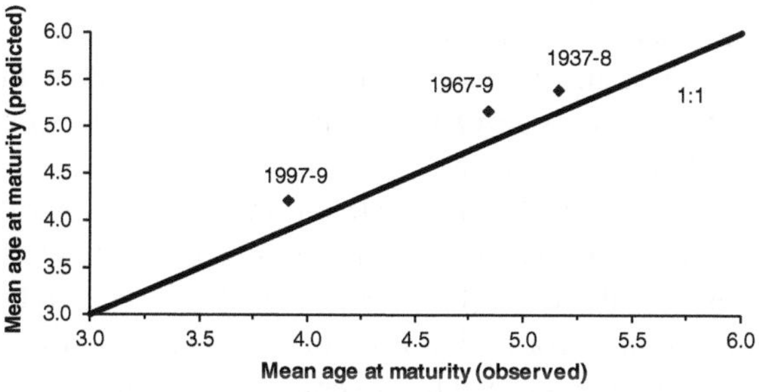

Fig. 9 Mean age at maturity for maximum egg production per female in relation to that observed during the three time periods

In this study, the catch from the net fishery was one of the main data sources (pre-1991) used to examine trends in run timing and age composition of salmon and, although this source may be biased (Moore et al., 1995), such data have been considered by others to accurately reflect changes within the population (Anon., 1994; Gudjonsson et al., 1995; Summers, 1995; MacLean et al., 2001; Baglinière et al., 2004). It is suggested that the observed change in the sea age composition is too large to be explained by sampling selectivity and it is unlikely that selection by the nets would affect smolt age. It is thus concluded that the gross changes in biological characteristics evident from the catch data reflect those of the population as a whole. There is the possibility that the results may reflect bias as a result of different people interpreting the scale features used to determine age, over the time period. However, for the Dee this was considered unlikely and that the observed change in the age composition of the smolt was not a consequence of differences in interpretation (Cragg-Hine et al., 2006).

The change in run timing for salmon is similar to that reported for other UK and Irish rivers (Gough et al., 1992; Anon., 1994; Youngson, 1995; Webb & Campbell, 2000; Milner et al., 2001; Youngson et al., 2002; Heddell-Cowie, 2005; Quinn et al., 2006), suggesting common causative factors. In the Dee the change appears to have taken place during the mid-1950s at the time when the scheme to regulate the flow in the Dee commenced but prior to the major construction of the reservoirs of Celyn (1964) and Brenig (1976). After the mid 1950s there is a steady increase in the proportions migrating from July onwards, while the decline in early running fish (March–May) continued to the early 1980s, they are now virtually absent. In the case of the latter this may in part relate to a decline in fishing effort in the early part of the season presumably due to a lack of fish. Similarly, the shift in sea age composition from a population dominated by multi-sea winter (MSW) salmon in the late 1930s and 1950s to one where one sea winter (1SW) fish now predominate is also evident from other river systems (Gee & Milner, 1980; Gough et al., 1992; Anon., 1994; Youngson, 1995).

Over the time period, there is evidence of an increase in freshwater growth rate which is likely to explain the decline in mean smolt age as the faster growing parr migrate to sea earlier (Metcalf et al., 1989; Økland et al., 1993). This change would appear to have started in the 1980s. The increase in growth rate may be related to a reduction in density-dependent processes (Gibson, 1993; Jenkins et al., 1999; Imre et al., 2005; Lobón-Cerviá, 2005), though other studies have found no relationship (Egglishaw & Shackley, 1977; Elliott, 1994). There is evidence from the catch data that the size of the population may have declined over the 60 years, assuming that catch reflects stock (Beaumont et al., 1991; Crozier & Kennedy, 2001; Gargan et al., 2001; Hansen, 2001; Whelan et al., 2001). The catch data suggest a gradual decline in the population from the 1960s to the early 1980s after which the population has remained relatively stable. The absence of synchrony between catch, being used as a surrogate for population size, and length at age one or two suggests other factors may be responsible.

Temperature is known to affect the timing and the size of alevins at emergence and their subsequent growth and survival (Elliott et al., 2000). Although the mean temperature in the Dee has increased by 0.2°C year^{-1} since the 1960s, it is considered unlikely that a change in emergence time could account for the increase in length at age one of $\sim$30 mm from ca. 50 mm in the late 1930s and 1960s to >80 mm during the 1990s.

Temperature also affects growth rate (Brett, 1979). Davidson & Hazlewood (2005) compared the back-calculated length at age for salmon with that predicted from the growth model of Elliott & Hurley (1997) and found for 0+ parr that prior to 1989 the growth model over estimated the back calculated mean lengths. After 1989 there was good agreement between the measured and predicted estimates of length at age one. For the 1+ parr the mean back-calculated lengths at the end of their second growing year were less than those predicted from the growth model for the year classes investigated 1979–1998. The growth model (Elliott & Hurley, 1997) assumes that fish are feeding on maximum rations. The fact that the growth model more accurately predicted the back-calculated length-at-age one after 1989 may indicate that food was no longer limiting growth. From the late 1970s there has been a change in land use, specifically agricultural intensification and may have resulted in enhanced invertebrate production and faster growth rate of juvenile salmonids. The lack of any association between observed and predicted length-at-age two (Davidson & Hazlewood, 2005) may reflect the increase in the proportion of 1-year-old smolts, so the 2-year-old fish sampled were the slower growing fraction of the population.

A decline in mean smolt age has been similarly observed in other British (Davidson & Hazlewood, 2005; J. Maclean, unpublished data) as well as in French populations of salmon (Baglinière et al., 2004). Davidson & Hazlewood (2005) reported an overall decline in the mean smolt age of 1SW and 2SW salmon on the Severn (England) and Wye (Wales) since the 1960s. The onset of this decline also appeared in the 1980s but was notably less marked on the Severn. On the North Esk (Scotland) there has similarly been a slow but steady decline in mean smolt age over the period 1971–2004 (J. Maclean, unpublished data). This decline in smolt age may effect reproductive success as egg size is smaller for S1 as opposed to S2 smolts of the same sea age and early survival (egg to swim-up) may also be lower (Moffett et al., 2006).

There appears to have been no systematic change in post-smolt growth of salmon over the 60-year period with growth rates in the late 1930s and 1960s being within the range observed currently. Since the early 1980s, the trend was for an increase in post-smolt growth up until the mid 1990s, since then there has been a decline. Davidson & Hazlewood (2005) examined post-smolt growth for a number of British and Irish rivers, between the late 1970s/early 1980s and the late 1990s, and showed a high degree of synchrony, suggesting the influence of common factors at sea. However, there are no similar long-term data sets with which to compare the findings on the Dee. The size at return has remained relatively stable for 1SW salmon but there is a suggestion that towards the end of the period the mean size of 2SW has increased. This is in contrast to the findings of Quinn et al. (2006) who showed a decline since the 1950s in the size of those salmon, considered to be predominantly grilse.

For some Atlantic salmon populations marine survival rates have been shown to be correlated with post-smolt growth (Friedland et al., 2000) while other have found no association (Crozier & Kennedy, 1999). This study suggested that overall mortality for salmon has increased over the 60 years and in particular between the late 1960s and the end of the 1990s. This would seem to indicate that post-smolt growth is not a good correlate/surrogate for marine mortality.

The increase in marine mortality (Potter et al., 2003a) has been considered either as a result of marine environmental changes (Anon., 1994; Summers, 1995; ICES, 2004) or as a result of selective exploitation of the older age classes (Gee & Milner, 1980; Consuegra et al., 2005). The situation could also reflect the selective loss of a particular component (MSW) of the stock through degradation of their freshwater habitat (Youngson, 1995). There is a tendency for the older sea age classes to spawn in the upper reaches of river catchments (Hawkins & Smith, 1986; Webb, 1989; Laughton & Smith, 1992). If the habitat has become degraded such that it is now unsuitable and even if restored the younger sea age classes (1SW salmon) could not spawn or reproduce successfully then the overall age at maturity would decline. Possible reasons for this include substrate size and/or the hydrological regime. Large fish bury their eggs deeper in the gravel (Jones, 1959) and if the depth that eggs are buried has an advantage by increasing their survival rate then this may explain why the upper catchment may be unsuitable for 1SW salmon.

It would seem unlikely that selective exploitation of the older age classes is the main factor affecting the change in age at maturity, as it is unclear as to how it might explain some of the long-term cycles in sea age composition (Martin & Mitchell, 1985; Summers, 1995). Exploitation will have had an influence on the overall level of mortality, and the increase in high seas fishing pressure off Greenland would result in fish being caught in the year(s) before they mature (Shearer, 1992). The effect of this would be to reduce the average size and age of the home water catch. Of the three time series, the increase in fishing pressure would mainly have affected the 1967–1969 sample. It has not been possible to quantify the change in exploitation. However, historically the home water level of exploitation is unlikely to be less than the level of 31.7% at the end of the 1990s (mean total net and rod exploitation 1997–1999), due to higher effort specifically in the net fishery. Thus the increase in mortality is unlikely to have been the result of an increase in exploitation. There is also the possibility that the change in age at maturity could simply reflect higher natural mortality in the second and subsequent years at sea, with no change in the maturation schedule. However, if this was the case then the model suggests that the selection pressure would still be for fish to mature early.

Some of this mortality may relate to the fact that smolts are migrating at a younger age and thus smaller size (Økland et al., 1993) resulting in a lower survival rate when compared to their larger conspecifics (Chadwick, 1987; Lundqvist et al., 1994), but this is not always the case (Dempson et al., 2003).

While it might be reasonable to assume European populations of salmon respond similarly to common marine factors this is less likely to be the situation in fresh water. Kennedy & Crozier (1995) did, however, report a significant positive correlation in egg to smolt survival between the rivers Bush and Burrishoole (Ireland) for 12 out of the 13 years, between 1973 and 1985. The European Common Agricultural Policy has resulted in large-scale intensification of agriculture during the 1980s, especially in the uplands. Hendry et al. (2003) reviewed the impact of changes in land use practice on juvenile salmon and though not quantified, the changes could result in an increase in mortality. In fact, fresh water mortality in the Dee might have been expected to have decreased over the time period as a result of density-dependent processes (Elliott, 2001). For example the net catch during the 1960s was three times that of the 1990s and if catch reflects stock size it would suggest a reduction in negative density-dependent factors. Also carrying capacity may have increased since the end of the 1980s, as a result of enrichment, further reducing density-dependent effects.

The impact of increased mortality could have been larger, if growth rate in freshwater had not improved. One of the consequences of the increase in growth rate and smolts migrating at an earlier age is to dampen the impact of the increase in mortality. There are about 40% more salmon spawners when their mean smolt age is 1.5 years compared with 2 years, assuming the same level of mortality and maturation schedule to that operating at the end of the 1990s. The impact on egg deposition may also have been higher than predicted. This is because a 1:1 sex ratio was assumed in the model for each age group and though no sex ratio data was available for the Dee data from other populations indicates that males dominate the 1SW and females the 2SW component in MSW populations (Environment Agency, 1998).

Of the processes that influence life histories, growth was observed to have increased, mortality estimated to have increased and there was no information on any changes in fecundity. Age at maturity has been found to decrease with an increase in growth rate (Alm, 1959). However, the change in growth rate has been taken into account and it would thus appear that the change in mortality has been the major influence on the change in the maturation schedule. The change in age at maturity can result from phenotypic plasticity or through genetic adaptation, to the change in environmental conditions. In this study it was not possible to determine the process but both have been proposed. Rago & Goodyear (1987) suggested that a phenotypic response could account for the change in age at maturity while Youngson (1995) and Consuegra et al. (2005) have suggested that the change reflects a loss of MSW genotypes. The close agreement between the observed and expected age at maturity suggests that age at maturity is being optimised. This provides a mechanism, which could explain the cycles in age at maturity. The model indicates that the fish are favoured to mature early when mortality is high and, when mortality reduces, the selection pressure will be for fish to mature later, if fitness is to be maximised. This is in agreement with the findings of Rago & Goodyear (1987) who showed that increased mortality (fishing pressure) reduced the age at which the fish were able to achieve their maximum reproductive potential. However, the findings are in contrast to those of Hutchings & Jones (1998) who reported that under conditions of low survival females are favoured to mature as 2SW fish while under high survival conditions maturation as grilse is favoured. The difference probably reflects the level of fitness used in the studies. In the present study the maximum number of eggs per females was used as the measure of fitness while Hutchings & Jones (1998) used the rate of growth of the population. In addition one of the consequences of a reduction in smolt age would be for S1 salmon to mature later than S2s for any specific level of mortality, if the number of eggs per females is to be maximised. There is certainly some evidence from the 1997 to 1999 data that this is the case. The main argument has focused on the selective advantage of delayed maturity for females. For males the benefits of large body size may be that they are more attractive, and better able to compete for access, to females such that reproductive success increases with body size as well as being, possibly better able to cope with the physical conditions (Holtby & Healey, 1990; Quinn, 2005).

The management of salmon is through biological reference points (Potter et al., 2003b) and separate reference points have been asked for different sea age groups (NASCO, 1999). If the change in sea age composition of salmon is a plastic response to an increase in mortality irrespective of when it occurs in the life cycle, then separate biological reference points for different sea age groups would not seem to be appropriate. If the aim is to increase the MSW component of the stock then the overall mortality needs to be reduced. Whether reducing exploitation would be enough to restore the MSW component to the Dee is dependent on what proportion of the total mortality is attributable to fishing. If the overall mortality can be reduced then there is evidence from other species that the population can respond quickly by delaying maturation (Reznick et al., 1990; Olsen et al., 2005). This assumes that that the observed change is due to phenotypic plasticity as opposed to genetic changes as the latter are likely to be more difficult to reverse. Clearly, the degree to which managers can influence the population life history depends on where the main mortality is taking place and its cause.

Acknowledgements We thank J. L. Baglinière, S. Barnard, L. Black, G. Mawle, N. Milner, M. Pawson, and G. Williams for their comments on this study and R. Wyatt for checking some of the outputs from the model using Winbugs. We gratefully appreciate the extremely constructive reviews of three anonymous referees. The views expressed are those of the authors and not necessarily those of the Environment Agency.

References

Alm, G., 1959. Connection between maturity, size, and age in fishes. Report of the Institute of Freshwater Research, Drottningholm 40: 5–145.

Anon., 1994. Run Timing of Salmon. MAFF Publications, London.

Anon., 2006. Annual Assessment of Salmon Stocks and Fisheries in England and Wales 2005. Environment Agency, Bristol; Centre for Environment, Fisheries and Aquaculture Science, Lowestoft.

APEM, 1998. An Investigation into the Impact of Flow Regulation on Fisheries in the River Dee. Environment Agency, Bristol.

Baglinière, J. L., L. Denais, E. Rivot, J. P. Porcher, E. Prévost, F. Marchand & V. Vauclin, 2004. Length and age structure modifications of the Atlantic salmon (*Salmo salar*) populations of Brittany and Lower Normandy from 1972 to 2002. Technical report, Institut National de la Recherche Agronomique and Conseil Supérieur de la Pêche, Rennes.

Beaumont, W. R. C., J. S. Welton & M. Ladle, 1991. Comparison of rod catch data and known numbers of Atlantic salmon (*Salmo salar*) recorded by a resistivity fish counter in a southern chalk stream. In Cowx, I. G. (ed.), Catch Effort Sampling Strategies, Their Application in Freshwater Fisheries Management. Fishing News Books, Oxford: 49–60.

Bennion, H., E. Shilland & P. Abbleby, 2003. An assessment of recent environmental change in Llyn Tegid using the sediment record. In Gritten, R. H., C. A. Duigan & H. Millband (eds), The Ecology, Conservation and Environmental History of the Largest Lake in Wales. University of Liverpool, Liverpool: 153–168.

Bisbal, G. A. & W. E. McConnaha, 1998. Consideration of ocean conditions in the management of salmon. Canadian Journal of Fish and Aquatic Sciences 55: 2178–2186.

Blackie, J. R. & M. D. Newson, 1986. The effects of forestry on the quantity and quality of runoff in upland Britain. In De L. G. Solbé, J. F. (ed.), Effects of Land Use on Fresh Waters Agriculture, Forestry, Mineral Exploitation, Urbanisation. Ellis Horwood Limited, Chichester: 398–412.

Brett, J. R., 1979. Environmental factors and growth. In Hoar, W. S., D. J. Randall & J. R. Brett (eds), Fish Physiology, Vol. 8. Academic Press, London: 599–675.

Chadwick, E. M. P., 1987. Causes of variable recruitment in a small Atlantic salmon stock. American Fisheries Society Symposium 1: 390–401.

Charlesworth, B., 1980. Evolution in Age-Structured Populations. Cambridge University Press, Cambridge.

Churchward, A. S. & P. Hickley, 1991. The Atlantic salmon fishery of the River Severn (UK). In Cowx, I. G. (ed.), Catch Effort Sampling Strategies, Their Application in Freshwater Fisheries Management. Fishing news Books, Oxford: 1–14.

Consuegra, S., C. García de Leániz, A. Serdio & A. Verspoor, 2005. Selective exploitation of early running fish may induce genetic and phenotypic changes in Atlantic salmon. Journal of Fish Biology 67(Supplement A): 129–145.

Cragg-Hine, D., D. C. Bradley & K. Hendry, 2006. Changes in salmon smolt ages in the Welsh River Dee over a sixty-six year period. Journal of Fish Biology 68: 1891–1895.

Crozier, W. W. & G. J. A. Kennedy, 1999. Relationships between marine growth and marine survival of one sea winter Atlantic salmon, *Salmon salar* L. from the River Bush, Northern Ireland. Fisheries Management and Ecology 6: 89–96.

Crozier, W. W. & G. J. A. Kennedy, 2001. Relationship between freshwater angling catch of Atlantic salmon and stock size in the River Bush, Northern Ireland. Journal of Fish Biology 58: 240–247.

Davidson, I. C., R. J. Cove, N. J. Milner & W. K. Purvis, 1996. Estimation of Atlantic salmon (*Salmo salar*, L.) and sea trout (*Salmo trutta*, L.) run size and angling exploitation on the Welsh Dee using mark-recapture and trap indices. In I. G. Cowx, (ed.), Stock Assessment in Inland Fisheries. Fishing News Books, Blackwell Science, Oxford: 293–307.

Davidson, I. C. & M. S. Hazlewood, 2005. Effect of climate change on salmon fisheries. Environment Agency Science Report W2-047/SR. Environment Agency, Bristol.

Davidson, I. C., M. S. Hazelwood & R. J. Cove, 2006. Predicted growth of juvenile trout and salmon in four rivers in England and Wales based on past and possible future temperature regimes linked to climate change. In Harris, G. S. & N. J. Milner (eds), Sea Trout: Biology, Conservation and Management. Proceedings of First International Sea Trout Symposium, Cardiff, July 2004. Fishing News Books, Blackwell Scientific Publications, Oxford: 401–416.

Dempson, J. B., C. C. Mullins, C. Bourgeois, M. F. O'Connell & D. G. Reddin, 2003. Perspectives on smolt production and marine survival of Newfoundland Atlantic salmon (*Salmo salar* L.) to smolt size and run timing. In Potter, E. C. E., N. Ó. Maoiléidigh & G. Chaput (eds), Marine Mortality of Atlantic Salmon, *Salmo salar* L: Methods and Measures. Department of Fisheries and Oceans Canadian Science Advisory Secretariat Research Document 2003/101, 27–43.

Egglishaw, H. J. & P. E. Shackley, 1977. Growth, survival and production of juvenile salmon and trout in a Scottish stream. Journal of Fish Biology 11: 647–672.

Elliott, J. M., 1994. Quantitative Ecology and the Brown Trout. Oxford University Press, Oxford.

Elliott, J. M., 2001. The relative role of density in the stock-recruitment relationship of salmonids. In Prévost, E., & G. Chaput (eds), Stock Recruitment and Reference Points Assessment and Management of Atlantic Salmon. INRA, Paris: 25–66.

Elliott, J. M. & M. A. Hurley, 1997. A functional model for maximum growth of Atlantic salmon parr, *Salmo salar*, from two populations in Northwest England. Functional Ecology 11: 592–603.

Elliott, J. M., M. A. Hurley & S. C. Maberly, 2000. The emergence period of sea trout fry in a Lake District stream correlates with the North Atlantic Oscillation. Journal of Fish Biology 56: 208–210.

Environment Agency, 1998. Salmon Action Plan Guidelines, Version 2. Environment Agency, Bristol.

Friedland, K. D., 1998. Ocean climate influences on critical Atlantic salmon (*Salmo salar*) life history events. Canadian Journal of Fish and Aquatic Sciences 55: 119–130.

Friedland, K. D., L. P. Hansen, D. A. Dunkley & J. C. Maclean, 2000. Linkage between ocean climate, post-smolt growth and survival of Atlantic salmon (*Salmo salar*, L.) in the North Sea area. ICES Journal of Marine Science 57: 419–429.

Friedland, K. D. & R. E. Hass, 1996. Marine post-smolt growth and age maturity of Atlantic salmon. Journal of Fish Biology 48: 1–15.

Friedland, K. D., D. G. Reddin & M. Castonguay, 2003. Ocean thermal conditions in the post-smolt nursery of North American Atlantic salmon. ICES Journal of Marine Science 60: 343–355.

García de Leániz, C., A. Serdio & S. Consuegra, 2001. Present status of Atlantic salmon in Cantabria. In García de Leániz, C., A. Serdio & S. Consuegra (eds), El Salmón, Joya de Nuestros Ríos. Consejería de Ganadería, Agricultura y Pesca, Santander: 55–82.
Gardner, M. L. G., 1976. A review of factors which may influence the sea-age and maturation of Atlantic salmon, *Salmo salar*. Journal of Fish Biology 9: 289–327.
Gargan, P., J. Stafford & N. Ó. Maoiléidigh, 2001. The relationship between salmon rod catch, stock size, rod exploitation and rod effort on the Erriff fishery western Ireland. In Shelton, R. (ed.), The Interpretation of Rod and Net Catch Data. The Atlantic Salmon Trust, Pitlochry: 68–75.
Gee, A. S. & N. J. Milner, 1980. Analysis of 70-year catch statistics for Atlantic salmon (*Salmo salar*) in the River Wye and implications for management of the stocks. Journal of Applied Ecology 17: 41–57.
George, A. F., 1984. Scottish salmon and grilse return-migration variations over 200 years. In Holden, A. H. (ed.), Proceedings of the Institute of Fisheries Management 15th Annual Study Course, Stirling University. Institute of Fisheries Management, Nottingham: 23–32.
George, A. F., 1991. Climate and the Salmon. Salmon, Trout and Sea-Trout. December 1991.
Gibson, R. J., 1993. The Atlantic salmon in freshwater: Spawning, rearing and production. Reviews in Fish Biology and Fisheries 3: 39–73.
Gough, P. G., A. J. Winstone & P. G. Hilder, 1992. Spring salmon a review of factors affecting the abundance and catch of spring salmon from the River Wye and elsewhere, and proposals for stock maintenance and enhancement. National Rivers Authority Welsh Region Technical Fisheries Report No. 2. National Rivers Authority, Bristol.
Gudjonsson, S., S. M. Einarsson, Th. Antonsson & G. Gudbergsson, 1995. Relation of grilse to salmon ratio to environment changes in several wild stocks of Atlantic salmon (*Salmo salar*) in Iceland. Canadian Journal of Fish and Aquatic Sciences 52: 1385–1398.
Hansen, L. P., 2001. Relationship between catches, rod exploitation and total run of Atlantic salmon in the River Drammenselv, Norway. In Shelton, R. (ed.), The Interpretation of Rod and Net Catch Data. The Atlantic Salmon Trust, Pitlochry: 85–90.
Hawkins, A. D. & G. W. Smith, 1986. Radio-tracking observations on Atlantic salmon ascending the Aberdeenshire Dee. Scottish Fisheries Research Report 36.
Heddell-Cowie, M., 2005. Importance of the River Teviot to Atlantic salmon, *Salmo salar*, rod catches in the River Tweed, Scotland. Fisheries Management and Ecology 12: 137–142.
Heller, M. D., 1992. pH and flow in the Afon Alwen and Afon Brenig below the reservoirs. National Rivers Authority, Technical Memorandum, EAU/92/TM05. National Rivers Authority, Cardiff.
Hendry, K., D. Cragg-Hine, M. O'Grady, H. Sambrook & A. Stephens, 2003. Management of habitat for rehabilitation and enhancement of salmonid stocks. Fisheries Research 62: 171–192.
Hodgson, P. B., 1993. The ecology and management of the coarse fish populations of the lower Welsh Dee. Ph.D. Thesis, University of Liverpool, Liverpool.
Hodgson, P. B., A. E. Weston & T. P. Westwood, 1980. Environmental factors influencing the management of the canalised estuarial reaches of the regulated River Dee. Paper presented at the Annual General Meeting of the Water Engineers and Scientists.
Holtby, L. B. & M. C. Healey, 1990. Sex-specific life-history tactics and risk-taking in coho salmon. Ecology 71: 678–690.
Hutchings, J. A. & M. E. B. Jones, 1998. Life history variation and growth rate thresholds for maturity in Atlantic salmon, *Salmo salar*. Canadian Journal of Fish and Aquatic Sciences 55(Supplement 1): 22–47.
ICES, 2004. Report of the working Group on North Atlantic salmon. Halifax, Canada, 29 March–8 April 2004. International Council for the Exploration of the Sea C. M.2004/ACFM: 20.
Imre, I., J. W. A. Grant & R. A. Cunjak, 2005. Density-dependent growth of young-of-the-year Atlantic salmon *Salmo salar* in Catamaran Brook, New Brunswick. Journal of Animal Ecology 74: 508–516.
Jenkins, T. M., Jr., S. Diehl, K. W. Kratz & S. D. Cooper, 1999. Effects of population density on individual growth of brown trout in streams. Ecology 80: 941–956.
Jones, J. W., 1939. Salmon of the Cheshire Dee, 1937 and 1938. Transactions of the Liverpool Biological Society 52: 19–79.
Jones, J. W., 1949. Studies of the scales of young salmon (*Salmo salar* L. (Juv.)) in relation to growth, migration and spawning. Fisheries Investigations Series 1, V.
Jones, J. W., 1950. Salmon of the Cheshire Dee. Fisheries Investigations Series 1, V, No. 3.
Jones, J. W., 1959. The Salmon. Collins, London.
Kennedy, G. J. A. & W. W. Crozier, 1995. Factors affecting recruitment success in salmonids. In Harper, D. M. & A. J. D. Ferguson (eds), The Ecological Basis for River Management. John Wiley & Sons Ltd., Chichester: 349–362.
Laird, L. M., 1976. Growth and movements of juvenile Atlantic salmon and brown trout in the Afon Mynach. Ph.D. Thesis, University of Liverpool, Liverpool.
Laughton, R. & G. W. Smith, 1992. The relationship between the date of river entry and the estimated spawning position of adult Atlantic salmon (*Salmo salar* L.) in two major Scottish east coast rivers. In Priede, I. G., & S. M. Swift (eds), Wildlife Telemetry: Remote Monitoring and Tracking of Animals. Ellis Horwood, New York: 423–433.
Lees, P. R., 1972. The salmonid fish of the rivers Dee and Clwyd. Ph.D. Thesis, University of Liverpool, Liverpool.
Lobón-Cerviá, J., 2005. Spatial and temporal variation in the influence of density dependence on growth of stream-living brown trout (*Salmo trutta*). Canadian Journal of Fish and Aquatic Sciences 62: 1231–1242.
Lundqvist, H., S. Mckinnell, H. Fängstam & I. Berglund, 1994. The effect of time, size and sex on recapture rates and yield after river releases of *Salmo salar* smolts. Aquaculture 121: 245–257.
MacLean, J. C., A. F. Youngson & R. J. Fryer, 2001. The use of rod catch data in assessing trends in relative abundance. In Shelton, R. (ed.), The Interpretation of Rod

and Net Catch Data. The Atlantic Salmon Trust, Pitlochry: 91–99.

Martin, J. H. A. & K. A. Mitchell, 1985. Influence of sea temperature upon the numbers of grilse and multi-sea-winter Atlantic salmon (*Salmo salar*) caught in the vicinity of the River Dee (Aberdeenshire). Canadian Journal of Fish and Aquatic Sciences 42: 1513–1521.

Mayall, S., 2003. Development of flow regulation in the River Dee catchment and the associated impacts on the level regime at Llyn Tegid. In Gritten, R. H., C. A. Duigan & H. Millband (eds), The Ecology, Conservation and Environmental History of the Largest Lake in Wales. University of Liverpool, Liverpool: 49–57.

Metcalfe, N. B., F. A. Huntingford, W. D. Graham & J. E. Thorpe, 1989. Early social status and the development of life-history strategies in Atlantic salmon. Proceedings of the Royal Society London B 236: 7–19.

Milner, N. J., I. C. Davidson, R. E. Evans, V. Locke & R. J. Wyatt, 2001. The use of rod catches to estimate salmon runs in England and Wales. In Shelton, R. (ed.), The Interpretation of Rod and Net Catch Data. The Atlantic Salmon Trust, Pitlochry: 46–67.

Milner, N. J., I. C. Davidson, R. J. Wyatt & M. W. Aprahamian, 2000. The application of biological targets to the management of salmon recreational fisheries in England and Wales. In Cowx, I. G. (ed.), Management and Ecology of River Fisheries. Fishing News Books, Oxford: 361–372.

Moffett, I. J. J., M. Allen, C. F. Flanagan, W. W. Crozier & G. J. A. Kennedy, 2006. Fecundity, egg size and early hatchery survival for wild Atlantic salmon, from the River Bush. Fisheries Management and Ecology 13: 73–79.

Moore, D. S., G. L. Chaput & P. R. Pickard, 1995. The effect of fisheries on the biological characteristics and survival of mature Atlantic salmon (*Salmo salar*) from the Miramichi river. In Chadwick, E. M. P. (ed.), Water, Science and the Public: The Miramichi Ecosystem. Special Publications of Canadian Journal of Fisheries and Aquatic Science 123: 229–247.

Moore, A. & C. P. Waring, 2001. The effects of a synthetic pyrethroid pesticide on some aspects of reproduction in Atlantic salmon (*Salmo salar* L.). Aquatic Toxicology 52: 1–12.

NASCO, 1999. Request for Scientific Advice from ICES. Report of the Sixteenth Annual Meeting of the Council, Annex 8. NASCO, Edinburgh.

Økland, F., D. Jonsson, A. J. Jensen & L. P. Hansen, 1993. Is there a threshold size regulating seaward migration of brown trout and Atlantic salmon. Journal of Fish Biology 42: 541–550.

Olsen, E. B., G. R. Lilly, M. Heino, M. J. Morgan, J. Brattey & U. Dieckmann, 2005. Assessing changes in age and size at maturation in collapsing populations of Atlantic cod (*Gadus morhua*). Canadian Journal of Fish and Aquatic Sciences 62: 811–823.

Pope, I. A., D. H. Mills & W. M. Shearer, 1961. The fecundity of the Atlantic salmon (*Salmo salar* Linn.). Freshwater and Salmon Fisheries Research, 26.

Potter, E. C. E., J. C. MacLean, R. J. Wyatt & R. N. B. Campbell, 2003b, Managing the exploitation of migratory salmonids. Fisheries Research 62: 127–142.

Potter, E. C. E., N. Ó. Maoiléidigh & G. Chaput, 2003a. Marine Mortality of Atlantic Salmon, *Salmo salar* L: Methods and Measures. Department of Fisheries and Oceans Canadian Science Advisory Secretariat Research Document 2003/101.

Prouzet, P., 1990. Stock characteristics of Atlantic salmon (*Salmo salar*) in France: A review. Aquatic Living Resources 3: 85–97.

Quinn, T. P., 2005. The Behaviour and Ecology of Pacific Salmon and Trout. University of Washington Press, Seattle.

Quinn, T. P., P. McGinnity & T. F. Cross, 2006. Long-term declines in body size and shifts in run timing of Atlantic salmon in Ireland. Journal of Fish Biology 68: 1713–1730.

Rago, P. J. & C. P. Goodyear, 1987. Recruitment mechanisms of striped bass and Atlantic salmon: Comparative liabilities of alternative life histories. American Fisheries Society Symposium 1: 402–416.

Reznick, D. N., H. Bryga & J. A. Endler, 1990. Experimentally induced life-history evolution in a natural population. Nature (London, U.K.) 346: 357–359.

Riddell, B. E., 1986. Assessment of selective fishing on the age at maturity in Atlantic salmon (*Salmo salar*): A genetic perspective. In Meerburg, D. J. (ed.), Salmonid Age at Maturity. Canadian special Publication of Fisheries and Aquatic Sciences 89: 102–109.

Roff, D. A., 1992. The Evolution of Life Histories. Chapman & Hall, London.

Russell, I. C. & E. C. E. Potter, 1996. Interception salmon fisheries – assessing their impact on national stocks. Proceedings of the Institute of Fisheries Management Study Course, Cardiff, Sept. 1993.

Shearer, W. M., 1992. The Atlantic Salmon Natural History, Exploitation and Future Management. Fishing news Books, Oxford.

Staurnes, M., L. P. Hansen, K. Fugelle & O. Haraldstad, 1996. Short-term exposure to acid water impairs osmoregulation, seawater tolerance, and subsequent marine survival of smolts of Atlantic salmon (*Salmo salar*). Canadian Journal of Fish and Aquatic Sciences 53: 1695–1704.

Stearns, S. C., 1992. The Evolution of Life Histories. Oxford University Press, Oxford.

Summers, D. W., 1995. Long-term changes in the sea-age at maturity and seasonal time of return of salmon, *Salmo salar* L., to Scottish Rivers. Fisheries Management and Ecology 2: 147–155.

Thomson, A. G. & R. M. Fuller, 2003. An investigation of land use change in the Llyn Tegid catchment using satellite remote sensing. In Gritten, R. H., C. A. Duigan & H. Millband (eds), The Ecology, Conservation and Environmental History of the Largest Lake in Wales. University of Liverpool, Liverpool: 169–187.

Turrell, W. R. & R. G. J. Shelton, 1993. Climatic change in the north-eastern Atlantic and its impacts on salmon stocks. In Mills, D. H. (ed.), Salmon in the Sea and New Enhancement Strategies. Fishing News Books, Oxford: 40–78.

Webb, J. H., 1989. The Behaviour of Adult Salmon (*Salmo salar* L.) in the River Tay as Determined by Radio Telemetry. Scottish Fisheries Research Report, 44.

Webb, J. H. & R. N. B. Campbell, 2000. Patterns of run timing in adult Atlantic salmon (*Salmo salar* L.) returning to Scottish rivers – some new perspectives and management implications. In Whoriskey, F. G., Jr. & K. B. Whelan (eds), Proceedings of the Fifth International salmon Symposium, Canada: 100–138.

Whelan, K. F., B. J. Whelan & G. Rogan, 2001. Catch as a predictor of salmon stock in the Burrishoole fishery, Co. Mayo, western Ireland. In Shelton, R. (ed.), The Interpretation of Rod and Net Catch Data. The Atlantic Salmon Trust, Pitlochry: 76–84.

Woolland, J. V., 1972. Studies on Salmonid Fishes in Llyn Tegid and the Welsh Dee. Ph.D. Thesis, University of Liverpool, Liverpool.

Youngson, A. F., 1995. Spring Salmon 1994 Bensinger-Liddell Memorial Fellowship. Atlantic Salmon Trust, Pitlochry.

Youngson, A. F., J. C. MacLean, & R. J. Fryer, 2002. Rod catch trends for early-running MSW salmon in Scottish rivers (1952–1997): Divergence among stock components. ICES Journal of Marine Science 59: 836–849.

Hydrobiologia (2008) 602:79–86
DOI 10.1007/s10750-008-9283-6

FISH AND DIADROMY IN EUROPE

Glass eel recruitment, *Anguilla anguilla* (L.), in a Mediterranean lagoon assessed by a glass eel trap: factors explaining the catches

A. J. Crivelli · N. Auphan · P. Chauvelon · A. Sandoz · J.-Y. Menella · G. Poizat

Abstract Although the colonisation of coastal rivers on the Atlantic and Mediterranean coast by glass eels, *Anguilla anguilla*, has been well studied and understood, the colonisation of lagoons by glass eels is much less known. For the first time in the Mediterranean region, the installation of a glass eel fish-pass in Grau de la Fourcade channels in the Rhône delta enabled us to determine which factors could explain the variations in the catches of glass eel entering the Vaccarès coastal lagoon system. Whatever be the procedure chosen, the results of the model were the same: the temperature, the cumulative water discharge from the channel in the 5 nights before the catch (freshwater lure) and time that the drainage pumps were working explained the glass eel catches in the fish-pass in the Grau de la Fourcade. The tide and the cumulative discharge from the channel for only 3 nights before the catch did not seem to have a significant role in explaining catches. These results show that it is important that the lagoons should continue to receive rainfall runoff from their watersheds so that their water levels are high in winter, and that there is a good colonisation by glass eels as a result of a freshwater lure effect, when strong north winds expel low salinity water to the sea.

Keywords Glass eel trap · Catches dynamics · Environmental factors

Guest editors: S. Dufour, E. Prévost, E. Rochard & P. Williot
Fish and diadromy in Europe (ecology, management, conservation)

A. J. Crivelli (✉) · P. Chauvelon · A. Sandoz · G. Poizat
Station Biologique de la Tour du Valat, Le Sambuc, Arles 13200, France
e-mail: a.crivelli@tourduvalat.org

N. Auphan · J.-Y.Menella
MRM (Association Migrateurs Rhône-Méditerranée), Zone Industrielle du Port Fluvial, Chemin des Ségonnaux, Arles 13200, France

Introduction

On the Atlantic coasts of France catches of glass eels (glass eels) are monitored every year, mainly from analyses of glass eel fisheries and to a lesser extent by means of fish-passes (Briand et al., 2003). In contrast, in the French Mediterranean region there is no annual monitoring of glass eels, since glass eel fishing is prohibited and no fish-passes have been installed as yet.

Glass eels use tides through selective tidal stream transport to move up estuaries along the Atlantic coast (McCleave & Kleckner, 1982; Elie & Rochard, 1994; Lambert, 1994) and to enter rivers around the Mediterranean (Gandolfi et al., 1984; Ciccotti et al., 1995). To guide their movements migrating glass eels use various environmental factors or clues (Harden-Jones, 1984) such as the temperature, river discharge, low salinity water and chemical factors in the water (see review by Elie & Rochard, 1994; Tongiorgi et al., 1986; Tosi et al., 1988, 1990). Physiological factors (body condition) might also be involved in the control of locomotor activity and then migratory behaviour of glass eels (Edeline et al., 2006).

In lagoons, only Finiger (1976) and Lecomte-Finiger & Razouls (1981) have attempted to determine which factors are involved in the migration of glass eels into Mediterranean coastal lagoons. They showed that the water discharge exiting, the associated drop in salinity and the wind were the main factors explaining the nocturnal migration of glass eels into the Bages-Sigean lagoon. In contrast to what is observed on the Atlantic coast, the glass eel catches were higher at low tide. The latter result and the low tidal amplitude observed in the Mediterranean (only about 20 cm) suggest that other triggering factors than those described in the Atlantic coast might determine glass eels' migration into lagoons of the Mediterranean region.

After a feasibility study covering the whole of Mediterranean France, including Corsica (Barral, 2001), a classification of sites that were favourable for the installation of glass eel fish-passes was drawn-up. The first site chosen was the Grau de la Fourcade in the Rhône delta. By means of the installation of this fish-pass, we attempted to determine which triggering factors were involved in the observed variations in catches of glass eels attempting to enter a Mediterranean coastal lagoon, the Vaccarès.

Study area

The Grau de la Fourcade is situated to the east of the town of Saintes-Marie de la Mer in the west of the Rhône delta (Fig. 1). This channel connects the sea to the lagoons closest to the coast, which themselves are connected to the Vaccarès lagoon, the whole system covering an area of 12,000 ha. The channel is blocked by a seawall that has13 sliding sluices fitted into it, so that the exchanges with the sea are entirely artificially managed. Data collected since 1993 on the management of this structure show that exchanges of water with the sea are dominated by outputs to the sea and the periods when sea water enters are limited (Chauvelon et al., 2003; Rosecchi et al., 2003). From this monitoring, it is also evident that the works are often closed from June to September and that there are rarely more than 7 sluices open. Beside this channel there is also an outlet that discharges the treated waste water from the town of Saintes-Marie de la Mer.

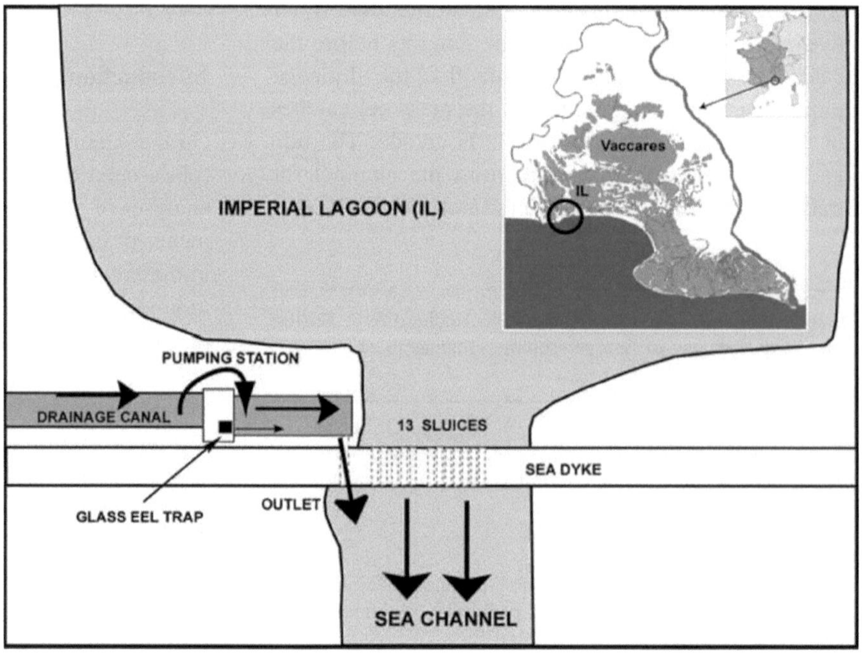

Fig. 1 The study area

The outlet is fitted with a drainage station composed of three Archimedes screws (2 with a pumping capacity of 1.5 m^3 s^{-1} and one of 1 m^3 s^{-1}) and a standby centrifugal pump. The outlet from the final storage lagoon of the treatment works communicates with the channel. When the sluices are closed, the discharge that can act as a lure consists almost entirely of the water from the outlet of the treatment works. This attracts glass eels, which then become trapped in a dead end, with no hope of reaching the Vaccarès system. This is precisely why a glass eel fish-pass was installed by the Fish-Pass Company (France) next to the waste water outlet pumps.

Materials and methods

The fish-pass

The fish-pass was installed in the autumn of 2003 and became operational in January 2004. Depending on the catches, the pass was visited every day, every other day or only once a week. The glass eels captured were weighed as a pooled sample, and six subsamples of 30 glass eels were also weighed (wet weight) at the fish-pass. Based on the mean of these sub-samples an estimate of the total number of fish caught was made. The following data were also collected: monitoring of the operation of the discharged waste water drainage station, the temperature of the discharged water with a temperature recorder and monitoring of the opening of the sluices in the seawall. In addition to our study, a hydraulic study of the exchanges between the sea and the lagoons provided hourly and daily data on the tide and the discharges exiting or entering through the works, in relation to the relative water levels on either side of the seawall and the opening of the sluices (Chauvelon et al., 2003; Rosecchi et al., 2003).

The freshwater lure

If the glass eels are to be able to detect the possibility of colonising an inland water body (river or lagoon), they must receive a signal indicating that such colonisation is possible, this signal being called the "lure" ("appel en mer" in French). It consists of low salinity, often turbid water coming either from a coastal river in flood, or from the outlet channel of a lagoon and which can form a plume mixing with the seawater, extending several kilometres from the coast in the case of large rivers. For the Grau de la Fourcade, we wanted to determine how far out to sea this lure could be detected by the glass eels in relation to the number of sluices open and the discharge from the channel. We used ten SPOT satellite images that covered various circumstances (from 0 to 7 sluices open; discharge from 0 to >500,000 m^3/day) to conduct this investigation. Because of the turbidity of the water, the dimensions of the plume of discharged water could be readily measured on the satellite images.

Data analysis

The following factors that are likely to explain the catches were analysed: the temperature of the water discharged through the fish-pass (°C), the cumulative volume discharged through the sluices for 3 and 5 nights before the day of sampling at time t (m^3 = the lure), the tide (slope of the variation in level on the sea side of the wall) and the time that the waste water drainage pumps were operating (minutes of operation/between two captures). As glass eels are only active at night, the values of these factors were calculated for the night (0800 pm–0700 am). For this analysis we only considered the period of high catches, i.e. from 7 January to 5 April 2004. All statistical procedures were performed within the Generalized Linear Model Framework (GLM, Statistica 6.0). Two problems were encountered with the database: (1) there were days with no catches and (2) when fish were caught, the catches did not always represent the catch for 24 h, since the fish-pass was not visited every day. Three analyses were therefore conducted: (a) using presence and absence data ($N = 50$), (b) by reducing the database to only those days that represented the catch for 24 h (i.e. each time the fish-pass was visited on two consecutive days; $N = 27$) and finally (c) for all the catches by calculating the catch per unit effort (CPUE, $N = 50$), i.e. the number caught per 24 h trapping. In the last case the mean of the environmental factors over the period in question was used. All the data were Log+1 transformed to fulfil the normality assumption. Model selection was based on the Akaike Information

Criterium (AIC; Burnham & Anderson, 1998). The minimum value of AIC indicates the best compromise between the fit of the model and the precision of parameter estimates. In the final model, the contribution of terms and their significance were checked by the χ^2 values of the Likelihood Type 3 test.

Results

The catches

From January 2004 to April 2004, the fish-pass was visited 50 times and 1,349,036 glass eels were captured in the Grau de la Fourcade fish-pass (Fig. 2), but from May 2004 to December 2004, only 30,000 glass eels were caught. Glass eels were therefore recorded in every month of the year.

The effect of environmental factors

Whatever the procedure chosen, the results of the model were the same (Table 1): the best model was the temperature (positively correlated), the cumulative discharge from the sluices for the preceding 5 nights (positively correlated) and the activity of the pumping station (negatively correlated). The lower the temperature and the longer the pumping station operated, the lower the number of glass eels captured;

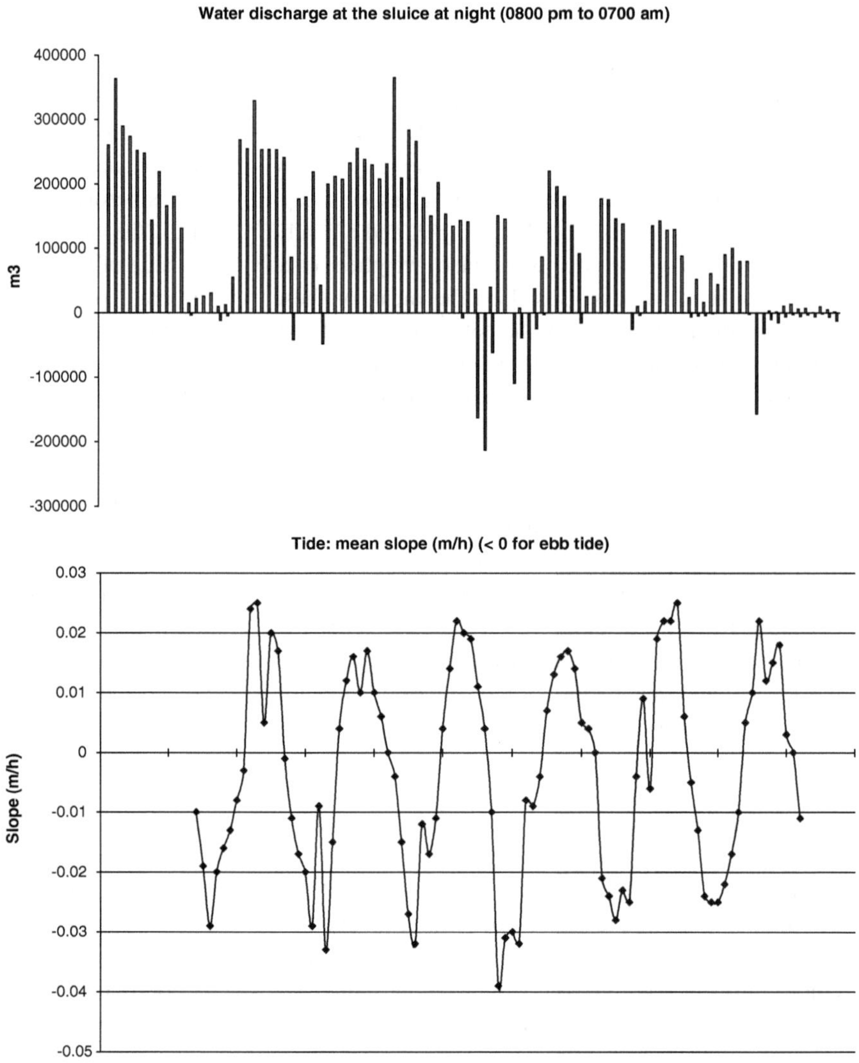

Fig. 2 Raw data taken into account in the analysis of glass eel catches with the exception of the drainage pumps. Positive data in the water discharge graph are water leaving the lagoon, negative data are water entering the lagoon

Fig. 2 continued

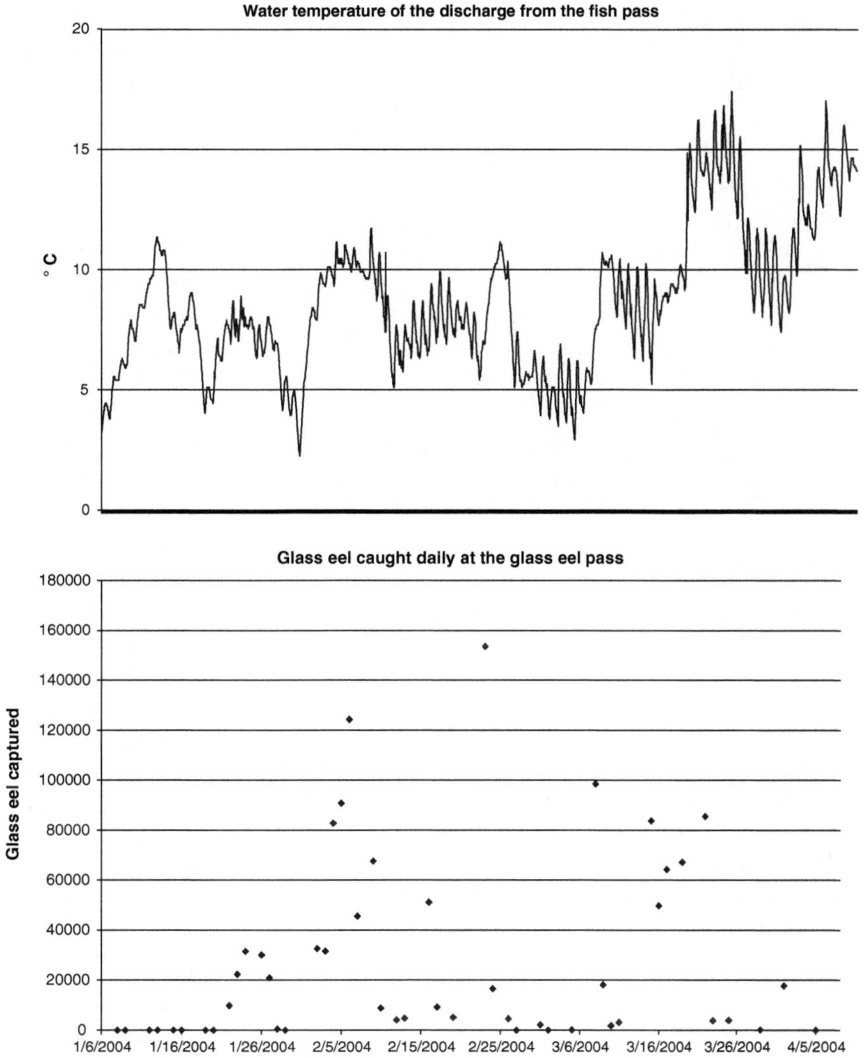

Table 1 Relative contribution of the different environmental factors to observed variations in glass eel catches

Treatment	Effect	d.f.	χ^2	*P*
Presence–absence	Temperature	1	4.93	<0.0263
	Discharge for 5 days	1	1.89	0.1686
	Pumps functioning	1	4.04	<0.044
CPUE	Temperature	1	13.59	<0.0002
	Discharge for 5 days	1	7.85	<0.0051
	Pumps functioning	1	11.29	<0.0007
24 h catches	Temperature	1	13.91	<0.0002
	Discharge for 5 days	1	6.80	<0.0091
	Pumps functioning	1	8.34	<0.0039

and the higher the cumulative sluice discharge for the preceding 5 nights, the higher the captures in terms of numbers.

The relations between the sluice discharge and the lure effect

Ten SPOT images acquired between 1995 and 2001 were used (Fig. 3). These images were taken on days when the number of sluices open varied 0–7 and to daily discharges varied between 0 and 550,000 m^3/day. The length of the plume in the sea was significantly correlated both with the number of sluices

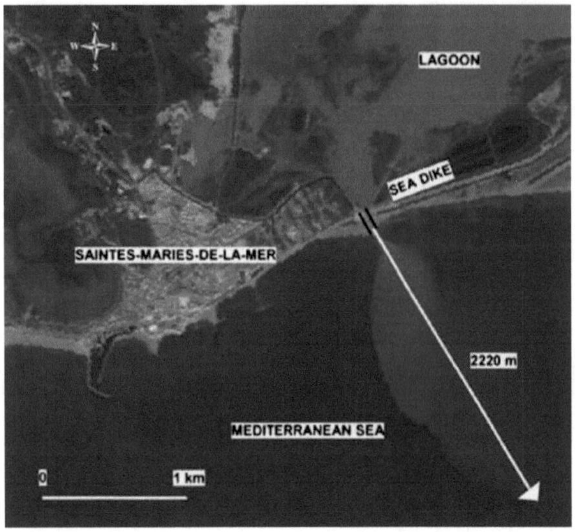

Fig. 3 Spot image from the 31st of March 2001. The connection between the lagoon and the sea is located between the dark lines. On that day, the length of the plume in the sea was 2,220 m

open (R^2: 0.9113) and with the discharge. The latter relation was better (R^2: 0.9234; Fig. 4). With 7 sluices open and a discharge of 550,000 m^3/day the length of the plume could reach up to 2 km, whereas it was only 300-m long with 2 sluices open and a discharge of 57,000 m^3/day. These results are valid for any given year as long as no change occurs in the connection with sea and at the sluices structure.

Discussion

Whatever be the procedure chosen, the results of the model were the same: the temperature, the

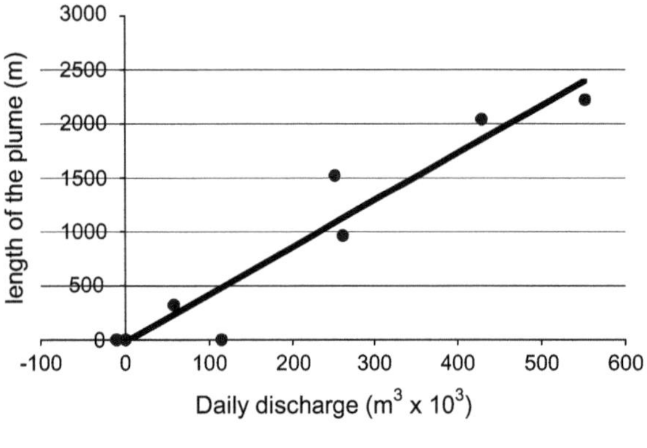

Fig. 4 Relation between the discharge from the Grau de la Fourcade and the length of the plume in the sea

cumulative discharge from the sluices in the 5 nights preceding the catch and the length of time that the drainage pumps were working explained the glass eel catches Grau de la Fourcade fish-trap. The tide and the cumulative discharge from the sluices during only 3 nights before the catch did not appear to play a significant role in explaining the catches.

With the exception of the role of tide, this result is remarkable, since it fits in well with the knowledge acquired by other studies conducted on glass eel catches elsewhere (Gascuel, 1986; Elie & Rochard, 1994; Martin, 1995; Jessop, 2003). It is in fact widely accepted that below a water temperature of 4–10°C (Tongiorgi et al., 1986; Elie & Rochard, 1994; Edeline et al., 2006), glass eels are inactive and are therefore not captured. Glass eels were only captured on two occasions when temperatures were about 5°C; all the other catches took place when the temperature was >6°C, the highest catches occurring with a temperature between 9.9 and 11.4°C, just below the threshold of 11–12°C at which a strong reduction in migratory activity of glass eels is observed (Jessop, 2003; Edeline et al., 2006).

The second most important factor was the cumulative discharge from the sluices during the 5 nights preceding the catch: this means that to have a good catch the freshwater lure that is produced by a strong discharge in the 5 days before the catch is essential. This confirms the great importance of the freshwater lure: low salinity water and surface water odours (Tosi et al., 1990; Tosi & Sola, 1993; Sola & Tongiorgi, 1996) guide migration (acting as clues) offshore, especially in the Mediterranean sea where tidal currents are too weak to orientate glass eels towards the outlet of lagoons systems. In terms of our glass eel catch data for 2004, it seems that a discharge of >200,000 m^3/day and at least 5 sluices open is necessary for very good catches.

Finally, the activity of the drainage pumps plays a mechanical role in the catches: the longer the pumps operate, the longer the catches are delayed because the current in the channel leading to the fish-pass ramp is too strong for the glass eels to be able to swim towards this ramp.

The tide did not play a significant role in explaining the catches in contrast to all other studies (Gascuel, 1986; Elie & Rochard, 1994; Martin, 1995; Jessop, 2003), including those in the Mediterranean region (Gandolfi et al., 1984; Ciccotti et al., 1995).

However, all the latter studies have been undertaken in river estuaries that have geomorphological and hydrological features quite different than those encountered in the connections between lagoon systems and the sea.

In conclusion, in the light of these results, it would appear that it is essential to let the lagoons receive the rainwater runoff from their watersheds, so that their level rises each winter and when there is a strong north wind (the "mistral"), low salinity water is driven south towards the sea and forms a strong lure. This is favourable for a good recruitment of glass eels in these lagoons. This assumption has been confirmed for the Vaccarès lagoon in a long-term study of glass eels' recruitment (1993–2005) that shows clearly that low glass eel recruitment occurred in drought years and good recruitment in wet years (Crivelli, unpublished data). The eel yields in shallow lagoon systems of the Mediterranean region have also been shown to increase greatly with increased quantities of water supplied by the watershed explaining again that recruitment is improved, the greater the freshwater lure between November and March (Crivelli et al., 1995). Such recruitment has a high economic significance, because eel fishing accounts for more than half of the income of the professional lagoon fisheries (Loste & Dusserre, 1996). Lastly, because many connections between the lagoon systems and the sea are equipped with sluices, the management policy for these sluices will be of prime importance, to obtain a good glass eel recruitment.

Acknowledgements This project was funded by the French "Ministère de l'Ecologie et du Développement Durable", by the "Conseil Supérieur de la Pêche", "Agence de l'Eau Rhône-Méditerranée-Corse", the MRM association (Migrateurs Rhône-Méditerranée), the "Fondation Sansouire", The "Conseil Régional Provence-Alpes-Côte d'Azur" and the "Conseil Général des Bouches-du-Rhône". We also thank the municipality of Saintes-Maries de la Mer that gave permission to build the fish-pass at the Grau de la Fourcade. Finally we thank Fish-Pass for its excellent design and installation of the fish-pass.

References

Barral, M. 2001. Etude préliminaire à la mise en place d' "un tableau de bord anguille" sur le bassin Rhône-Méditerranée-Corse, campagne d'études 2000. MRM, Fiche signalétiques des différents obstacles expertisés, 86 pp + annexes.

Briand, C., D. Fatin, G. Fontenelle & E. Feunteun, 2003. Estuarine and fluvial recruitment of the European glass eel, *Anguilla anguilla*, in an exploited Atlantic estuary. Fisheries Management and Ecology 10: 377–384.

Burnham, K. P. & D. R. Anderson, 1998. Model Selection and Inference: A Practical Information-Theoretic Approach. Springer Verlag, New York.

Chauvelon, P., M. G. Tournoud & A. Sandoz, 2003. Integrated hydrological modelling of a managed coastal Mediterranean wetland (Rhone delta, France): initial calibration. Hydrology & Earth System Sciences 7: 123–131.

Ciccotti, E., T. Ricci, M. Scardi, E. Fresi & S. Cataudella, 1995. Intraseasonal characterization of glass eel migration in the River Tiber: space and time dynamics. Journal of Fish Biology 47: 248–255.

Crivelli, A. J., M.-C. Ximenes, B. Gout, G. Lasserre, P. Freon & T. Do Chi, 1995. Causes and effects of terrestrial runoff and riverine outflow on brackish/coastal marine fisheries ecosystems in the northern Mediterranean region. FAO Fisheries Technical Paper No. 349: 59–88.

Edeline, E., P. Lambert, C. Rigaud & P. Elie, 2006. Effects of body condition and water temperature on *Anguilla anguilla* glass eel migratory behavior. Journal of Experimental Marine Biology and Ecology 331: 217–225.

Elie, P. & E. Rochard, 1994. Migration des civelles d'anguilles (*Anguilla anguilla* L.) dans les estuaires, modalités du phénomène et caractéristiques des individus. Bulletin Français de Pêche et de Pisciculture 335: 81–98.

Finiger, R., 1976. Contribution à l'étude biologique et écologique des civelles (*Anguilla anguilla* LINNE 1758) lors de leur pénétration dans un étang Méditerranéen. I. Recrutement et Biométrie au cours d'un cycle annuel. Vie et Milieu 26: 123–144.

Gandolfi, G., M. Pesaro & P. Tongiorgi, 1984. Environmental factors affecting the ascent of glass eels, *Anguilla anguilla* (L.), into the Arno River. Oebalia 10: 17–35.

Gascuel, D., 1986. Fow carried and active swimming migration of the glass eel (*Anguilla anguilla*) in the tidal area of a small estuary on the French Atlantic coast. Helgolander Meeresunters 40: 321–326.

Harden-Jones, F. R., 1984. A view from the ocean. In McCleave, J. D., G. P. Arnold, J. J. Dodson & W. H. Neill (eds), Mechanisms of Migration of Fishes. Plenum Press, New York: 1–26.

Jessop, B. M., 2003. Annual variability in the effects of water temperature, discharge, and tidal stage on the migration of American eel elvers from estuary to river. American Fisheries Society Symposium 33: 3–16.

Lambert, P., 1994. Synthèse des concepts de modélisation du phénomène de migration des civelles d'*Anguilla anguilla* (L., 1758) en estuaire. Bulletin Français de Pêche et de Pisciculture 335: 99–110.

Lecomte-Finiger, R. & C. Razouls, 1981. Influence des facteurs hydrologiques et météorologiques sur la migration anadrome des civelles dans le Golfe du Lion. Cahier du Laboratoire de Montereau 12: 13–16.

Loste, C. & K. Dusserre, 1996. La pêche sur l'étang de Bages-Sigean. Report of the CEPRALMAR (Centre d'Etudes et de Promotion des Activités Lagunaires et Maritimes), Montpellier.

Martin, M. H., 1995. The effect of temperature, river flow, and tidal cycles on the onset of glass eel and elver migration into freshwater in the American eel. Journal of Fish Biology 46: 891–902.

McCleave, J. D. & R. C. Kleckner, 1982. Selective tidal stream transport in the estuarine migration of glass eels of the American eel (*Anguilla rostrata*). Journal du Conseil International pour l'Exploration de la Mer 40: 262–271.

Rosecchi E., P. Chauvelon, G. Poizat & A. J. Crivelli, 2003. Conséquences de la variabilité hydro-saline d'un complexe lagunaire méditerranéen, induite par la gestion hydraulique et les contraintes climatiques, sur ses peuplements piscicoles: le cas du système Vaccarès. Programme National Liteau-98 du M.E.D.D. Rapport final. Tour du Valat, 61 pp. + annexes.

Sola, C. & P. Tongiorgi, 1996. The effect of salinity on the chemotaxis of glass eels, *Anguilla anguilla*, to organic earthy and green odorants. Environmental Biology of Fishes 47: 213–218.

Tongiorgi, P., L. Tosi & M. Balsamo, 1986. Thermal preferences in upstream migrating glass-eels of *Anguilla anguilla*. Journal of Fish Biology 28: 501–510.

Tosi, L. & C. Sola, 1993. Role of Geosmin, a typical inland water odour, in guiding glass eel *Anguilla anguilla* (L.) migration. Ethology 95: 177–185.

Tosi, L., L. Sala, C. Sola, A. Spampanato & P. Tongiorgi, 1988. Experimental analysis of the thermal and salinity preferences of glass-eels, *Anguilla anguilla* (L.), before and during the upstream migration. Journal of Fish Biology 33: 721–733.

Tosi, L., A. Spampanato, C. Sola & P. Tongiorgi, 1990. Relation of water odour, salinity and temperature to ascent of glass-eels, *Anguilla anguilla* (L.): a laboratory study. Journal of Fish Biology 36: 327–340.

Hydrobiologia (2008) 602:87–98
DOI 10.1007/s10750-008-9284-5

FISH AND DIADROMY IN EUROPE

Impact of glass eel fishery on by-catch fish species: a quantitative assessment

E. Gisbert · M. A. López

Abstract The present study evaluated the effect of the glass eel fishery on by-catch ichthyofauna in the Ebro River delta by investigating the primary responsible factors of discard mortality, in order to assess the adequacy of fishing and sorting procedures, and assess the ecological impact of this traditional activity. Results revealed that glass eel fishery had a negative impact on by-catch ichthyofauna mainly composed of mugilid fry and small-size estuarine species. Data showed that between 10 and 69% of incidental species died as a consequence of glass eel capture and sorting procedures. Differences in by-catch fish mortality rates depended on the glass eel sorting time, the total weight of catch, the discard ratio, by-catch fish size, and by-catch fish species composition. The total weight of the catch and the discard ratio were two of the main parameters affecting discarded fish mortality, since both had a direct effect on the period of time required for glass eel sorting during which by-catch fish were exposed to air and suffered from very aggressive sorting procedures that caused severe external and internal injuries. By-catch fish measuring less than 40 mm L_T were more susceptible to glass eel fishery procedures than those larger (>60 mm L_T), independent of the species considered or their stage of development. This fishery, similar to other ones, should strive to minimization the amount of discards and the mortality of discarded fish, by improving the selectivity of the fish trap and developing new sorting and selection procedures.

Keywords Glass eel fishery · By-catch · Discard mortality · Ebro River delta · *Anguilla anguilla* · Grey mullet fry

E. Gisbert and M. A. López have contributed equally to this study.

Guest editors: S. Dufour, E. Prévost, E. Rochard & P. Williot
Fish and diadromy in Europe (ecology, management, conservation)

E. Gisbert (✉)
Unitat de Cultius Experimentals (UCE), Centre d'Aqüicultura, Institut de Recerca i Tecnologia Agroalimentaries (IRTA), Aptat. Correus 200, 43540 Sant Carles de la Rapita, Tarragona, Spain
e-mail: enric.gisbert@irta.es

M. A. López
Institut de Recerca i Tecnologia Agroalimentaries (IRTA), Aptat. Correus 200, 43540 Sant Carles de la Rapita, Tarragona, Spain

Introduction

Postlarvae of the European eel (*Anguilla anguilla*, L.), known as glass eels, are caught by professional fishermen along most coastal waters of the Atlantic Ocean, from Scandinavia to Morocco, and rivers of the Baltic and Mediterranean seas (Moriarty & Dekker, 1997). This fishery target glass eels during their onshore migration into estuarine and freshwater environments after several years of migration from the spawning grounds in the Sargasso Sea. A progressive reduction of eel captures has been extensively reported in the last ten years (Moriarty & Dekker, 1997; Dekker, 2002, 2005; Ringuet et al., 2002), overlapping with an increasing demand for glass eels worldwide. This global over-exploitation of eels coupled with long-term climatic changes might have also contributed to the historical decline of the natural population of this catadromous species (EIFAC/ICES, 2003; Dekker, 2002, 2005; Wirth & Bernatchez, 2003).

The main European glass eel fishery is located in France, capturing ca. 82% of glass eels in Europe (Castelnaud et al., 1994; Castelnaud, 2002). Other important glass eel fisheries are also found along the Atlantic coasts of Portugal, Spain, Morocco and the Bristol Channel in the United Kingdom (Ringuet et al., 2002; Dekker, 2003). Minor fisheries exist in the Mediterranean Sea. Different nets are used for the glass eel fishery, depending on the country and their fishery traditions. These are hand-held or ship-based, which are moved manually or are fixed, and include trawls, stow nets and fyke nets (Ringuet et al., 2002).

The Ebro River delta (Western Mediterranean Spain) is the fourth largest delta in the Mediterranean Sea, occupying a triangular coastal platform of 320 km^2 with an average altitude of 0.5 m over the sea level. Freshwater outflows reach marine environments, directly from the river (mean annual flow = 400 $m^3 s^{-1}$) and through a net of drainage channels connecting rice fields and coastal lagoons with two bays located at both sides of the Delta (Fig. 1). These estuarine environments are one of the main sites for glass eel capture in the Mediterranean coast of the Iberian Peninsula, where glass eels are captured by means of a traditional fixed and non-selective fish trap known in the Catalan language as a "*bussó*" (Fig. 2). In this area, like in most European coastal waters, catches have steadily declined since the eighties (Fig. 3) and the fishery is now tightly regulated by the regional government. The fishing season extends form from November to March and fishing operations are allowed only from dusk to dawn (Fernández & Pech, 2000). Fishing licences (1 fishing trap per fisherman) are given by the administration by raffle (381 fishing licenses in the 2004–2005 season). As a consequence of the dramatic reduction in glass eel captures (Fig. 3) and their very high value (600 euros/kg in local markets), illegal captures are

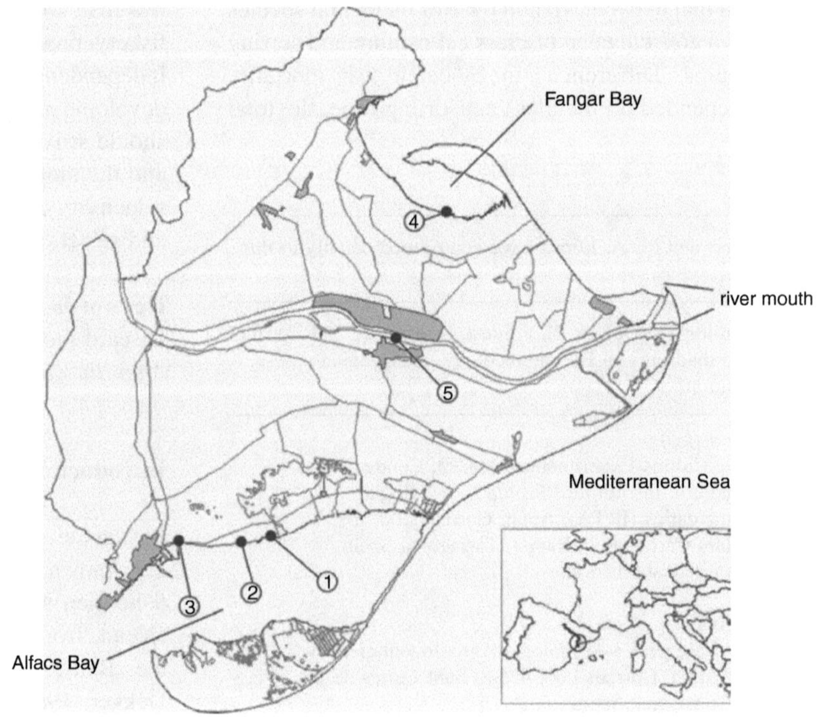

Fig. 1 Location of the study area and monitored glass eel fishing sites in the Ebro River delta

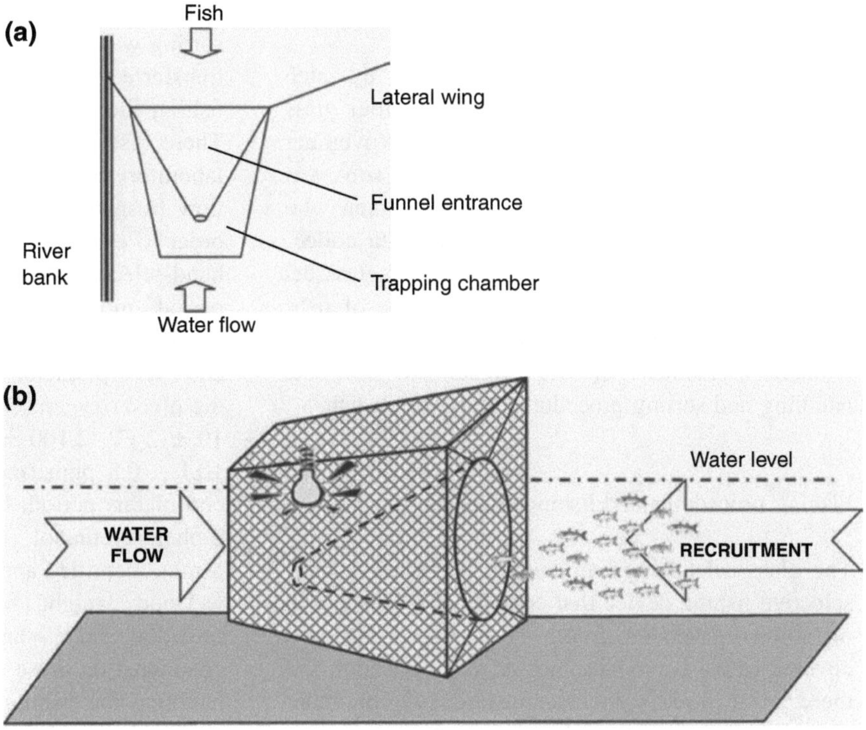

Fig. 2 Glass eel fishing trap ("*bussó*") used in the glass eel fishery of the Ebro Delta River. (**a**) Schematic aerial view of the fishing trap, (**b**) principle of glass eel fishing method

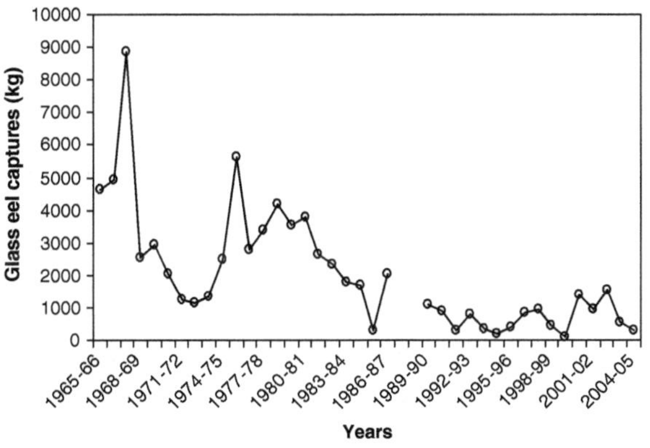

Fig. 3 Time-series of glass eel catches by the Sant Pere fishermen guild in the Ebro River delta. These data did not represent the total of glass eel captures in the area, but were the only available time-series data (data courtesy of the Ebro Delta Natural Parc, PNDE)

very common, which makes difficult to exactly know the magnitude of captures in the area.

This fishery is performed by professional fishermen using artesanal and rudimentary sorting methods, however not a great deal of care is taken in relation to glass eel manipulation and health (Fernández & Pech, 2000; Ringuet et al., 2002), and even less to by-catch fish. The captured fish are therefore exposed to high levels of stress during their capture and handling, exacerbated by air exposure, hypoxia and skin lesions. These factors, which normally do not affect the glass eel, might have a great impact on the health and survival of discarded fish species. Discarded fish mortality is a critical problem in the management of worldwide fisheries (Davis, 2002) that might produce effects on fish size distributions, species composition, and ecosystem diversity (Pauly et al., 2001; Christensen et al., 2003). Despite this, discard mortality rates in specific fisheries, such as that of the glass eel in the Western Mediterranean area, are rarely known and there exist very few field studies on this topic (Guerault et al., 1989; Robin, 1990).

The objectives of the present study were to evaluate the effect of the glass eel fishery on by-catch ichthyofauna by examining the primary responsible factors of discard mortality in order to assess the adequacy of fishing and sorting procedures, and assess the ecological impact of this traditional activity. This information might be the first step for the implementation of guidelines in the glass eel fishery management and the development of environmentally friendly fishing methods.

Materials and methods

The impact of the glass eel fishery on by-catch ichthyofauna was assessed by monitoring four glass eel fishing sites in the Ebro Delta from November 2004 to February 2005. Each fishing site was managed by authorized professional fishermen who agreed in collaborating with the study. Data collection was obtained by direct observation of glass eel fishing practices without direct interference of fishermen's procedures, which allowed us to conduct a realistic assessment of the effects of glass eel handling and sorting procedures on by-catch fish.

Fishing procedures and fishing site selection

The glass eel fishing trap (*bussó*) is a passive non-selective fishing device that consists of a trapezoidal metallic or wooden frame (100 × 80 × 100 cm) covered with a 1-mm mesh net. At the trap's entrance, there is a funnel directed inwards (7 cm inner diameter) that prevents fish escapement (Fig. 2a). A light placed inside the trap attracts fish during night hours. The principle of glass eel capture is the migration of glass eels from brackish to freshwater environments through a salinity gradient (Fig. 2b).

Fishing traps were placed along both riverbanks and drainage channels at dusk. At dawn after a fishing period of ca. 14 h, the fish trap was removed from the water and its content was emptied in a 30 × 50 cm square rigid sieve with a mesh of 3–5 mm inner diameter. Once in the sieve, the catch was sorted by forcing the glass eels to pass through the mesh by strong diagonal hand movements and using the fingertips to push fishes against the sieve mesh. The principle of this sorting procedure was that glass eels pass through the mesh due to their anguilliform movements, while the rest of the by-catch fish got retained in the sieve. Sorting times may vary depending on the quantity of glass eels and by-catch fish captured, and the presence of invertebrates and litter in the sieve. After fish sorting, glass eels were kept in a polystyrene box and transferred to the wholesaler, while the discarded by-catch fish were dumped into the water, whereas in the present study we kept those fish in order to evaluate their survival after glass eel handling and sorting procedures.

Once fish traps were emptied and glass eel hand sorting was finished, all the discarded live fish were transferred by road (5–25 min trip, depending on the fishing site) in a 200-l aerated tank to the laboratory. There, discarded fish were progressively acclimated to laboratory conditions and kept for ten days in an open-flow freshwater 500-l tank (flow rate: 5 l min^{-1}) in order to evaluate the effect of glass eel capture and hand-selection methods on their survival. During this period, mortality was recorded and water quality monitored regularly. Water temperature, conductivity and pH (pHmeter 507, Crison Instruments) and dissolved oxygen (OXI330, Crison Instruments) were 10 ± 2°C, $2{,}100 \pm 200$ μS cm^{-1}, 7.5 ± 0.3 and 10.1 ± 0.5 ppm (mean ± SD), respectively. At the end of this period, live fish were anaesthetised with 2-phenoxyethanol (0.3 ml l^{-1}) for a maximum of 5 min, identified and their total length (L_T, mm) and wet body weight (W_B, mg) measured. Species identification and L_T and W_B measurements were also conducted on those fish that died during the tasks of handling the fishing trap and glass eel sorting, and during the ten-day period that they were kept in our facilities. These data allowed us to evaluate the non-target fish species size classes were more vulnerable to glass eel fishery.

Fishing area selection was established according to the wide range of estuarine environments of the study area: the Southern (Alfacs) and Northern (Fangar) Bays of the Ebro River delta and the main river course (Fig. 1, Table 1). Their selection depended on the agreement of professional fishermen to participate in the study, which resulted in that not all fishing sites could be studied. It should be noted that the fishing site located in the main river course (Fs 5) was abandoned during the course of the study due to its very low yield; such low captures were explained by the fact that this fishing site was located near the upper limit of riverine glass eel fisheries. After discarding the above-mentioned fishing site, a new fishing site was selected in the Southern bay coastline (Fs 3).

Variable definition

The effect of the glass eel fishery on discarded ichthyofauna was independently evaluated at each fishing point by evaluating the primary responsible

Table 1 Water quality parameters (means ± SD) from different fishing sites monitored during 2004–2005 glass eel fishery

	Temperature (°C)	Conductivity (μS cm^{-1})	Dissolved oxygen (ppm)	pH	Turbidity (FAU)
Fs 1 $n = 6$	7.0 ± 1.7	2,906 ± 1,362	11.7 ± 1.4	7.6 ± 0.2	5.0 ± 3.8
Fs 2 $n = 4$	6.3 ± 1.8	5,040 ± 2,295	10.2 ± 0.6	7.9 ± 0.5	13.0 ± 9.5
Fs 3 $n = 3$	8.9 ± 2.9	2,712 ± 1,350	8.7 ± 1.3	7.6 ± 0.2	33.8 ± 7.5
Fs 4 $n = 5$	7.1 ± 1.8	3,166 ± 2,167	10.1 ± 0.4	7.9 ± 0.4	18.0 ± 10.4

Temperature, conductivity, dissolved oxygen, pH and turbidity were measured on the same days as the fishing site was monitored

factors for discarded fish mortality. The total mortality (M_T, %) of by-catch fish was calculated as:

$$M_T(\%) = M_{ST} + M_{LT};$$

where short term mortality (M_{ST}) was computed as the addition of the mortality derived from fish trap handling (M_H, %) and that associated to glass eel sorting procedures (M_S, %). M_{LT} (%) was defined as the long term mortality, representing the percentage of discarded fish that died during the ten days posterior to their capture. This would represent the rate of discarded fishes that would die after they were returned to the water by fishermen, once glass eel sorting was over.

In addition, other variables, such as the glass eel and by-catch fish captured biomass (B, g), fish size (L_T), glass eel hand sorting times (T, min), and species composition were recorded. The above-mentioned data were used to calculate the discard ratio (D, %) which was defined as the ratio of discarded fish to total captured biomass (Alverson et al., 1994).

Statistical analysis

Total mortality percentage (arcsine $\sqrt{M_T}$ transformed) and glass eel hand-sorting time (log T-transformed) from different fishing sites were statistically compared by means of a one way ANOVA (Sokal & Rohlf, 1995) in order to check if different sites and fishermen procedures had an effect on non-target fish survival. Variables (M_H, M_S, M_{LT}, T) were checked for normality by means of the Kolmogorov–Smirnov test and correlated by means of the non-parametric Spearman Rank Order correlation due to the non normality of most variables (SigmaStat, 2004). Discarded fish mortality was correlated to fish size (L_T) by means of lineal regression (Sokal & Rohlf, 1995).

The estimation of the global impact of the glass eel fishery in the area was conducted under two different approaches. The first one was based on the strong correlation between glass eel captures and by-catch fish biomass (Table 4), which was used to estimate (mean ± SE) by linear regression the total of by-catch fish captures in the area according to reported glass eel yields. These data were used to estimate the impact of the glass eel fishery (discarded fish mortality) by means of average field M_T data (36.5%, Table 3). The second approximation to estimate the total mortality of by-catch fish during the glass eel fishery season was performed using the following data: (1) 25 and 75% percentiles of the daily by-catch fish biomass per fishing trap (percentiles were chosen since this variable did not show a normal distribution; Kolmogorov–Smirnov test, $P < 0.001$); (2) the mean M_T value; (3) the number of fishing licences given by the administration (1 licence per fishing trap); and (4) length of fishing season (November–March: 107 labour days). This estimation method is generally considered a conservative assessment and probably underestimates the true magnitude of the effects of the glass eel fishery. In general, based on past empirical data and studies, estimates of losses based on countable dead fish are conservative (AFS, 1992). Under both scenarios, the total species-specific mortality, in number of specimens for those species that dominated by-catch biomass (*Liza ramada*, *L. aurata* and *Mugil cephalus*) was estimated in the studied area according to their abundance in field captures, percentage of mortality for each species and mean body weight.

Table 2 Occurrence of by-catch fish species captured in glass eel fishing traps

	Fishing sites				
	Fs 1 % (*n*)	Fs 2 % (*n*)	Fs 3 % (*n*)	Fs 4 % (*n*)	Total % (*n*)
Chelon labrosus					
Juvenile (>100 mm L_T)	0.1 (1)	0.1 (34)	0.6 (10)	–	0.2 (45)
Liza aurata					
Fry (30–40 mm L_T)	31.3 (313)	0.1 (37)	0.3 (6)	0.4 (5)	8.0 (361)
Juvenile (>100 mm L_T)	1.2 (12)	>0.1 (4)	0.1 (1)	–	0.4 (17)
L. ramada					
Fry (15–25 mm L_T)	52.7 (528)	91.0 (23,756)	95.1 (1,719)	85.2 (983)	81.0 (26,986)
Juvenile (>70 mm L_T)	1.2 (12)	0.4 (105)	1.3 (23)	0.1 (1)	0.7 (141)
L. saliens					
Fry (30–40 mm L_T)	–	>0.1 (9)	–	0.1 (1)	>0.1 (9)
Juvenile (>60 mm L_T)	0.8 (8)	0.1 (35)	0.1 (1)		0.3 (45)
Mugil cephalus					
Fry (30–40 m L_T)	0.5 (5)	3.1 (815)	0.7 (13)	0.5 (6)	1.2 (839)
Juvenile (>100 L_T)	–	0.3 (81)	0.3 (6)	–	0.2 (87)
Gambusia holbrooki					
<20 mm L_T	0.5 (5)	1.2 (317)	0.1 (2)	0.1 (1)	0.5 (325)
30–50 mm L_T	3.3 (33)	1.8 (476)	0.3 (5)	0.3 (3)	1.4 (517)
Atherina boyeri					
Fry (<20 mm L_T)	–	>0.1 (3)	–	0.6 (7)	0.2 (10)
Juvenile/adult (30–70 mm L_T)	1.0 (10)	>0.1 (10)	–	0.5 (6)	0.4 (26)
Pomatoschistus microps					
Fry (<20 mm L_T)	1.0 (10)	0.3 (73)	0.1 (2)	0.2 (2)	0.4 (87)
Juvenile /adult (30–60 mm L_T)	6.4 (64)	0.1 (31)	0.4 (8)	2.6 (30)	2.4 (133)
Syngnathus abaster					
Juvenile (40–80 mm L_T)	–	>0.1 (7)	–	0.2 (2)	0.1 (9)
Dicentrarchus labrax					
Fry (20–30 mm L_T)	–	>0.1 (11)	–	–	>0.1 (11)
Aphanius iberus					
Juvenile/adult (20–40 mm L_T)	–	0.1 (21)	–	0.2 (2)	0.1 (23)
Salaria fluviatilis					
Adult (40–60 mm L_T)	–	>0.1 (4)	–	–	>0.1 (4)
Barbus graellsii					
Juvenile (50–80 mm L_T)	–	0.1 (30)	–	–	0.1 (30)
Carassius sp.					
Juvenile (60–80 mm L_T)	–	0.1 (14)	–	–	0.1 (14)
Scardinius erythrophthalmus					
Juvenile (70–100 mm L_T)	–	>0.1 (5)	–	–	>0.1 (5)
Pseudorasbora parva					
Fry (20–40 mm L_T)	–	0.1 (18)	0.6 (10)	0.2 (2)	0.3 (30)
Juvenile (>60 mm L_T)	–	0.8 (202)	0.1 (1)	8.9 (103)	3.3 (306)
Alburnus alburnus					
Adult (90–100 mm L_T)	-	-	-	0.2 (2)	<0.1 (2)

Table 3 Summary statistics of fishing performance and mortality rates at studied fishing sites

	$B_{(\text{glass eel})}$ (g)	$B_{(\text{by-catch fish})}$ (g)	D (%)	M_T (%)	M_H (%)	M_S (%)	M_{LT} (%)	T (min)
Fs 1	128	78.3	43.3	37.5	2.8	13.6	16.6	5.8
$n = 6$	(46–300)	(60–150)	(16.7–56.6)	(10.1–44.9)	(0.4–4.7)	(2.3–29.6)	(6.1–32.4)	(5–7)
Fs 2	3.8	46.3	93.5	37.9	1.7	16.1	19.9	3.5
$n = 4$	(1–10)	(20–65)	(85.7–97.6)	(25.2–52.9)	(0.2–3.6)	(9.1–25.0)	(6.2–41.6)	(2–4)
Fs 3	0.8	196.7	96.8	32.2	5.2	16.6	13.4	4
$n = 3$	(0.5–1)	(20–550)	(95.2–99.8)	(10.0–56.8)	(1.1–7.3)	(3.6–23.3)	(4.0–26.2)	(3–5)
Fs 4	53	561	88.8	37.5	1.9	32.2	3.6	13.5
$n = 5$	(5–118)	(189–1,688)	(79.1–93.5)	(14.4–68.8)	(1–2.4)	(11.4–40.2)	(7.8–1.2)	(7–20)

All data are shown as the mean and value range. Data were obtained from all monitored fishing sites and at different dates

Variables: B, Glass eel and by-catch fish biomass; D, discard ratio; M_T, total discard mortality; M_H, mortality of fish due to fish trap handling; M_S, by-catch mortality of derived from glass eel handmade sorting procedures; M_{LT}, mortality of discarded fish after a ten-day long period: T, length in time of glass eel selection

In this study we used indistinctly the terms: discarded fish, by-catch, incidental catch and non-target fish species, since these terms are synonyms in this kind of fishery where only the target species (glass eels) are retained. The specific means and uses and abuses of this nomenclature are deeply described in Alverson's et al. review (1994).

Results

Non-target species composition

A total of 17 by-catch species were captured during glass eel fishing season. Species composition was dominated by grey mullets (Osteichthyes, Mugilidae), which represented 90% of the total discarded fish individuals. Among them *L. ramada*, *L. aurata* and *M. cephalus* fry were the most abundant species. Species composition from different glass eel fishing sites is presented in Table 2. By-catch fish biomass was statistically correlated with the quantity of glass eel captured ($r = 0.65$; $P = 0.003$; $n = 18$, Table 4).

Mortality of by-catch fish species

The total mortality rate (M_T) of discarded fish varied between 14.4 and 68.8% of total by-catch fish (Table 3). Although no statistically significant differences were recorded comparing mean M_T values from each fishing site (ANOVA, $P > 0.05$). The contribution of M_H, M_S and M_{LT} to M_T values varied depending on fishing sites and dates (ANOVA, $P < 0.05$). Mortality derived from fish trap handling (M_H) ranged from 0.2 to 7.3% of by-catch fish and was negatively correlated to the by-catch fish biomass ($r = -0.53$, $P = 0.023$; $n = 18$; Table 4). During the emptying of the fish trap, some by-catch fish adhered to the walls getting trapped inside it and died of anoxia, while some others got crushed against the rigid structures of the trap. Mortality derived from hand-sorting of glass eels (M_S) varied from 2.3 to 40.2% of by-catch fish. Hand-sorting times recorded from different fishing sites were similar (ANOVA, $P > 0.05$) ranging from 2 to 20 min, although this wide range of recorded values varied depending on the quantity of glass eels ($r = 0.79$, $P < 0.001$; $n = 18$; Table 4) and by-catch fish biomass ($r = 0.70$, $P < 0.001$; $n = 18$; Table 4), and the presence of invertebrates (*Gammarus* sp.) and macroalgae litter in the sieve while sorting. The mortality of discarded by-catch fish when they were returned into the water (M_{LT}) ranged between 1.2 and 41.6% of by-catch fish depending on the fishing sites and dates. M_{LT} was negatively correlated to glass eel ($r = -0.48$, $P = 0.042$; $n = 18$; Table 4) and by-catch fish biomass ($r = -0.54$, $P = 0.021$; $n = 18$; Table 4), and the length of hand-sorting time for glass eels ($r = -0.59$, $P = 0.010$; $n = 18$; Table 4).

Table 4 Spearman Rank order correlations between the biomass (B) of captured glass eels and by-catch fish, mortality of fish due to fish trap handling (M_H), by-catch mortality of derived from glass eel handmade sorting procedures (M_S), mortality of discarded fish after a five-day long period (M_{LT}) and length in time of glass eel selection (T)

	$B_{(by\text{-}catch\ fish)}$	M_H	M_S	M_{LT}	M_T	T
$B_{(glass\ eel)}$	**0.649**	−0.364	−0.036	**−0.483**	−0.281	**0.794**
	0.003	*0.134*	*0.882*	*0.042*	*0.255*	*<0.001*
$B_{(by\text{-}catch\ fish)}$		**−0.532**	0.189	**−0.536**	−0.078	**0.702**
		0.023	*0.446*	*0.021*	*0.754*	*<0.001*
M_H			0.154	0.162	0.209	−0.286
			0.535	*0.514*	*0.397*	*0.244*
M_S				−0.321	**0.598**	0.191
				0.189	*0.009*	*0.441*
M_{LT}					0.447	**−0.587**
					0.062	*0.010*
M_T						−0.121
						0.626

Correlations were performed using data from all monitored fishing sites and at different dates ($n = 18$). In bold letters are presented those correlations that were statistically significant ($P < 0.05$) and in italics their significance level

Although total mortality was similar from different fishing sites and dates (Table 3), the rate of mortality varied amongst by-catch species (ANOVA; $P < 0.05$). Species mortality was directly associated to the abundance of non-target fish species (Table 4 and Fig. 4) and their size (Fig. 5). The grey mullets were the most affected species, as most of the captured *L. ramada* (10–25 mm L_T), *L. aurata* (20–40 mm L_T) and *M. cephalus* (20–40 mm L_T) fry died ($M_T = 97.2$, and 100 and 70.0%, respectively), as a consequence of trapping and selecting glass eels. Fishing had a lower impact on the other grey mullet species, such as *L. saliens* (40–60 mm L_T) and *C. labrosus* (80–100 mm L_T) ($M_T = 42.4$ and 15.9%, respectively). Total mortality of other species, such as *A. boyeri* (20–40 mm L_T) and *P. microps* (20–40 mm L_T), and *G. holbrooki* (20–40 mm L_T) and the endangered *A. iberus* (20–40 mm L_T), was 85.0, 82.9, 72.6 and 70.0%, respectively. Fishing had a lower impact on larger species, such as *Carassius*

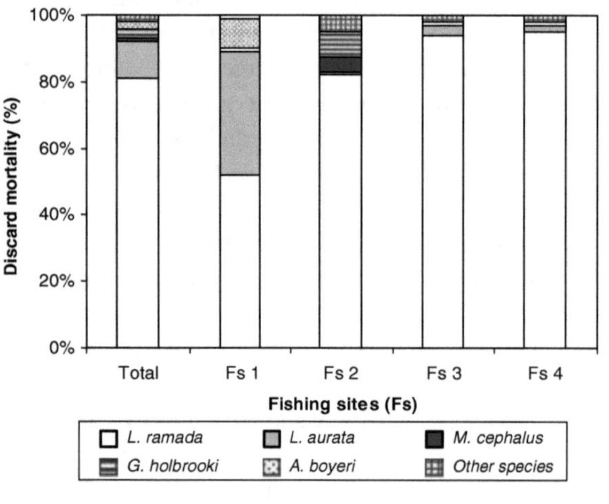

Fig. 4 Discard mortality rate (%) of by-catch fish species captured in glass eel fishing traps

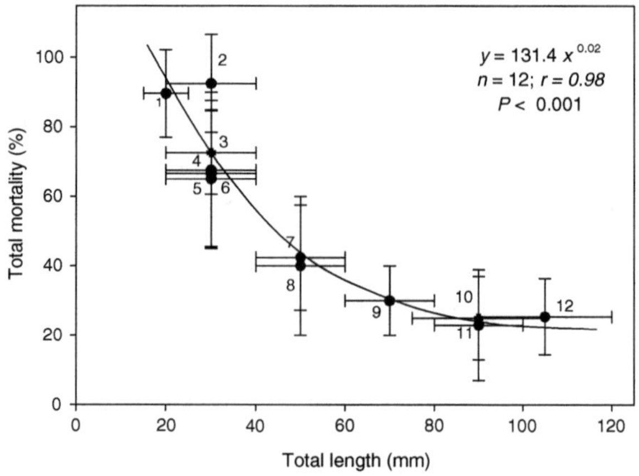

Fig. 5 Regression between discard mortality rate (%) and fish size (L_T, mm). Abbreviations: 1, *L. ramada*; 2, *L. aurata*; 3, *G. holbrooki*; 4, *M. cephalus*; 5, *P. microps*; 6, *L. ibera*; 7, *P. parva* ; 8, *L. saliens*; 9, *C. auratus*; 10, *S. erythrophthalmus*; 11, *C. labrosus*; 12, *A. alburnus*

sp. (60–80 mm L_T; $M_T = 30\%$), *S. erythrophthalmus* (70–100 mm L_T; $M_T = 25\%$) and *A. alburnus* (90–120 mm L_T; $M_T = 25\%$).

Impact of glass eel fishery

The glass eel catch in the Ebro River delta recorded during 2004–2005 fishing season was 1 t. According to present results, we estimated that the capture of 1 t glass eels would represent the accidental capture of 8.6 ± 2.0 t of other fish species [$B_{\text{by-catch fish}} = -208.4 + (8.8\ B_{\text{glass eels}})$; a standard error = 2.03, $R = 0.77$, $P < 0.001$; $n = 18$)]. Approximately 36.5% (3.1 ± 1.8 t) died as a consequence of the glass eel fishing and sorting procedures being used. The estimated number of specimens of *L. ramada*, *L. aurata* and *M. cephalus* fry that died during glass eel fishing season were $23 \times 10^6 \pm 13 \times 10^6$, $13 \times 10^6 \pm 7.2 \times 10^6$ and $3.3 \times 10^6 \pm 2.1 \times 10^6$, respectively.

When considering the second method for estimating total mortality originating from the glass eel fishery in the Ebro River delta, two different scenarios might be considered depending on the chosen daily by-catch fish biomass (25 and 75% percentile). Thus, the most conservative by-catch fish mortality estimation ($B_{\text{by-catch fish}}$ 25% percentile = 60 g) showed that a total of 2.5 t of by-catch fish were captured in the area during the studied season (6.4 kg per fishing trap), from which 0.9 t died after glass eel capture and sorting operations. The second scenario in which by-catch fish mortality assessment was conducted using 75% percentile of by-catch fish biomass ($B_{\text{by-catch fish}}$ 75% percentile = 755 g) estimated that 11.2 t of by-catch fish died during the glass eel fishing season from a total capture of 30.8 t of by-catch fish (81 kg per fishing trap and season). As grey mullets were the dominant group of fish species in glass eel by-catch, we estimated the quantity of individuals of *L. ramada*, *L. aurata* and *M. cephalus* fry that died as a consequence of the glass eel fishery. The two former scenarios were also considered for estimating species-specific mortality. The number of specimens of *L. ramada* fry that died during the glass eel fishing season ranged from 6.6×10^6 to 83.1×10^6, *L. aurata* and *M. cephalus* were also affected by the glass eel fishery but to a lesser extent, thus the number of dead fry for both species ranged between 0.4×10^6 and 4.9×10^6 and between 0.3×10^6 and 3.4×10^6, respectively.

Discussion

Present results showed that in the Ebro Delta during glass eel fishery season (November–March) a total of 17 non-target fish species were caught in glass eel fish traps, representing 43% of the fish species of the area (Demestre et al., 1977). It should be noted that although the information of by-catches and discards obtained from the glass eel fishery were a valuable tool for estimating the impact of this kind of traditional activity, these data should not to be considered as a study of the ichthyology of the Ebro Delta, since captures were only obtained in estuarine environments that did not represent the totality of habitats of this area (350 km^2). According to the ecological classification developed by Demestre et al. (1977), euryhaline-marine species (*L. ramada*, *L. aurata*, *L. saliens*, *M. cephalus*, *C. labrosus*, *P. microps*, *A. boyeri* and *S. abaster*) were the most abundant species (94%) in glass eel traps, while euryhaline-limnetic species, such as *G. holbrooki*, *A. iberus* and *Carassius* sp., represented 3% of total captures. The dominance of eurihaline-marine species in the glass eel fishery by-catch and discards was attributed to the location of fishing sites, since monitored glass eel traps were situated in the network of drainage channels connecting coastal lagoons with the estuary and sea. Although stenohaline-limnetic species (*S. fluviatilis*, *A. alburnus*, *P. parva*, *B. graellsii* and *S. erythrophthalmus*) do not normally penetrate into estuarine environments, their presence in glass eel traps was common (3%) and it was attributed to the forced drainage of irrigation water and the species contained within by pumping from the delta's continental part into the estuary.

As Davis (2002) reviewed, stress and mortality in discarded fish from towed- or fixed-gear fisheries often results from several classes of interacting stressors. These included capture stressors (net entrainment, mesh passage, crushing and wounding, etc.), fishing conditions (water and atmospheric conditions, anoxia, handling procedures, exposure to air, etc.), and biological attributes (behaviour, size and species). Most of the variability reported here on the mortality rate of the by-catch species was explained by the glass eel sorting time, the total weight of catch, the discard ratio, by-catch fish size, and by-catch fish species composition. The total weight of the catch and the discard ratio were two of the main parameters affecting discarded fish mortality, since both had a direct effect on the glass eel sorting time. This task is conducted on the river or channel bank during which the catch is exposed to air. Stress is induced in fish due to a combination of several stressors of the by-catch process, including capture, environmental factors and handling. Exposure to air is an integral part of fish capture, and increases any stress and mortality that may result from capture (Davis et al., 2001; Davis, 2002). In the glass eel fishery, air exposure times were a function of handling and sorting times, and ranged from 2 min for small catches, or catches with a small amount of by-catch fish, to 20 min for larger catches or when the biomass of the by-catch fish was larger than that of glass eels. The occurrence of large amounts of litter in the sieve also increased sorting time. In particular, the longest recorded times in the present study corresponded to those catches in which there were large amounts of macroalgae (*Ulva* sp.) or amphipods (*Gammarus* sp.). The presence of the remains of macroalgae were due to changes in water flow direction within the drainage channel during night hours, while amphipod species were attracted to the fishing trap due to the presence of light inside the fishing device.

Not only was the time that by-catch fish were exposed to air critical for discarded fish survival, but also the artesanal sorting procedure itself caused a lot of physical damage (e.g. scale loss, skin abrasions, superficial haematomas and internal lesions) to by-catch fish, that often turned into bacterial and fungal infections (15% incidence). Laboratory experimental tests with *L. ramada* and *L. aurata* fry revealed that by-catch fish mortality rates derived from fish trap handling and air exposure was magnified by the stress and physical injuries suffered during the glass eel sorting period (López & Gisbert, unpublished data).

Present results revealed that smaller fish were more susceptible to be damaged when caught and sorted than larger fish. Field data showed that, by-catch fish measuring less than 40 mm L_T were more susceptible to glass eel fishery procedures than those larger than 60 mm L_T, independently on the species considered or their stage of development. Adult mature *P. microps* females (30–40 mm L_T) and *G. holbrooki* (30–40 mm L_T) specimens were as susceptible to glass eel sorting and air exposure as grey mullets fry (15–40 mm L_T). These findings suggested that fish size rather than the stage of development (fry vs. adult) was the key factor that would explain different by-catch fish mortality patterns between species during glass eel fishery procedures, confirming previous studies on discards from trawl fisheries that reported that smaller fish are more sensitive to capture stressors than larger ones (Sangster et al., 1996; Davis & Parker, 2004; Davis, 2005).

The assessment of the global impact of the glass eel fishery in the Ebro River delta revealed the important effect of this activity in the area studied whichever approximation to the problem that was considered. The most affected by-catch species were *L. ramada* and *L. aurata* fry, which recruitment season overlapped with that of glass eels (Gisbert et al., 1995). The glass eel fishery also had a negative effect on the rest of the by-catch species, but its effect on the cyprinodontiform *A. iberus*, which was one of the most abundant species in these kinds of habitats, and nowadays is an endangered fish species, should be highlighted. Although habitat loss, contamination and trophic competition with other species have been considered as the key factors for the critical conservation status of this species (Doadrio, 2002; Rincón et al., 2002), present results revealed that glass eel fishing might also have been another important factor affecting *A. iberus* populations. However, the authors would like to point out that this study represents a first approach to this problem in the area, as fishing site selection was sometimes constrained by fishermen's reluctance to collaborate in the study, which made it difficult to have a larger number of sampling sites and limiting somewhat the robustness of the data. Nevertheless, this approximation on discarded fish mortality derived from the glass eel fishery is of value and showed this kind of fishery might have a direct effect on by-catch species. However, it should also be noted, another factor affecting the underestimation of the effects of the glass eel fishery on discard mortality is the poaching of glass eels, which is a common practice among fisherman and farmers in the area. There is no direct data on the incidence of illegal fishing, but we estimate this would represent ca. 20–25% of the total legal fishing licenses granted by the administration.

The overall impact of glass eel fishery might be considered at the population, trophic (food chain) and ecosystem level (Alverson et al., 1994). Firstly, it might have an important effect on the species composition and dominance, and population dynamics of by-catch species. If discards are significant in numbers and consist largely of under-sized individuals (i.e. high juvenile mortality), which was the case in the present study, then a reduction in yield-per-recruit and spawning potential would be expected. However, in practice it might be difficult to demonstrate the impact of discarding on a population, because of the lack of detailed data on many fisheries to separate out this factor from other causes of mortality (Mac-Donald et al., 1994). The high incidence of by-catch fish mortality among grey mullet fry might disrupt the ecological equilibrium in estuarine habitats where glass eel fishing was conducted, since mugilid fish are known to enhance small phytoplankton in freshwater macrophyte-free environments due to zooplankton depletion (Cardona et al., 2001).

These results revealed the dramatic effect of the glass eel fishery on by-catch fish species and suggested the importance of developing environmentally friendly fishing methods for minimizing the amount of discards and discarded fish mortality. These objectives might be achieved by the improvement of the artesanal sorting procedures conducted by fishermen, such as reducing glass eel handling and sorting times (length of time of air exposure), and implementing non-aggressive sorting methods that would improve by-catch fish welfare and reduce discarded fish mortality. However, these fishing procedure modifications would not completely avoid discard mortality, since glass eel traps are non-selective fishing devices. Thus, fishing trap design modifications might have to be considered in order to reduce the entrance of by-catch fish into the fishing trap to the minimum possible threshold, without affecting glass eel capture. These modifications would substantially reduce the amount of by-catch fish biomass, discard rates and sorting times.

In conclusion, the capture and return of non-targeted fish to the water during glass eel fishing season produced an important source of mortality in estuarine environments. Sorting time and procedures were the key factors that underlied by-catch fish mortality in the glass eel fishery. This fishery, similar to other ones, should strive for a minimization of the amount of discards and the mortality of discarded fish, by improving the selectivity of the fish trap and developing new sorting and selection procedures.

Acknowledgements This study was funded by 2004ACOM-0047 by Generalitat de Catalunya (Spain). E. Gisbert was funded by Programa Ramón y Cajal (Ministerio de Educación y Ciencia, Spain). Authors are grateful to all professional fishermen, especially Angel Sánchez, who collaborated in the study and to the Parc Natural del Delta de l'Ebre (Generalitat de Catalunya) and their staff for coordination tasks with fishermen. We would also like to express our gratitude to Maria Garcia for their excellent technical help during the glass eel fishing season, and Karl Andree, Josep M. Olmo and Luis Cardona for reviewing the manuscript.

References

Alverson, D. L., M. H. Freeberg, J. G. Pope & S. A. Murawski, 1994. A global assessment of fisheries bycatch and discards. FAO Fisheries Technical Paper. No. 339. FAO, Rome.

American Fisheries Society, 1992. Investigation and valuation of fish kills. AFS Special Publication 24, AS. Bethesda, MD.

Cardona, L., P. Royo & X. Torras, 2001. Effects of leaping grey mullet *Liza saliens* (Osteichthyes, Mugilidae) in the macrophyte beds of oligohaline Mediterranean coastal lagoons. Hydrobiologia 462: 233–240.

Castelnaud, G., 2002. Caractéristiques de la pêcherie civellière du golfe de Gascogne. Contrat européen n° 99/023. IFREMER, CEMAGREF, Groupement de Bordeaux, Bordeaux.

Castelnaud, G., D. Guerault, Y. Desaunay & P. Elie, 1994. Yield and abundance of the glass eel (*Anguilla anguilla* L.) at the beginning of the nineties. Bulletin Français de la Pêche et de la Pisciculture 335: 263–287.

Christensen, V., S. Guenette, J. J. Heymans, C. J. Esaters, R. Watson, D. Zeller & D. Pauly, 2003. Hundred-year decline of North Atlantic predatory fishes. Fish and Fisheries 4: 1–24.

Davis, M. W., 2002. Key principles for understanding fish bycatch discard mortality. Canadian Journal of Fisheries and Aquatic Sciences 59: 1834–1843.

Davis, M. W., 2005. Behaviour impairment in captured and released sablefish: ecological consequences and possible substitute measures for delayed discard mortality. Journal of Fish Biology 66: 245–265.

Davis, M. W., B. L. Olla & C. B. Schreck, 2001. Stress induced by hooking, net towing, elevated sea water temperature and air in sablefish: lack of concordance between mortality and physiological measures of stress. Journal of Fish Biology 58: 1–15.

Davis, M. W. & S. J. Parker, 2004. Fish size and exposure to air: potential effects on behavioural impairment and mortality rates in discarded fish. North American Journal of Fisheries Management 24: 518–524.

Dekker, W., 2002. Slipping through our hands, Population dynamics of the European eel. Doctoral Thesis, University of Amsterdam, Amsterdam.

Dekker, W., 2003. On the distribution of the European eel (*Anguilla anguilla*) and its fisheries. Canadian Journal of Fisheries and Aquatic Sciences 60: 787–799.

Dekker, W. (ed.), 2005. Report of the Workshop on National Data Collection for the European Eel, 6–8 September 2005, Sånga Säby, Stockholm, Sweden.

Demestre, M., A. Roig, A. de Sostoa & F. J. de Sostoa, 1977. Contribution to the study of the continental ichthyofauna of the Ebre Delta (NE of Iberian Peninsula). Treballs de l'Institut Català d'Història Natural 8: 145–226.

Doadrio, I. (ed.), 2002. Atlas y libro rojo de los peces continentales de España. Ministerio de Medio Ambiente, Madrid.

EIFAC/ICES, 2003. Report of the Thirteenth Session of the Joint EIFAC/ICES Working Group on Eels. EIFAC Occasional Paper. No. 36. FAO, Rome.

Fernández, V. & A. Pech, 2000. L'angula al delta de l'Ebre. Revista Catalunya Rural i Agrària 65: 22–23.

Gisbert, E., L. Cardona & F. Castelló, 1995. Evidence of competition between Mediterranean mullet fry. Journal of Fish Biology 47: 414–420.

Guerault, D., Y. Desaunay, P. Bellois & P. Grellier, 1989. The Glass-eel Fishery in the Loire Estuary (1987): estimate of by-catches and effects on resources. EIFAC Working Party on Eel, Porto.

MacDonald, D. S., J. G. Pope, N. Daanand & J. D. Reynolds, 1994. Impact of fishing on non-target species. Report to the Commission of the European Communities. MAFF Directorate of Fisheries, R.I.V.O and The University of East Anglia.

Moriarty, C. & W. Dekker (eds.), 1997. Management of the European Eel. Fisheries Bulletin (Dublin) 15: 1–110.

Pauly, D., M. L. Palomares, R. Froese, P. Saa, M. Vakily, D. Preikshot & S. Wallace, 2001. Fishing down Canadian aquatic food webs. Canadian Journal of Fisheries and Aquatic Sciences 58: 51–62.

Rincón, P. A., A. M. Correas, F. Morcillo, P. Risuen & J. Lobon-Cervia, 2002. Interaction between the introduced eastern mosquitofish and two autochthonous Spanish toothcarps. Journal of Fish Biology 61: 1560–1585.

Ringuet, S., F. Muto & C. Raymakers, 2002. Eel: their harvest and trade in Europe and Asia. Traffic Bulletin 19: 1–27.

Robin, J. P., 1990. By-catch of juvenile flounder, *Platichthys flesus* L. 1758, in the glass-eel fishery of the Loire Estuary, France. Aquatic Living Resources 3: 75–83.

Sangster, G. I., K. Lehman & M. Breen, 1996. Commercial fishing experiments to assess the survival of haddock and whiting after escape from four sizes of diamond mesh cod-ends. Fisheries Research 25: 323–345.

SigmaStat, 2004. User's Manual SigmaStat, Statistical Software, Version 3.10 for Windows, Systat Software, Inc.

Sokal, R. R. & F. J. Rohlf, 1995. Biometry, 3rd edn. W. H. Freeman and Co, New York.

Wirth, T. & L. Bernatchez, 2003. Decline of North Atlantic eels: a fatal synergy? Proceedings of the Royal Society of London B 270: 681–688.

Hydrobiologia (2008) 602:99–109
DOI 10.1007/s10750-008-9281-8

Mercury contamination and life history traits of Allis shad *Alosa alosa* (Linnaeus, 1758) and Twaite shad *Alosa fallax* (Lacépède, 1803) in the Gironde estuary (South West France)

Aude Lochet · Régine Maury-Brachet · Claire Poirier · Javier Tomás · Michel Lahaye · Miran Aprahamian · Eric Rochard

Abstract Mercury concentration [Hg] was assessed in 20 adult Allis shad *Alosa alosa* (54–59 cm) and 20 adult Twaite shad *Alosa fallax* (36–44 cm) collected during their spawning migration in the Dordogne and the Garonne rivers (France). [Hg] was measured in the gills, dorsal muscle, liver and kidney. Twaite shad exhibited higher [Hg] than Allis shad. Median [Hg] values were $[Hg]_{Gills} = 0.33$ µg g^{-1} (dw), $[Hg]_{Muscle} = 1.22$ µg g^{-1}, $[Hg]_{Liver} = 1.99$ µg g^{-1}, $[Hg]_{Kidney} = 1.93$ µg g^{-1} for Twaite shad and $[Hg]_{Gills} = 0.06$ µg g^{-1}, $[Hg]_{Muscle} = 0.20$ µg g^{-1}, $[Hg]_{Liver} = 1.18$ µg g^{-1}, $[Hg]_{Kidney} = 1.08$ µg g^{-1} for Allis shad. In order to understand such differences, we investigated some life history traits of the two species: migratory history, age (3–6 years), size at age (an expression of growth) and the number of spawning events (0–2 events). The difference in estuarine residence time between the juveniles of both species, which was assumed to influence [Hg], was investigated using otolith Sr:Ca ratio. The microchemical analysis revealed a significant difference in the residence time of juveniles in the estuary (medians are 21 d and 10 d for Twaite shad and Allis shad, respectively) but this residence time seems too short to influence [Hg] (Allis shad: $r_{spearman} < 0.128$; Twaite shad: $r_{spearman} < -0.340$). As both species show the same age structure, the influence of age on [Hg] was negligible. The literature shows that the differences in growth and in the number of spawning events reported in our study are in favour of a higher [Hg] for Twaite shad than for Allis shad. Although trophic status was not investigated here, the literature reveals that it is another factor that could produce higher [Hg] in Twaite shad, since its diet includes higher trophic levels than Allis shad.

Guest editors: S. Dufour, E. Prévost, E. Rochard & P. Williot
Fish and diadromy in Europe (ecology, management, conservation)

A. Lochet (✉) · E. Rochard
Cemagref, Groupement de Bordeaux, Estuarine ecosystems and Diadromous Fish Research Unit, 50 avenue de Verdun, 33612 Cestas Cedex, France
e-mail: aude.lochet@bordeaux.cemagref.fr

R. Maury-Brachet · C. Poirier
Laboratoire d'Ecophysiologie et Ecotoxicologie des Systèmes Aquatiques (LEESA), UMR CNRS 5805, Université Bordeaux 1, Place du Docteur Peyneau, 33120 Arcachon, France

J. Tomás
Departamento de Recursos Naturales, CSIC/UIB-Institut Mediterrani d'Estudis Avançats, Miguel Marqués 21, 07190 Esporles, Illes Balears, Spain

M. Lahaye
Centre de Caractérisation des Matériaux Avancés (CeCaMA), 87 avenue du Docteur Schweitzer, 33608 Pessac Cedex, France

M. Aprahamian
Environment Agency, National Fisheries Technical Team, Richard Fairclough House, Knutsford Road, Warrington WA4 1HG, UK

Keywords Allis shad · Twaite shad · Mercury · Sr:Ca · Life traits · Estuary

Introduction

Mercury (Hg) is of increasing concern in aquatic environmental quality (Slemr & Langer, 1992) as well as in human health (Frery et al., 2001; Cordier et al., 2002; Zahir et al., 2005) since it bioccumulates in food webs, mainly in the form of methylmercury (Boudou & Ribeyre, 1997; Power et al., 2002). In the Gironde estuary, which is the largest estuary in Western Europe covering 635 km^2 at high tide (Salomon, 2002) and the second in terms of fish species richness (Lobry et al., 2003), there is no known local point source of mercury. Although total estimated quantities of Hg are relatively low in comparison to other European estuaries, methylation rates in the Gironde estuary are higher due to high turbidities, underoxygenation of the water column during the summer period, and the long water residence time (Tseng et al., 2001), which leads to significant bioaccumulation. In the Gironde system, [Hg] is higher in the estuary than in the adjacent coastal area: in coastal waters, total mercury values are between 2.5 and 27.6 pmol l^{-1}, whereas in the Gironde estuary, the values range from 21.8 to 103.2 pmol l^{-1} (Cossa & Noël, 1987). In an initial study carried out in this estuary, Twaite shad *Alosa fallax* (Lacépède, 1803) exhibited, surprisingly, the highest Hg concentration of the eight fish species under study (Durrieu et al., 2005a). [Hg] in Twaite shad muscle was 1.35 ± 0.34 μg g^{-1} dry weight (d.w.), approximately half the safety limit set by the World Health Organization (WHO, 1990). In the Gironde system, Twaite shad live in sympatry with Allis shad *Alosa alosa* (Linnaeus, 1758) which supports a large fishery (Castelnaud et al., 2001) but for which [Hg] is unknown. [Hg] in fish depends on the complex interaction between numerous factors like age and size of fish (Mol et al., 2001; Roulet & Maury-Brachet, 2001), trophic position (Power et al., 2002; Ikingura & Akagi, 2003), and the movements of individuals among water bodies of different quality (Riget et al., 2000). Twaite and Allis shad are anadromous species. Although life history characteristics such as age and size at maturity, iteroparity rate and feeding habits of both species are well documented (Aprahamian et al. 2003; Baglinière et al. 2003), the time juveniles spend in the estuary is not well known. Regarding the differences in [Hg] between the estuary and the coastal area, one can hypothesize that [Hg] in fish depends on the time juveniles spend in the estuary. According to Taverny (1991), juveniles are observed in the estuary from 3 to 6 months after hatching for Allis shad and for nearly 2 years for Twaite shad. However, Martin Vandembulcke (1999) presented evidence of shorter residence times. In order to estimate precisely the residence time of juveniles in the estuary, an individual tracking, based on deducing information of juvenile migratory history from otoliths of adult fish, is used. The objectives of this study are (i) to examine the levels of [Hg] in Allis and Twaite shad adults from the Gironde estuary, (ii) to investigate the relationship of these concentrations to several life traits: residency time of juveniles in the estuary, age, size and the number of spawning events. For the investigation of these relationships, we developed an approach based on otolith microstructure, otolith microchemistry and scale structure.

Material and methods

Study area

The Gironde estuary, located in the south-west of France, is the result of the confluence of the Garonne and the Dordogne rivers, 75 km from the sea (Fig. 1). It varies in width from 2 km at the confluence to 11 km at its widest part. The substrate is mainly composed of mud and sand, with a high turbidity level in the water column (about 1 g l^{-1} at surface and 10 g l^{-1} near the bottom). The Gironde estuary is contaminated by several metals (cadmium, lead, copper and mercury) (Durrieu et al., 2005a).

Fish and tissue sampling

Twenty mature Allis shad and 20 mature Twaite shad were sampled in both rivers (Fig. 1). They were caught during their spawning migration, between 23 April and 5 May 2003 for Allis shad and from 12

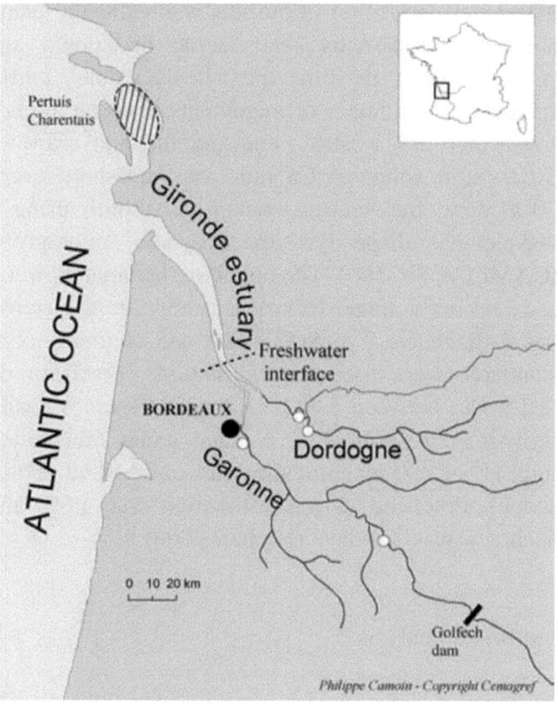

Fig. 1 Sampling locations of Allis shad and Twaite shad adults in the Dordogne and the Garonne rivers (empty circles), juveniles of Allis shad at the Golfech dam (black rectangle) and subadults of Allis shad in the Pertuis Charentais (striped area)

to 26 May 2003 for Twaite shad (Table 1), by commercial fishermen using a trammel net (stretched mesh: 110 mm). Total length, fresh body weight and sex were recorded for each individual. Eight scales per fish were taken from the standard area for scale sampling defined for the adult fish of both species, cleaned in water using a bath of a very dilute solvent (sodium hypochlorite 0.5%) and placed between two glass slides for examination (Baglinière et al., 2001; Gazeau et al., 2001). Scales were used to assess individual age and the number of spawning events. Individual age was estimated by counting annulus of winter ring (described as a white, clear line running parallel to the border of the scale showing concentric thin striae whose width increases from the centre outwards) (Baglinière et al., 2001). Spawning marks arise from the resorption of scales during freshwater spawning migration. They appear as a drastic change in the structure of the scale (Baglinière et al., 2001). In order to estimate estuarine residence time before seaward migration, sagittal otoliths were extracted and cleaned, then stored in a desiccation room before microchemical analysis. For the analysis, a transverse section of otoliths was prepared according to the recommendations of Secor et al. (1992). Each otolith was embedded sulcus side-down in Araldite® resin before being sectioned transversally (with an Isomet® saw). The resulting section was attached to a glass slide with thermoplastic glue (CrystalBond®), ground with wet sand paper (grit 1200 and 4000) and polished on one side (polishing cloth with alumina paste of 1 and 0.3 μm) until reaching the primordium. The preparation was then turned out, affixed again to a slide with the polished face down, ground and polished to remove extra material until the core area was reached.

Samples of gill arch, liver, kidney and skeletal muscle from the dorsal surface were taken using stainless steel scalpels and plastic instruments, to minimize mercury contamination. Samples were frozen (−20°C) until analysis for mercury concentration.

Table 1 Total length (TL) and fresh body weight (W) of the Allis shad and Twaite shad adults used in this study

Species	Sex	Present study			Other studies	
		n	TL (cm)	*W* (g)	TL (cm)[a]	*W* (g)[a]
Allis shad	Female	10	59 (58–60)	2,212 (2,111–2,368)	57	2,015
	Male	10	54 (53–55)	1,560 (1,541–1,650)	52	1,510
	Total	20	57 (54–59)	1,883 (1,562–2,198)		
Twaite shad	Female	10	44 (43–45)	772 (623–797)	44	726
	Male	10	36 (35–37)	334 (309–374)	38	411
	Total	20	40 (36–44)	451 (343–771)		

[a] Data from Martin Vandembulcke (1999) for Allis shad and Taverny (1991) for Twaite shad

Biometrics of individuals of both species coming from other studies in the Gironde estuary are also presented. Values are median (25–75th percentiles) for the present study and mean for the other studies

Mercury determination

Total mercury concentration (inorganic + organic mercury) was measured by flameless atomic absorption spectrometry (Durrieu et al., 2005a). Analysis was carried out automatically after drying organ samples by thermal decomposition at 750°C, under an oxygen flow (AMA 254, Prague, Czech Republic). The detection limit was 0.01 ng Hg. The validity of the analytical method was checked during each series of measurements against the same three standard biological reference materials.

Metal concentration in the different organs is expressed in terms of dry weight (48 h at 45°C) and in $\mu g\ g^{-1}$.

Estuarine residence time of juveniles

In many river/estuarine systems, including the Gironde, the ratio of strontium (Sr) to calcium (Ca) (Sr:Ca) is positively related to water salinity (Limburg, 1995; Secor et al., 1995; Daverat et al., 2005). Thus, the migratory history of fish can be reconstructed by measuring Sr:Ca ratio along transects on the otolith, that correspond to different times in the life of the fish (Kafemann et al., 2000; Limburg, 2001). In order to identify the otolith Sr:Ca ratios corresponding to the juvenile estuarine phase, young-of-the-year and subadults of Allis shad were collected (Lochet, 2006). Sr:Ca ratio was measured along transects on the otoliths of 9 Allis shad young-of-the-year sampled upstream from the Golfech dam in the Garonne River (Fig. 1). This sampling location ensures that all the Sr:Ca measurements ($n = 195$ values) on the otoliths represent freshwater values. The 95th percentiles of these measurements were used to define the threshold of the entrance in the estuary: 1.00×10^{-3} (expressed as weight percentage). In order to estimate the threshold of seaward exit, Allis shad subadults ($n = 10$) were sampled in offshore area (Pertuis Charentais) (Fig. 1). As the time they had spent in this area was unknown, Sr:Ca ratio at the outermost edge of the otoliths was used. The 5th percentiles of these measurements ($n = 10$ values) were used to estimate the threshold of seaward exit: 1.87×10^{-3}. Thus, we considered that, for both species, the estuarine phase ranged from 1.00×10^{-3} to 1.87×10^{-3}. As daily growth increment deposition in otoliths was validated using hatchery-reared Allis shad larvae of known age (Lochet, 2006), the time spent in the estuary corresponds to the number of increments between the last freshwater Sr:Ca ratio value and the first seawater Sr:Ca ratio value. Sr:Ca ratio was measured every 20 μ along the ventral axis of the otolith using a wavelength dispersive spectroscopic microprobe (CAMECA SX-100). The operating parameters were: accelerating voltage, 15 keV; current, 20nA; electron beam diameter, 8 μ. SrFCl (Sr) and andradite (Ca) standards were used for calibrations. Precision on Sr:Ca was between 2 and 3% for both species. Daily growth increments were counted under transmitted light at $\times$1,000 magnification (Leica DMRA2 microscope). Precision of age estimation was 4.6% and accuracy was 3.9 days (Lochet, 2006).

Statistical analysis

As most of the variables were not normally distributed (Kolmogorov-Smirnov one sample test), we used non-parametric statistics. Values are median with the range corresponding to the 25th and 75th percentiles. Spearman rank-order correlation coefficient r_s was used to measure the association between [Hg] and the estuarine residence time of juveniles (Siegel & Castellan, 1988). Differences in [Hg] among fish of different age, size and number of times the fish spawned were investigated using either Kruskal–Wallis or Mann–Whitney U one-way analysis of variance by ranks (Siegel & Castellan, 1988). The relationship between [Hg] and size was examined by comparing [Hg] between a group of "small" fish (total length less than the median size of the sample, $n = 10$) and a group of "large" fish (total length greater than the median size of the sample, $n = 10$). Statistics were performed using Systat 10©. A significance level of 0.05 was used.

Results

Biometric characteristics of adults

Total length ranged from 54 to 59 cm for Allis shad and from 36 to 44 cm for Twaite shad. Fresh body weight ranged from 1,562 to 2,198 g for Allis shad and from 343 to 771 g for Twaite shad (Table 1). Allis shad were significantly larger in terms of both length and weight when compared with Twaite shad (Mann–Whitney U, $P < 0.001$). For both species, females were significantly larger than males (Mann–Whitney U, $P < 0.001$).

Mercury contamination

For Allis shad, [Hg] in the gill arch, kidney, liver and muscle was not significantly different between the Garonne and Dordogne rivers (Mann–Whitney U, $P > 0.30$). This was also true for Twaite shad with the exception of muscle ($[Hg]_{Dordogne} > [Hg]_{Garonne}$, Mann–Whitney U, $P = 0.03$) (Table 2). Consequently, data from both rivers were pooled. For each organ, Twaite shad were significantly more contaminated than Allis shad (Mann–Whitney U, $P < 0.001$) (Table 2). [Hg] in liver and kidney for some Twaite shad reached or were above the safety limit defined by the WHO. For Allis shad, there was no significant difference in [Hg] between sexes (Mann–Whitney U, $P > 0.174$) with the exception of liver for which males were more contaminated than females (Mann–Whitney U, $P < 0.01$) (Table 2). For Twaite shad, females were significantly more contaminated than males (Mann–Whitney U, $P < 0.01$), except for liver for which there was no significant difference between sexes (Mann–Whitney U, $P = 0.36$) (Table 2).

[Hg] and estuarine residence time of juveniles before seaward migration

Juvenile migratory history was assessed from otoliths of 19 Allis shad adults and 19 Twaite shad adults. Twaite shad spend significantly more time in the estuary than Allis shad (Mann–Whitney U, $P = 0.01$): median values were 10 days for Allis shad and 21 days for Twaite shad (Fig. 2). For both species, the correlation between the time juveniles spent in the estuary and [Hg] in each organ was not significant (for each organ of Allis shad: $r_s < 0.128$, $P > 0.214$; for each organ of Twaite shad: $r_s < -0.340$, $P > 0.092$).

[Hg] and age

Ages ranged from 4 to 6 years for Allis shad and from 3 to 6 years for Twaite shad (Fig. 3). Age structure was not significantly different between species (Mann–Whitney U, $P = 0.92$). [Hg] was not significantly different between Allis shad of different ages (Kruskal-Wallis, $P > 0.20$, for each organ) (Table 2). A similar conclusion was drawn for

Table 2 Mercury concentrations in gills, muscle, liver and kidney for mature Allis shad ($n = 20$) and Twaite shad ($n = 20$) by sampling location (Gar, Garonne, $n = 10$; Dor, Dordogne,$n = 10$, for both species), sex (F, Females, $n = 10$; M, Males, $n = 10$, for both species), age (Allis shad: 4 years $n = 8$, 5 years $n = 9$, 6 years $n = 3$; Twaite shad: 4 years $n = 7$, 5 years $n = 6$, 6 years $n = 4$), size (Small $n = 10$; Large $n = 10$ for both species) and the number of spawnings (Twaite shad: 0 spawning $n = 2$, 1 spawning $n = 13$, 2 spawnings $n = 3$). Values are median. Whatever the tissue, the safety limit defined by the World Health Organization is 2.5 μg g^{-1} (dw)

	Sampling location			Sex		Age (year)			Size		Number of spawnings		
	Gar	Dor	Gar + Dor	F	M	4	5	6	Small	Large	0	1	2
Allis shad													
Gills	0.05	0.05	0.05	0.05	0.06	0.05	0.05	0.06	0.06	0.05			
Muscle	0.22	0.18	0.20	0.21	0.18	0.21	0.20	0.21	0.18	0.21			
Liver	1.42	1.10	1.18	0.97	1.58	1.04	1.58	1.05	1.56	0.97			
Kidney	0.98	1.15	1.08	0.94	1.26	0.97	1.08	1.24	1.16	0.99			
Twaite shad													
Gills	0.30	0.41	0.33	0.44	0.24	0.43	0.41	0.33	0.24	0.44	0.35	0.35	0.39
Muscle	1.09	1.38	1.22	1.40	1.01	1.37	1.40	1.05	1.01	1.40	1.12	1.29	1.06
Liver	2.03	1.99	1.99	2.08	1.93	2.50	2.31	1.78	1.93	2.08	1.84	2.17	1.87
Kidney	1.93	1.90	1.92	2.15	1.67	2.01	2.16	1.79	1.67	2.15	1.86	2.03	1.73

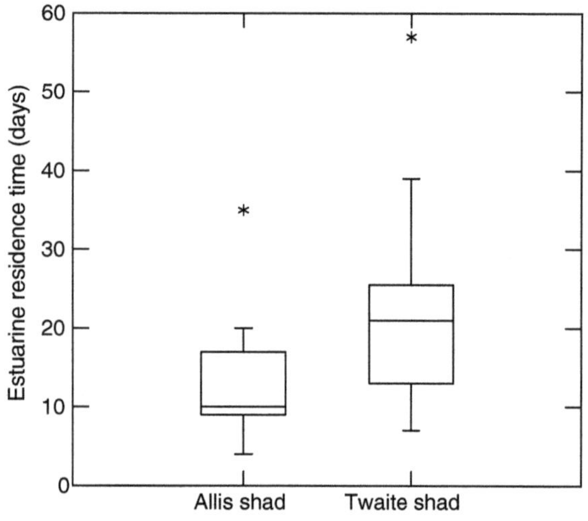

Fig. 2 Time spent in the Gironde estuary (in days) before juvenile seaward migration, for both species. The length of each box plot shows the range within which the central 50% of the values fall, with the box edges at the 25th and 75th percentiles. The whiskers show the range of values between the 5th percentiles and the 95th percentiles

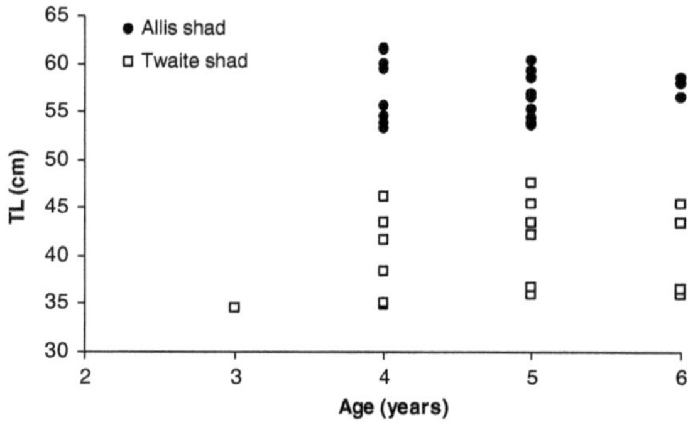

Fig. 3 Growth expressed as total length (TL) by age of Allis shad (filled circle) and Twaite shad (empty square)

Twaite shad except for $[Hg]_{muscle}$ (5-year-old fish > 6-year-old fish, Mann–Whitney U, $P = 0.01$) and $[Hg]_{liver}$ (6-year old fish < all of the other age groups, Mann–Whitney U, $P < 0.03$) (Table 2).

[Hg] and size, growth

As both species exhibit a strong linear relation between weight and length ($r_s = 0.88$ and 0.91 for Allis shad and Twaite shad respectively; $P < 0.001$), only the relation between [Hg] and length is shown. For Allis shad, [Hg] was not significantly different between small and large fish, except for liver (Table 2). For Twaite shad, [Hg] was higher in large individuals than in small ones (Table 2) except for liver. When we compare the size of fish at a given age, which is a way to compare their growth, it appears that growth is faster for Allis shad than for Twaite shad (for each age group, Mann–Whitney U, $P < 0.034$) (Fig. 3).

[Hg] and spawning history

All Allis shad were virgin fish, spawning for the first time, while most of the Twaite shad (88.9%) exhibited evidence of previous spawning with one (72.2%) or two (16.7%) spawning marks on their scales indicating reproduction in 2001 and 2002. [Hg] was not significantly different between fish that showed different numbers of spawning events (Kruskal–Wallis, $P > 0.09$ for each organ) (Table 2).

Discussion

This study provides the first comparison of Hg contamination in Allis shad and Twaite shad. It reveals drastic differences in [Hg] among these closely related, sympatric, *Alosa* species with Twaite shad being significantly much more contaminated than Allis shad. Compared with other species (Table 3), Twaite shad from either the Gironde or the Severn estuary exhibit the highest [Hg]. On the contrary, Allis shad exhibit the lowest concentrations, similar to those measured in sardine *Sardina pilchardus* (Walbaum, 1792) or herring *Clupea harengus* (Linnaeus, 1758).

The age range and biometric characteristics of the spawners of Allis and Twaite shad sampled are similar to those reported by Taverny (1991) and Martin Vandembulke (1999) for *Alosa* spp. from the Gironde-Dordogne-Garonne system. In this study, we focused on the time juveniles spent in the estuary before seaward migration as it was expected to be a potential factor discriminating the ecology of these two species. Although otolith Sr:Ca ratio analysis showed a significant difference in the estuarine residence time between the juveniles of both species, the time length seems much too short to influence

Table 3 Mercury concentrations in Allis shad and Twaite shad in the Gironde estuary compared with mercury concentrations of other species from the Gironde estuary, Twaite shad from the Severn estuary and marine clupeid fish species

Species	Location	Period	Gills	Muscle	Liver	Kidney	Sources
European eel (*Anguilla anguilla*)	Gironde estuary	2001	0.12	0.78	1.07	1.02	Durrieu et al. (2005a)
Bass (*Dicentrarchus labrax*)			0.08	0.54	0.59	0.37	
Meagre (*Argyrosomus regius*)			0.27	0.61	0.37	0.63	
Flounder (*Platichthys flesus*)			0.14	0.59	0.56	0.62	
Thin-lipped grey mullet (*Liza ramada*)			0.07	0.50	2.55	4.57	
Sole (*Solea vulgaris*)			0.11	0.26	0.27	0.42	
Canary drum (*Umbrina canariensis*)			0.21	0.32	0.46	0.59	
Twaite shad (*Alosa fallax*)			0.60	1.18	2.62	2.47	
Twaite shad (*Alosa fallax*)	Gironde estuary	2003	0.33	1.22	1.99	1.93	This study
Allis shad (*Alosa alosa*)			0.06	0.20	1.18	1.08	
Twaite shad (*Alosa fallax*)	Severn estuary	1985–1986		1.08[a]	2.60[a]	2.12[a]	Aprahamian. pers.com.
Sardine (*Sardina pilchardus*)	Atlantic ocean			0.21[b]			Cabañero et al. (2005)
Herring (*Clupea harengus*)	Atlantic ocean			0.20[b]			Cossa et al. (1990)

[a] Dry weight data calculated from fresh weight data according to the relation: 2.5 μg g^{-1} dry weight or 0.5 μg g^{-1} wet weight (WHO, 1990)

[b] Values are mean

Values are median except for sardine and herring

mercury concentrations observed in the mature adults. The same is true for the freshwater phase: although early life occurs in freshwater (either the Garonne or the Dordogne), it is a short time phase in comparison with the whole life cycle of fish. The freshwater phase is from 54 to 124 days for Allis shad and from 7 to 84 days for Twaite shad (Lochet, 2006). If we consider the marine part of the migratory history of both species, the distribution at sea is not the same: Twaite shad have a more coastal distribution than Allis shad (Taverny & Elie, 2001a). Depending on the water quality of their marine habitats, the accumulation of mercury of both species may be different. Our study reveals estuarine residence time of juveniles much lower than previously thought (Taverny, 1991; Martin Vandembulcke, 1999). This discrepancy mainly relies on the use of different approaches. Taverny (1991) and Martin Vandembulcke (1999) sampled juveniles in the estuary, at different points in time. Such an approach reflects the population pattern of seaward migration, but cannot inform us about the individual patterns. The Sr:Ca ratio approach is an individual a posteriori tracking that mainly depends on our ability to interpret properly the otolith information (Campana & Thorrold, 2001). Our experiments to identify the estuarine phase on the otolith and to validate daily growth increments make this Sr:Ca ratio approach reliable. The same kind of discrepancy was found for *Alosa sapidissima* (Wilson, 1811) seaward migration. Seaward migration was considered to be a late-season phenomenon based on capture of juveniles (e.g. O'Leary & Kynard, 1986) while individual tracking using otoliths revealed that seaward migration begins as early as June (Limburg, 1995, 1996).

Considering the other life history traits taken into account in this study, the absence of a significant difference in age structure between the two species supports the idea that age is not an important factor explaining the different level of contamination between the two species. Our results for Twaite shad are consistent with an accumulation of [Hg] as fish get larger (Baeyens et al., 2003; Agusa et al., 2004) and is reinforced by the lower [Hg] in smaller Twaite shad collected by Durrieu et al. (2005a) (e.g. $[Hg]_{muscle}$ of 0.4 μg g^{-1} for a 30 cm TL fish). The narrower length range in Allis shad probably explains the absence of Hg accumulation with length.

The difference of growth between Allis shad and Twaite shad is in favour of a higher [Hg] for the latter species. Indeed, for a given period, fast growing fish produce more tissues than slow growing ones, leading to an effect of "dilution" of the pollutants by the tissue. Such a phenomenon has been reported for Artic Charr (*Salvelinus alpinus*) (Hammar et al., 1993) or walleye (*Sander vitreus*) (Simoneau et al., 2005).

The number of spawning events, which is rarely cited as a factor influencing [Hg], could act in at least two ways. Firstly, spawning events are costly in terms of energy. This implies that fish will have to compensate metabolic needs by feeding extensively (Aprahamian, 1989; Trudel et al., 2000). Secondly, Niimi (1983) reported a 6–22% increase in [Hg] following the deposition of eggs in different species. This increase was related to differences in relative [Hg] between fish and eggs and to the body weight loss associated with spawning. Hence, higher [Hg] can be expected in an iteroparous species, like Twaite shad, than in a semelparous one like Allis shad. A significant accumulation of Hg through multiple spawnings probably depends on the number of spawning events performed during a fish's life. In the Severn estuary, Twaite shad spawn up to seven times (Aprahamian et al., 2003) and exhibit a significant correlation between [Hg] and the number of spawning events (Aprahamian, unpublished). In the Gironde estuary, Twaite shad rarely spawn more than three times (Taverny, 1991). The low number of spawning events of Twaite shad in the Gironde estuary could explain why no effect is detected.

Although not demonstrated in this study, feeding habits are of great importance in the exposure to contaminants. It is known that mercury is accumulated by fish via two pathways: directly from the water through the gills and from their diet, with a much more importance of uptake from diet than from water (Hall et al., 1997; Harris & Bodaly, 1998). As mercury (especially the methylmercury form) accumulates through the trophic food web (Boudou & Ribeyre, 1997; Watras et al., 1998; Simon & Boudou, 2001), piscivorous fish are more contaminated than non piscivorous fish. As an example, Durrieu et al. (2005b) showed that $[Hg]_{muscle}$ in French Guiana piscivorous fish was 715 times higher than in herbivorous fish. A literature survey reveals marked differences in the diet of Allis shad and Twaite shad, especially at the subadult and adult stages (Table 4).

Table 4 Diet of Allis and Twaite shad at different stages and in different locations

Prey items	Larvae and juveniles in river		Juveniles in estuary		Subadults and adults at sea	
	Allis shad [1, 2]	Twaite shad [3, 4]	Allis shad [4, 5]	Twaite shad [3, 6, 7]	Allis shad [2, 8, 9]	Twaite shad [2, 8, 9]
Insects (Ephemeroptera. Diptera. Trichoptera)	++	++	++			
Malacostraca						+
Mysids			++	++	+	
Amphipods			++	++		
Isopods			++	+		
Decapoda					++	
Euphausids					++	+
Copepod			++	++	++	
Cladocera (e.g. Daphnia)	++			+	++	
Fish (mainly anchovy and sprat)				++	+	++
Polychaete worms					++	

[1] Cassou-Leins & Cassou-Leins (1981); [2]Cemagref. unpublished data; [3]Aprahamian (1989); [4]Sabatié (1993); [5]Taverny (1991); [6]Oesmann & Thiel (2001); [7]Pasquaud, pers.com; [8]Taverny & Elie (2001b); [9]Maitland & Lyle (2005)

++ indicates predominant food items cited by the authors and + indicates prey of lesser importance

Allis shad is mainly planktivorous whereas Twaite shad is more piscivorous (Taverny & Elie, 2001b; Maitland & Lyle, 2005; Cemagref source, unpublished data and S. Pasquaud, pers. com.). This difference should promote a higher [Hg] for the latter species.

As far as we know, no health concerns have been reported from eating Allis shad or Twaite shad. Even if Twaite shad is more contaminated than Allis shad, the level of contamination in muscle does not reach the safety limit set by the WHO. However, because of mercury bioaccumulation, excessive consumption is not suitable, particularly for Twaite shad.

Acknowledgements We thank Stéphanie Pasquaud (Cemagref, Bordeaux) who provided data on shad prey in the Gironde estuary, Stéphanie Bourrachot and Frank Poulain for stomach content analysis and Marina Daubas for age reading. Three anonymous referees and Karin Limburg (university of Syracuse, New York) provided comments that improved the manuscript.

References

Agusa, T., T. Kunito, S. Tanabe, M. Pourkazemi & D. G. Aubrey, 2004. Concentrations of trace elements in muscle of sturgeons in the Caspian Sea. Marine Pollution Bulletin 49: 789–800.

Aprahamian, M. W., 1989. The diet of juvenile and adult Twaite shad *Alosa fallax fallax* (Lacépède) from the rivers Severn and Wye (Britain). Hydrobiologia 179: 173–182.

Aprahamian, M. W., J. L. Baglinière, M. R. Sabatié, P. Alexandrino, R. Thiel & C. D. Aprahamian, 2003. Biology, status, and conservation of the anadromous Atlantic Twaite shad *Alosa fallax fallax*. In Limburg, K. E. & J. R. Waldman (eds), Biodiversity, Status, and Conservation of the World's Shads. American Fisheries Society, Bethesda, Maryland: 103–124.

Baeyens, W., M. Leermakers, T. Papina, A. Saprykin, N. Brion, J. Noyen, M. De Gieter, M. Elskens & L. Goeyens, 2003. Bioconcentration and biomagnification of mercury and methylmercury in North Sea and Scheldt estuary fish. Archives of Environmental Contamination and Toxicology 45: 498–508.

Baglinière, J. L., M. R. Sabatié, M. W. Aprahamian, P. J. Alexandrino, C. D. Aprahamian, C. A. Assis, J. J. Cassou-Leins, M. Le Corre, C. Mennesson-Boisneau, D. Martin-Vandembulcke, E. Rochard & C. Teixeira, 2001. Guide pour l'interprétation des écailles et l'estimation de l'âge chez les aloses (*Alosa* sp.) de la façade Atlantique-Est et de la Méditerranée-Ouest. Bulletin Français de la Pêche et de la Pisciculture 357/360: 485–530.

Baglinière, J. L., M. R. Sabatié, E. Rochard, P. Alexandrino & M. W. Aprahamian, 2003. The Allis shad *Alosa alosa*: biology, ecology, range and status of populations. In Limburg, K. E. & J. R. Waldman (eds), Biodiversity, Status, and Conservation of the World's Shads. American Fisheries Society, Bethesda, Maryland: 85–102.

Boudou, A. & F. Ribeyre, 1997. Mercury in the food web: accumulation and transfer mechanisms. In Sigel, A. & H. Sigel (eds), Metal ions in Biological Systems. Mercury and its effects on Environment and Biology. Marcel Dekker, New York: 289–319.

Cabañero, A. I., C. Carvalho, Y. Madrid, C. Batoréu & C. Camára, 2005. Quantification and speciation of mercury and selenium in fish samples of high consumption in Spain and Portugal. Biological Trace Element Research 103: 17–35.

Campana, S. E. & S. R. Thorrold, 2001. Otoliths, increments, and elements: keys to a comprehensive understanding of fish populations? Canadian Journal of Fisheries and Aquatic Sciences 58: 30–38.

Cassou-Leins, F. & J. J. Cassou-Leins 1981. Recherches sur la biologie et l'halieutique des migrateurs de la Garonne et principalement de l'alose *Alosa alosa* L. Ph.D thesis, Institut National Polytechnique, Toulouse, 382p.

Castelnaud, G., E. Rochard & Y. Le Gat, 2001. Analyse de la tendance de l'abondance de l'alose *Alosa alosa* en Gironde à partir de l'estimation d'indicateurs halieutiques sur la période 1977–1998. Bulletin Français de la Pêche et de la Pisciculture 362/363: 989–1015.

Cordier, S., M. Garel, L. Mandereau, H. Morcel, P. Doineau, S. Gosme-Seguret, D. Josse, R. White & C. Amiel-Tison, 2002. Neurodevelopmental investigations among methylmercury-exposed children in French Guiana. Environmental Research 89: 1–11.

Cossa, D. & J. Noël, 1987. Concentrations of mercury in near shore surface waters of the bay of Biscay and in the Gironde estuary. Marine Chemistry 20: 389–396.

Cossa, D., D. Auger, B. Averty, M. Luçon, P. Masselin, J. Noël & J. Sanjuan, 1990. Niveaux de concentration en métaux, métalloïdes et composés organochlorés dans les produits de la pêche côtière française. Rapport IFREMER, 59p.

Daverat, F., J. Tomás, M. Lahaye, M. Palmer & P. Elie, 2005. Tracking continental habitat shifts of eels using Sr/Ca ratios: validation and application to the coastal, estuarine and riverine eels of the Gironde-Garonne-Dordogne watershed. Marine and Freshwater Research 56: 619–627.

Durrieu, G., R. Maury-Brachet, M. Girardin, E. Rochard & A. Boudou, 2005a. Contamination by heavy metals (Cd, Zn, Cu, and Hg) of eight fish species in the Gironde estuary (France). Estuaries 28: 581–591.

Durrieu, G., R. Maury-Brachet & A. Boudou, 2005b. Goldmining and mercury contamination of the piscivorous fish *Hoplias aimara* in French Guiana (Amazon basin). Ecotoxicology and Environmental Safety 60: 315–323.

Frery, N., R. Maury-Brachet, E. Maillot, M. Deheeger, B. Merone & A. Boudou, 2001. Goldmining activities and mercury contamination of native amerindian communities in French Guiana: key role of fish in dietary uptake. Environmental Health Perspectives 109: 1–5.

Gazeau, C., T. Lefrançois & E. Rochard, 2001. Mise en place d'une démarche qualité dans le cadre de l'échantillonnage d'une population de grande alose (*Alosa alosa*). Bulletin Français de la Pêche et de la Pisciculture 362/363: 1017–1024.

Hall, B. D., R. A. Bodaly, R. J. P. Fudge, J. W. M. Rudd & D. M. Rosenberg, 1997. Food as the dominant pathway of methylmercury uptake by fish. Water, Air, and Soil Pollution 100: 13–24.

Hammar, J., P. Larsson & M. Klavins, 1993. Accumulation of persistent pollutants in normal and dwarfed Artic charr (*Salvelinus alpinus* sp. complex). Canadian Journal of Fisheries and Aquatic Sciences 50: 2574–2580.

Harris, R. C. & R. A. Bodaly, 1998. Temperature, growth and dietary effects on fish mercury dynamics in two Ontario lakes. Biogeochemistry 40: 175–187.

Ikingura, J. R. & H. Akagi, 2003. Total mercury and methylmercury levels in fish from hydroelectric reservoirs in Tanzania. The Science of the Total Environment 304: 355–368.

Kafemann, R., S. Adlerstein & R. Neukamm, 2000. Variation in otolith strontium and calcium ratios as an indicator of life-history strategies of freshwater fish species within a brackish water system. Fisheries Research 46: 313–325.

Limburg, K. E., 1995. Otolith strontium traces environmental history of subyearling American shad *Alosa sapidissima*. Marine Ecology Progress Series 119: 25–35.

Limburg, K. E., 1996. Growth and migration of 0-year American shad (*Alosa sapidissima*) in the Hudson River estuary: otolith microstructural analysis. Canadian Journal of Fisheries and Aquatic Sciences 53: 220–238.

Limburg, K. E., 2001. Through the gauntlet again: demographic restructuring of American shad by migration. Ecology 82: 1584–1596.

Lobry, J., L. Mourand, E. Rochard & P. Elie, 2003. Structure of the Gironde estuarine fish assemblages: a comparison of European estuaries perspective. Aquatic Living Resources 16: 47–58.

Lochet, A., 2006. Dévalaison des juvéniles et tactiques gagnantes chez la grande alose *Alosa alosa* et l'alose feinte *Alosa fallax*: apports de la microchimie et de la microstructure des otolithes. Ph.D thesis, University of Bordeaux 1, Bordeaux, 208p.

Maitland, P. S. & A. A. Lyle, 2005. Ecology of Allis Shad *Alosa alosa* and Twaite Shad *Alosa fallax* in the Solway Firth, Scotland. Hydrobiologia 534: 205–221.

Martin Vandembulcke, D., 1999. Dynamique de population de la grande alose (*Alosa alosa*, L. 1758) dans le bassin versant Gironde-Dordogne-Garonne (France): analyse et prévision par modélisation. Ph.D thesis, INP-ENSAT, Toulouse, 114p.

Mol, J. H., J. S. Ramlal, C. Lietar & M. Verloo, 2001. Mercury contamination in freshwater, estuarine, and marine fishes in relation to small-scale gold mining in Suriname, South America. Environmental Research Section A 86: 183–197.

Niimi, A. J., 1983. Biological and toxicological effects of environmental contaminants in fish and their eggs. Canadian Journal of Fisheries and Aquatic Sciences 40: 306–312.

Oesmann, S. & R. Thiel, 2001. Feeding of juvenile Twaite shad (*Alosa fallax* Lacépède, 1803) in the Elbe estuary. Bulletin Français de la Pêche et de la Pisciculture 362/363: 785–800.

O'Leary, J. A. & B. Kynard, 1986. Behavior, length and sex ratio of seaward migrating juvenile American shad and blueback herring in the Connecticut River. Transactions of the American Fisheries Society 115: 529–536.

Power, M., G. M. Klein, K. R. R. A. Guiguer & K. H. Kwan, 2002. Mercury accumulation in the fish community of a sub-Arctic lake in relation to trophic position and carbon sources. Journal of Applied Ecology 39: 819–830.

Riget, F., G. Asmund & P. Aastrup, 2000. Mercury in Arctic char (*Salvelinus alpinus*) populations from Greenland. The Science of the Total Environment 245: 161–172.

Roulet, M. & R. Maury-Brachet, 2001. Le mercure dans les organismes aquatiques amazoniens (Annexe 4). In IRD (ed.), Le mercure en Amazonie. Paris: 204–271.

Sabatié, R., 1993. Recherches sur l'écologie et la biologie des aloses au Maroc (*Alosa alosa* Linné, 1758 et *Alosa fallax* Lacépède, 1803): exploitation et taxonomie des populations atlantiques, bioécologie des aloses de l'Ouest Sebou. Ph.D thesis, University of Bretagne Occidentale, Brest, 326p.

Salomon, J. N., 2002. L'inondation dans la basse vallée de la Garonne et l'estuaire de la Gironde lors de la "tempête du siècle" (27–28 décembre 1999). Géomorphologie : relief, processus, environnement 2: 127–134.

Secor, D. H., J. Mark Dean, & E. H. Laban, 1992. Otolith removal and preparation for microstructural examination. In Stevenson, D. K. & S. E. Campana (eds), Otolith microstructure examination and analysis. Canadian Special Publication of Fisheries and Aquatic Sciences 17: 19–57.

Secor, D. H., A. Henderson-Arzapalo & P. M. Piccoli, 1995. Can otolith microchemistry chart patterns of migration and habitat utilization in anadromous fishes? Journal of Experimental Marine Biology and Ecology 192: 15–33.

Siegel, S. & N. J. Castellan, 1988. Nonparametric statistics for the behavioral sciences. 2nd edn. McGraw-Hill, New York.

Simon, O. & A. Boudou, 2001. Direct and trophic contamination of the herbivorous carp *Ctenopharyngodon idella* by inorganic mercury and methylmercury. Ecotoxicology and Environmental Safety 50: 48–59.

Simoneau, M., M. Lucotte, S. Garceau & D. Laliberté, 2005. Fish growth rates modulate mercury concentrations in walleye (*Sander vitreus*) from eastern Canadian lakes. Environmental Research 98: 73–82.

Slemr, F. & E. Langer, 1992. Increase in global atmospheric concentrations of mercury inferred from measurements over the Atlantic Ocean. Nature 355: 434–437.

Taverny, C., 1991. Contribution à la connaissance de la dynamique des populations d'aloses (*Alosa alosa* et *Alosa fallax*) dans le système fluvio-estuarien de la Gironde: pêche, biologie et écologie. Etude particulière de la dévalaison et de l'impact des activités humaines. Ph.D thesis, University of Bordeaux 1, Bordeaux, 568p.

Taverny, C. & P. Elie, 2001a. Répartition spatio-temporelle de la grande alose *Alosa alosa* (Linné, 1766) et de l'alose feinte *Alosa fallax* (Lacépède, 1803) dans le golfe de Gascogne. Bulletin Français de la Pêche et de la Pisciculture 362/363: 801–821.

Taverny, C. & P. Elie, 2001b. Régime alimentaire de la grande alose *Alosa alosa* (Linné, 1766) et de l'alose feinte *Alosa fallax* (Lacépède, 1803) dans le golfe de Gascogne. Bulletin Français de la Pêche et de la Pisciculture 362–363: 837–852.

Trudel, M., A. Tremblay, R. Schetagne & J. B. Rasmussen, 2000. Estimating food consumption rates of fish using a mercury mass balance model. Canadian Journal of Fisheries and Aquatic Sciences 57: 414–428.

Tseng, C. M., D. Amouroux, G. Abril, E. Tessier, H. Etcheber & O. F. Donard, 2001. Speciation of mercury in a fluid mud profile of a highly turbid macrotidal estuary (Gironde, France). Environmental Science and Technology 35: 2627–2633.

Watras, C. J., R. C. Back, S. Halvorsen, R. J. M. Hudson, K. A. Morrison & S. P. Wente, 1998. Bioaccumulation of mercury in pelagic freshwater food webs. The Science of the Total Environment 219: 183–208.

WHO (World Health Organization), 1990. Environmental health criteria, 101. Methylmercury. WHO/ICPS, Geneva.

Zahir, F., S. J. Rizwi, S. K. Haq & R. H. Khan, 2005. Low dose mercury toxicity and human health. Environmental Toxicology and Pharmacology 20: 351–360.

Hydrobiologia (2008) 602:111–127
DOI 10.1007/s10750-008-9282-7

OpenAccess

FISH AND DIADROMY IN EUROPE

Upstream passage problems for wild Atlantic salmon (*Salmo salar* L.) in a regulated river and its effect on the population

H. Lundqvist · P. Rivinoja · K. Leonardsson · S. McKinnell

Abstract Due to hydropower development, the upstream migration of wild anadromous salmon and brown trout is impaired in many European rivers, causing negative effects on the long-term survival of natural salmonid populations. This study identified problems for Atlantic salmon during upstream migration in a regulated river in northern Sweden, Umeälven (mean flow: 430 m^3 s^{-1}). Tagging from 1995 to 2005 involved radio tags (n = 503), PIT tags (n = 1574) and Carlin tags (n = 573) to study the spawning migration of salmon from the coast past the regulated section of the river to a fish ladder at the dam/spillway 32 km upriver. The results demonstrate that migration success from the coast to the fish ladder varied between 0% and 47% among years, indicating an average loss of 70% of potential spawners. Discharge from the turbines attracted the salmon away from the bypass route. Echo-sounding in the turbine outlet showed that salmon were normally found at 1–4 m depths. They responded with upstream and/or downstream movements depending on flow changes; increased spill in the bypass channel attracted salmon to the bypass. Once in the bypass channel, salmon could be delayed and had difficulties passing the first rapid at high spills. Additional hindrances to upstream migration were found at rapids and the area of the fish ladder, located further upstream in the regulated river section. The average migration duration was 44 days from the estuary to the top of the fish ladder, with large variation among individuals within years. Modelling the salmon population dynamics showed a potential population increase of 500% in 10 years if the overall migration success could be improved from the current 30% to levels near 75%. Consequently improved migration facilities at the regulated river section should be implemented to achieve a long-term sustainability of these threatened anadromous salmonids.

Keywords Hydropower station · Tag · Migration success · Bypass channel · Fish ladder · Population model

Guest editors: S. Dufour, E. Prévost, E. Rochard & P. Williot
Fish and diadromy in Europe (ecology, management, conservation)

H. Lundqvist (✉) · P. Rivinoja · K. Leonardsson
Department of Wildlife, Fish, and Environmental Studies, SLU (Swedish University of Agricultural Sciences), Umeå 901 83, Sweden
e-mail: hans.lundqvist@vfm.slu.se

K. Leonardsson
Department of Ecology and Environmental Science, Umeå University, Umeå 901 87, Sweden

S. McKinnell
North Pacific Marine Science Organization, P. O. Box 6000, Sidney, BC, Canada V8L 4B2

Introduction

Natural salmonid populations have been eliminated or substantially reduced in many regulated rivers

(Eriksson & Eriksson, 1993; NRC, 1996). In a regulated river where water is diverted to turbines, changes to the natural flow of the river affect the fish migration. Consequently, problems can arise for adult fish on their way to spawning areas (Arnekleiv & Kraabøl, 1996; Rivinoja et al., 2001; Karppinen et al., 2002). Furthermore, turbines and dams cause elevated mortalities for downstream migrating smolts (Montén, 1985; Coutant & Whitney, 2000) and kelts (Scruton et al., 2002). Even if various fishways and guidance devices are constructed to maintain migration possibilities (Clay, 1995), their ability to attract and permit rapid and safe passage of fish (Katopodis, 1990) varies considerably. Upstream migrants can encounter problems in flow-controlled areas where they must find a way past turbine outlets to bypass channels where water volumes are relatively low compared to the main river (Arnekleiv & Kraabøl, 1996; Quinn et al., 1997; Thorstad et al., 2003). Similarly, large variations in river flow or intermittent spills from dams can hinder the upstream migration (Rivinoja et al., 2001). Problems can also arise in the vicinity of fish ladders, where ladder attraction and passage flows might be ineffective in ensuring high success at upstream migration (Bjornn & Peery, 1992). At the same time, environmental factors such as discharge volume and water temperature can affect the migration of fish in complex ways (Banks, 1969; Northcote, 1998). Discharge is one of the most important factors for attracting upstream migrants to the entrances of fishways, while adequate flows and water velocity within the fishway then secure upstream passage (Larinier, 1998; Williams, 1998).

In this paper, the results obtained from a ten-year study of adult Atlantic salmon (*Salmo salar* L.) passing through the flow regulated lower part of the River Umeälven are summarised. The migration behaviour of salmon at various obstacles in the regulated river is described with a focus on the migratory performance at different flows. A population model was used to estimate the consequences of improved escapements to the spawning grounds. Since anthropogenic impacts such as river regulation tend to diminish anadromous fish abundance, we highlight the long-term solutions to these migration problems so that viable populations can be maintained in future. Our results can be applied to other regulated river systems where bypasses are used to provide a migratory route for salmonids.

Materials and methods

Salmon and the River Umeälven study area

The rivers Umeälven and Vindelälven originate in parallel valleys with their headwaters in the mountains close to the Norwegian border, c. 450 km from the Bothnian Bay (Fig. 1). The Umeälven is dammed for hydroelectric power production throughout its length, so the passage of anadromous fish in this river is blocked by the first dam, Stornorrfors. The Vindelälven merges with the Umeälven 12 km above (64°N,

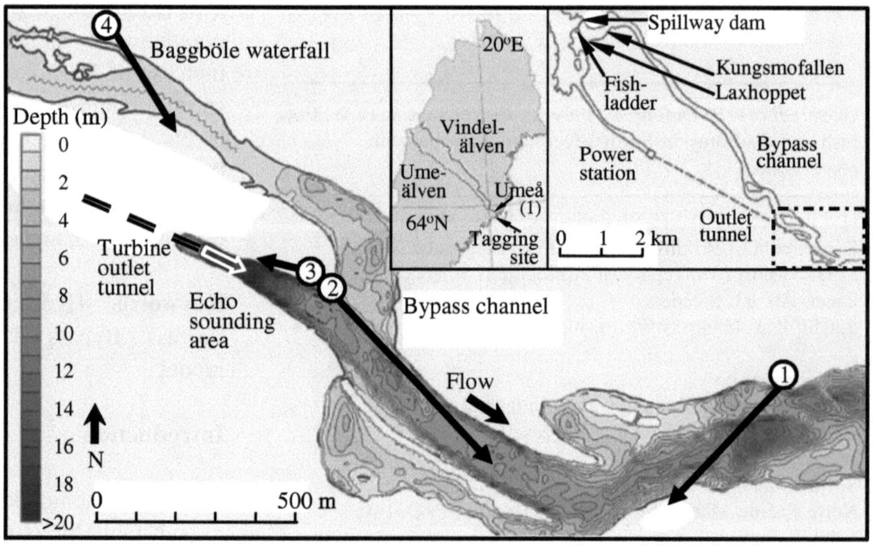

Fig. 1 The regulated area that upstream migrating fish meet in the River Umeälven is the confluence of the turbine outlet and the bypass channel. Archival receivers, their location (open circles with numbers) and main reading transects (black arrows) were positioned for observation of individually radio-tagged salmon. From 1995 to 2003 the first archival receiver was situated in Umeå (1). The echo-sounding area at the turbine outlet is indicated by the white arrow

20°E) Stornorrfors. Anadromous Atlantic salmon and brown trout (*Salmo trutta* L.) gain access to the Vindelälven by way of a fish ladder at Stornorrfors, located 32 km upstream from the coast (Fig. 1). A hatchery immediately below the dam attempts to compensate for lost production of wild fish from the Umeälven by annually releasing c. 80,000 salmon and 20,000 sea trout smolts with their adipose fin removed to distinguish them from wild fish. At the top of the ladder, all migrating fish are trapped, counted, weighed, sexed and identified as either of wild (adipose fin intact) or hatchery origin (adipose fin removed). About 70% of the salmon at the ladder is of wild origin. These fish are released above the dam to continue their migration upstream to spawning areas in Vindelälven. Total annual counts have varied from 250 to 6065 salmon between 1974 and 2005, totalling 52,671 over the years (Fig. 2).

Migrating wild salmon in this river system generally enter the coastal areas in May and after holding for a period of time, start their upriver migration in June. In this report, upstream migrants are those adult salmon that ascended the river as far as the confluence area where turbine discharge mixes with water in the river, the first migration obstacle in the part of the river that is regulated (described below). In total, the river rises about 75 m from the sea level to the top of the fish ladder. To successfully reach to natural spawning areas in River Vindelälven, the salmon must pass the following sections of the regulated river (Fig. 1):

(1) The lower section of the river which has slow-flowing water and extends from the coast up to the city of Umeå. The movements of radio-tagged salmon in the uppermost part of this section were covered with an automatic receiver, located 17–21 km upstream from the coast.

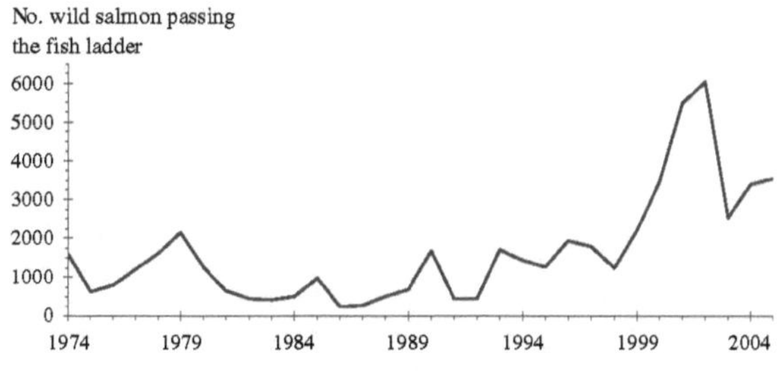

Fig. 2 Annual numbers of wild salmon, from 1974 to 2005, released past the fish ladder in River Umeälven to continue spawning migration to River Vindelälven (n = 52,671)

(2) The middle section includes the confluence area. This section has relatively homogeneous fast-flowing water that extends from the first receiver in the lower section up to the confluence of the turbine discharge and the natural riverbed that is used as a bypass channel, 22–23 km upriver. The turbine outlet features a 250-m long, 20–40-m deep channel at the end of the submerged turbine tunnel.

(3) The bypass channel is 8 km in length with a total fall of 70 m, and includes rapids and the fish ladder. The first rapid, Baggböle (height of 7.0 m), is located 1 km upstream of the entrance to the bypass. The next impediments, in upstream order, are N. Kungsmofallet (height of 2.5 m), then Ö. Kungsmofallet (height of 5.4 m) and finally Laxhoppet (height of c. 4.2 m), located 29–31 km from the coast. The fish ladder at the base of the dam (the location of a second receiver) is 240 m long, constructed of 65 ascending pools with associated weirs and orifices and has a total climb of 18 m.

The Stornorrfors power station (four Francis-turbines) has a maximum capacity of c. 1000 $m^3\ s^{-1}$, and legislation requires minimum spills to the bypass of 10 $m^3\ s^{-1}$ from 20 May to 15 June and 15–50 $m^3\ s^{-1}$ from 15 June to 1 October. Large spills can occur during periods of extreme discharge from melting snow during the spring- and mountain floods, and no water is released into the bypass from 1 October to 20 May. The fish ladder flow is maintained at c. 1 $m^3\ s^{-1}$, and spill flows of up to 19 $m^3\ s^{-1}$ act as an auxiliary source of attraction water (to the ladder). Bypass channel flows during the salmon migration period from 20 May to 1 October (measured by the power station company Vattenfall AB) during the years of this study varied from relatively low volumes with an average of 23 $m^3\ s^{-1}$ in 2003 (max flow: 85 $m^3\ s^{-1}$) to a maximum of 2022 $m^3\ s^{-1}$ in 1995 (average: 182 $m^3\ s^{-1}$). Average turbine flows during the study were 569 $m^3\ s^{-1}$, lowest in 1996 (297 $m^3\ s^{-1}$) and highest in 2001 (806 $m^3\ s^{-1}$). Bypass flows were experimentally altered from normal levels during the salmon upstream migration periods in 2001–2005. In 2001 artificial freshets of 70–120 $m^3\ s^{-1}$ were released from the dam for about 30 h three times and in 2002 80 $m^3\ s^{-1}$ were released for 78 h on two occasions, while 50 $m^3\ s^{-1}$ were spilled for 75–102 h three times in 2003. In 2004 a major flood resulted in spills of up to 1600 $m^3 s^{-1}$ in the middle of the salmon migration period (8–21 July), while spills thereafter were altered in July–August from the normal levels by releasing 50 $m^3 s^{-1}$ for 8 h at nights and 20 $m^3 s^{-1}$ for the remainder of the day. In 2005 the spill flow amounts were modified according to a model by Leonardsson et al. (2005), which aimed to optimise the upstream migration of salmon past the confluence area and the first rapid in the bypass channel while minimising the losses of power production due to the increased spills. This resulted in 90 $m^3 s^{-1}$ released during one 60-h period and 80 $m^3 s^{-1}$ was released for 156 h during the subsequent event.

The ambient river temperature during the study years ranged from 8 to 10°C in early June with a peak of 20–23°C in July–August. Thereafter it dropped slowly to about 4–6°C by October (data from Vattenfall AB). In 2001, hourly variation in water temperature was measured in the lower part of the bypass channel and at the turbine outlet (Onset-TidBit temperature loggers). The data showed only minor differences between the two locations, and the mean daily water temperature at the outlet was on average c. 0.2–0.3°C lower than in the bypass. Nevertheless daily temperature differences were greater in the bypass channel (2–3°C) than in the turbine outlet (0.5–1°C).

Tagging and tracking of salmon

During the nine-year period between 1995 and 2005, various tags were used to study salmon migrating upstream after their capture in a hoop-net at the mouth of River Umeälven, 63°41′36″ N 20°19′45″ E (Fig. 1). The earliest tagging occurred on 3 June (1996) and the latest on 29 August (1996). In total, 2650 salmon were tagged (Table 1) with either external radio tags, gastric radio tags, passive integrated transponder (PIT) tags or external Carlin-tags. The handling, tagging and genetic sampling of fish followed Rivinoja (2005) who, together with references therein, reported that these tags are unlikely to affect the swimming performance of adult salmon. The annual number of tagged fish varied from a minimum of 30 radio tags in 1995 to a maximum of 573 Carlin tags in 1996. The total lengths (L_T) of tagged salmon ranged from 39 to 116 cm. Annual mean sizes varied between years ($P < 0.05$, d.f. = 8, 2648, ANOVA)

Table 1 Yearly data on adult Atlantic salmon tagged and released at the mouth of the River Umeälven

	1995	1996	1997	1999	2001	2002	2003	2004	2005	Total
Date of tagging	30/6–17/8	3/6–29/8	24/6–3/7	16/6–13/7	25/6–4/7	6/6–26/7	23/6–8/8	16/6–30/7	13/6–25/7	
Size (cm)	63 (49–92)	86 (48–112)	89 (69–109)	84 (63–105)	86 (71–105)	66 (39–116)	77 (46–106)	80 (47–112)	87 (55–116)	
Wild:Hatchery	30:0	484:89	55:25	60:0	70:0	493:0	391:0	503:0	450:0	
Female:Male	20:10	387:186	74:6	34:26	60:10	126:367	226:165	263:240	235:215	
External radio tag	30		80	60	70	14	6			260
Gastric radio tag						69	58	60	56	243
PIT tag						410	327	443	394	1574
Carlin tag		573								573
Total	30	573	80	60	70	493	391	503	450	2650

from a minimum of 63 cm (S.D. ± 12.5) in 1995 to a maximum of 89 cm (S.D. ± 9.9) in 1997. Larger salmon (mostly females) arrived earlier in the lower river and at the fish ladder than smaller salmon, as was reported by McKinnell et al. (1994). Beginning in 1999, radio-tagged salmon were tagged with a PIT tag and a small cut or puncture on the adipose fin (Rivinoja et al., 2006), so that tagged fish could be identified at the ladder even if a salmon had lost its radio tag. Genetic analyses of the radio-tagged fish that passed the first archival receiver indicated that all fish belonged to the River Vindelälven population which has a uniquely high frequency of a particular composite haplotype (Vasemägi et al., 2005).

Radio-tagged fish were frequently located in the regulated part of the river using manual receivers (ATS R2100, Televilt RX8910) from a boat or from the shore. The exact positions of radio tags were needed to determine both the positions of the salmon and to ascertain whether a tag had become detached from its host. Automatic archival receivers (LOTEK SRX_400 with 4 or 9-element Yagi-antennas) were used at the confluence area and further up- and downriver (Fig. 1). From 1995 to 2003 the first archival receiver was located in Umeå c. 6 km downstream of the confluence area, while in 2004 and 2005, it was moved c. 4 km upriver of its previous position to study a narrower area (Fig. 1). The fish ladder and adjacent rapids were covered with receivers in 2001, 2002, 2004 and 2005. Since the average loss of external radio tags over the years 1995–2001 was 22%, gastric-implanted tags were used after 2001 to reduce tag losses. The average loss of radio tags was lowered to about 7%.

Echo-sounding in the turbine outlet area

Echo-sounding as described by Lilja (2004) was performed with a hydro-acoustic split-beam echo-sounder (Simrad EY60, GPT 200 kHz) at the turbine outlet area (Fig. 1). The positional data and directional movements of adult salmon were recorded from a boat in 2004 and 2005. The equipment was connected to a computer running ER60 software that recorded data together with a GPS (Geographical Positioning System, U2 SIRF Star II, WAAS-EGNOS). The equipment was calibrated at the top of the fish ladder using salmon of known size. At the turbine outlet area, echo-sounding was carried out on various dates during the migration using various transducer angles and depths, but mainly at a compass heading of 200° from a position located at 63°50'8.6″ N 20°7'32.6″ E. At each recording, the transducer depth and angle was noted, and the data were analysed with the post-processing software Sonar 5 (Balk & Lindem, 2004).

Modelling effects of power station losses on population dynamics

A model with basic data (Table 2) from ICES (2001), together with values from Rivinoja et al. (2005) and own unpublished data (Department of Aquaculture, SLU), was used to predict how improved upstream migration success of adult spawners past the regulated river section might affect future escapements of the spawning stock to River Vindelälven. The following assumptions were made:

- Eggs hatch and juveniles remain in the river for 2–3 years with survival p_X, where X is age.
- The probability of smolting at age 2+ is given by p_S while the rest smoltify as 3+ (dependent on local adaptations and river-specific growth conditions).
- Age 2+ and 3+ smolts migrate seawards and are exposed to reduced survival at the power station (p_T) and due to natural mortality (p_3) during the migration to the Baltic, where the single-survival stages are multiplied to estimate the real product outcome.
- The probability of survival at sea decreases substantially with increasing age due to the Baltic fishery, which was assumed to remain unchanged during the time frame modelled.
- The probability of returning to the river (p_{AX}) for spawning, given that the individual is alive, increases with age.
- The probability of reaching and ascending the fish ladder is given by p_U.
- The unsuccessful fraction $(1 - p_U)$ returns to the sea without spawning, but may return the following year.
- All kelts die at the power station during their seaward migration.

With the above model formulation, the outcome of improving smolt survival during their seaward

Table 2 Description of parameters in the Leslie matrix and their numerical values

Parameter	Numerical value	Description
Probability of survival		
p_0	0.025–0.125	During the first year after hatching
p_1	0.40–0.60	In the river from ages 1 to 2
p_2	0.40–0.60	In the river from ages 2 to 3
p_3	0.06–0.80	During the smoltification year, includes seaward migration and the first season in the sea. Assumed same for the 2- and 3-year-old smolts
p_4	0.60–0.80	In the sea between ages 4 and 5
p_5	0.05–0.15	In the sea between ages 5 and 6
p_6	0.025–0.075	In the sea between ages 6 and 7
p_7	0.025–0.075	In the sea each year after age 7
Proportion of		
$p_{A(4)}$	0.005–0.015	4-year females returning to river
$p_{A(5)}$	0.10–0.30	5-year females returning to river
$p_{A(6)}$	0.80–1.00	6-year females returning to river. Older females are assumed to always aim for a return migration to the river
p_R	**0** (0.01–0.05 vs. 0.05–0.15)	Spawners returning to the sea could become possible with a downstream bypass. This parameter also adjusts for reduced maturation size at following spawning occasion
p_S	0.40–0.60	Smoltified at age 2[a]
p_T	**0.75** (1.0, 1.09)	Smolts surviving passage of turbines or a downstream bypass[c]
p_U	**0.3** (0.5, 0.75)	Returning salmon that passes the fish ladder[d]
No. of eggs per female of age and weight[b]		
F_4	1260 ± 10%	4, W = 2.1 kg
F_5	2640 ± 10%	5, W = 4.4 kg
F_6	5220 ± 10%	6, W = 8.7 kg
F_7	9600 ± 10%	7, W = 16 kg

When a range of parameter values is presented, the parameter values were assigned to the numerical values following a uniform random distribution. The numerical values for parameter p_3 were solved to have a stable initial population size, given all the other parameter values. The bold values denote the observed survival or proportion with existing migration possibilities. (Notes from: [a]ICES, 2001; [b]Lundqvist et al., 1994; [c]Rivinoja, 2005, [d]this report.) The hypothetical values in note 3 and 4 adjusts for the survival/return rate improvement

migration as well as improving conditions for the upstream migration of adults could be predicted. The mathematical formulation of the model, where N_t is the size (age)-structured population vector containing the number of females in each size-class at time t and L is the Leslie matrix containing the survival and fecundity data is:

$$N_{t+1} = \mathrm{L} \cdot N_t$$

The matrix formulation becomes:

$$\mathrm{L} = \begin{bmatrix} 0 & 0 & 0 & 0 & p_{A(4)}p_U F_4 & p_{A(5)}p_U F_5 & p_{A(6)}p_U F_6 & p_U F_7 \\ p_0 & 0 & 0 & 0 & 0 & 0 & 0 & 0 \\ 0 & p_1 & 0 & 0 & 0 & 0 & 0 & 0 \\ 0 & 0 & (1-p_s)p_2 & 0 & 0 & 0 & 0 & 0 \\ 0 & 0 & p_S p_T p_3 & p_T p_3 & 0 & 0 & 0 & 0 \\ 0 & 0 & 0 & 0 & (1-p_U p_{A(4)})p_4 & 0 & 0 & 0 \\ 0 & 0 & 0 & 0 & 0 & (1-p_U p_{A(5)})p_5 & 0 & 0 \\ 0 & 0 & 0 & 0 & 0 & 0 & (1-p_U p_{A(6)})p_6 & (1-p_U)p_7 \end{bmatrix}$$

The robustness of the results was checked using a range of parameter values for all parameters except p_3 and p_U. The other parameter values were varied 10,000 times, within the ranges given in Table 2, by assuming that all parameter values within the ranges were equiprobable. For each run, the real-valued eigenvectors were scaled to become the initial population vectors yielding a non-growing population consisting of 1000 successfully reproducing females, approximately the number of females that have passed the fish ladder per year in recent years. By increasing the return probability (p_U) and predicted number of fish passing the ladder, the effects on the salmon population size of a hypothetical improvement in upstream migration success were evaluated. The model assumes density-independent growth for the population, justified by recognition of that the present population is far below its potential carrying capacity (ICES, 2001; 2005). Nevertheless, predictions beyond 15–20 years should be considered with caution since density-dependent effects were not included in the model. Mathematica ver. 5.2 (Wolfram Research, Inc. 2005) was used for the calculations.

Results

Salmon entry from the coast to the river

An average of 83% (range: 73–93% between years, Table 3) of all 478 radio-tagged wild fish migrated relatively quickly to the first receiver located 17–21 km upriver (Fig. 1). Individual migration durations in this section ranged from 0.5 to 80 days, and upstream migrations occurred both in day time and at the night (which is not dark at this latitude in summer). All fish that passed the first receiver reached the confluence area where a majority stayed for several days. Although 19 tagged fish were recaptured by fishermen at the coast of the Bothnian Bay, the fate of most of the 17% of radio-tagged individuals that were never registered in the river remains unknown. Four radio tags were lost near the tagging site.

Migration in the confluence area

In the confluence area between the power station outlet and the bypass channel, salmon generally

Table 3 Yearly data for individually registered radio-, Carlin- and PIT-tagged wild adult Atlantic salmon in the River Umeälven

Year	1995	1996	1997	1999	2001	2002	2003	2004	2005	Total
Tagged (*n*)	30	484	55	60	70	493	391	503	450	2650
Passed 1st archival receiver	73%	83%[a]	84%	83%	79%	78%	83%	93%	80%	83%
Days to 1st archival receiver	4 (1–28)	–	3 (0.7–20)	3 (0.8–15)	4 (0.6–12)	3 (0.5–47)	4 (0.7–80)	4 (0.7–29)	4 (0.6–26)	4 (0.6–80)
Passed ladder[b]	0%	18%	26%	34%	18%	47%	35%	14%	47%	30%[c]
Days to ladder	–	52 (10–95)	52 (27–77)	44 (22–81)	45 (31–61)	44 (11–91)	46 (9–88)	44 (14–91)	39 (14–101)	44 (9–101)

[a] Estimated average for radio-tagged fish for all years

[b] Amount based on fish that entered the river (passed the first archival receiver, numbered 1 in Fig. 1).

[c] The calculated weighed average takes into account the number of fish tagged each year

followed the large flows from the turbine outlet. During periods with high turbine discharge and low bypass flow, fish were attracted from the bypass channel, delaying their upstream migration. Most salmon that reached this area spent a relatively long time (min–max: 1–82 days, mean: 12 days, median: 9 days) in the high flows below the turbine outlet before ascending into the bypass channel. Salmon showed mainly three migratory responses in relation to the flow regimes:

(1) located in the turbine outlet, moving up- and downstream in the main stem depending on turbine flows in search for an upstream route. Some of these fish entered the bypass channel, yet stayed only for a limited time which caused unsuccessful advancement for further upstream migration,
(2) entered the bypass channel and held positions for a relatively long time below the first rapid at Baggböle without passing upriver. Increasing spill flows into the bypass channel generally attracted the salmon upstream, yet fish in the bypass responded to both increasing and decreasing spill flows by exhibiting downstream movements,
(3) entered the bypass channel and moved continuously upstream. Successful upstream passages of these rapids generally occurred at spill flows $<200\ m^3\ s^{-1}$.

These migratory responses were consistent for all years when telemetry was used to understand the positions of radio-tagged salmon. A four-year evaluation (adequate data obtained in 1997, 1999, 2001 and 2002) of up- and downstream movements of radio-tagged salmon in the main stem of the river, demonstrated that many salmon that were recorded in the tunnel outlet area had directed downstream movements and were registered on the receiver located in Umeå c. 6 km downstream the turbine outlet. On average, about 40% of all radio-tagged salmon that reached the confluence area in the years of this study (106 of a total of 268), 26%, 27%, 53% and 48% in 1997, 1999, 2001 and 2002, respectively, returned downstream at least three times and were registered on the receiver in large numbers over the whole 24-h period. For example, in 2001, the 55 salmon that began their upstream migration passed over the downstream receiver 174 times with a maximum of 11 detections recorded for each of two. In general, most downstream movements were observed c. 7 h after the turbine discharge had decreased, while some of the registrations took place when the discharge increased. In 1997 wild (n = 11) and hatchery salmon (n = 13) showed similar up- and downstream movements at the receiver. Combined data indicated that the number of registrations at the downstream receiver decreased in late summer (beginning in August), yet no relationship between the number of registrations in this area versus salmon sex or size, nor the date of tagging was found (Fig. 3). Nevertheless, a comparable behavioural pattern, where fish responds to rapid changes in flow regimes, was observed in all the years and we coin the expression "yo-yo migration" for this type of salmon behaviour.

Echo-sounding in the turbine outlet area

At the confluence area, both tagged and untagged salmon were observed to enter the turbine outlet repeatedly, indicating that the salmon were searching for an upstream route. Echo-sounding in this area revealed that the salmon were predominantly found near the surface (1–4 m depths). An event on 5th August 2004 demonstrated these yo-yo migrations, with about 70% of all movements directed downstream (dotted grey line in Fig. 4, left). Salmon at this area were also observed to dive to the bottom at depths of up to 40 m and also swim back and forth over the whole channel width (Fig. 4, right).

Salmon migrations from the confluence area to entry of the bypass channel

Data collected from 1997 to 2003 showed that salmon generally spend a long time in the confluence area before ascending the bypass channel; however, most fish in the confluence area responded to increased spill flows and moved quickly into the bypass. Fish reached the first rapid at Baggböle, immediately upstream of the entrance to the bypass, after an average over all years of c. 13 days (median = 10 days) after tagging. On average, they passed this section of the river after c. 25 days (median = 14 days). Detailed modelling of the relationship between bypass flow and the proportion of upstream migrating salmon passing the rapid at Baggböle is presented by Leonardsson et al. (2005). Data from

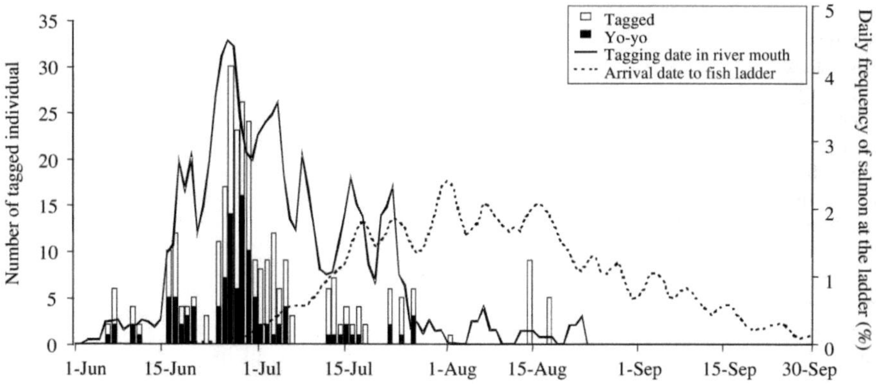

Fig. 3 Number of wild radio-tagged salmon showing "yo-yo movements" between the confluence area and the archival receiver in Umeå, 6 km downstream, in relation to tagging day. A total of 106 out of 268 radio-tagged salmon passed the archival receiver at least three times, resulting in 1 992 registrations in the four years (97, 99, 01 and 02). The total number of wild salmon tagged per day is illustrated by the black line, while the dashed line shows the relative frequency of salmon arrival date to the ladder

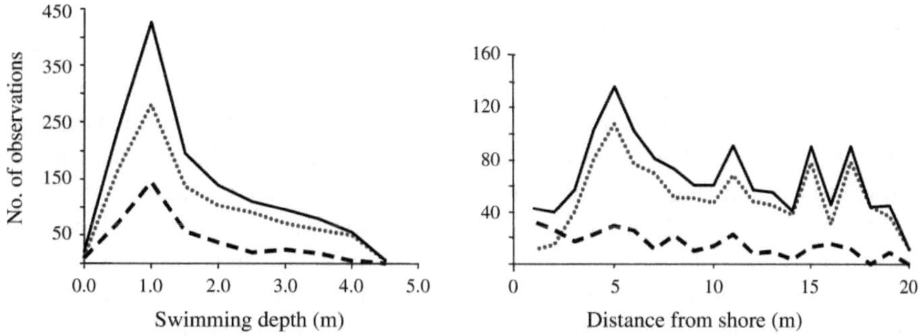

Fig. 4 Salmon in the turbine outlet channel were mainly <4 m deep (left figure) and showed up- and downstream movements over the whole channel width (right figure). The solid black line show total observed movements, the dotted grey line downstream movements and the dashed black line upstream movements (recorded on 5th August 2004)

2001 and 2003 illustrate the typical migration responses of radio-tagged salmon at the confluence area in relation to flows (Figs. 5, 6). In 2001, flows in the bypass channel were increased in the weekend to 50 $m^3 s^{-1}$, a normal discharge pattern at this dam. At the first period (30 June–2 July) only 2 of the 21 salmon located in the confluence area successfully ascended to the bypass channel, but when the bypass flow was increased to 200 $m^3 s^{-1}$ (7–8 July) the fraction increased to 13 of 28 (Fig. 5). For 10 days (20–30 July) when excess flows were spilled (on average 160 $m^3 s^{-1}$), numerous upstream ($n = 36$) and downstream ($n = 16$) movements of radio-tagged salmon were observed in the lower part of the bypass channel and also passages ($n = 11$) of the waterfall were observed. Similarly, data from 2003 (Fig. 6) show that upstream migration to the bypass increased with the amount of spill, and in addition, that turbine flows below 200 $m^3 s^{-1}$ facilitated salmon bypass ascent. Most salmon entered the bypass during periods of reduced ambient light at the night time hours (Fig. 6).

Overall responses to flows (Fig. 7) confirmed that increased spill flows and lower turbine flows generally attracted salmon to the bypass. Yet, at the same time, the waterfall Baggböle near the bypass entrance could also hinder the upstream migration as salmon seemed to hold and even move downstream from the bypass if the spill flows exceeded 150–200 $m^3 s^{-1}$ (see also Leonardsson et al., 2005). By correlating bypass flows with fish responses in the area just below this waterfall, different discharges were evaluated to see how flows could initiate or hinder fish migrations. Over the salmon migration period in 1997 the mean spill flow when salmon successfully entered and passed the bypass channel and the rapid was

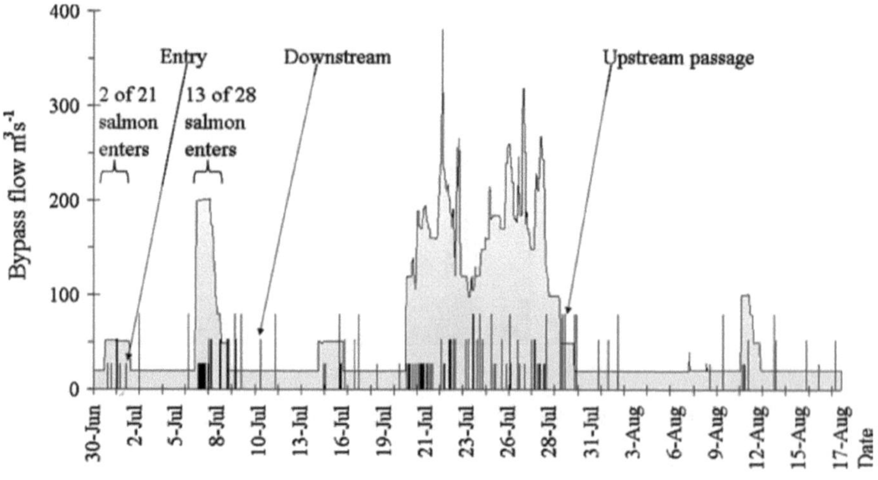

Fig. 5 Registrations ($n = 121$) of directional salmon movements (black bars) and bypass flow (grey area) from 30th June to 17th August 2001. The short bars indicate salmon entry from the confluence area to the bypass, the intermediate bars downstream migration from the bypass and the long bars salmon upstream passage of the first rapid in the bypass. The majority of registrations took place during the high spill flow period in late July

150 $m^3 s^{-1}$ (Rivinoja et al. 2001), while flows higher than these might cause unsuccessful passage regardless of the turbine discharge. From an event in 2002 (25 June–12 July), the migration at flows of 50 and 80 $m^3 s^{-1}$ showed how salmon responded to the increased flows within a few hours by moving upstream into the bypass and then stopped their upstream migration at the rapid (Fig. 8). A decreased discharge from 50 or 80 $m^3 s^{-1}$ to 20 $m^3 s^{-1}$ caused 50% of the salmon to pass the rapid and move upriver (7 out of 14 fish), while the remaining fish returned downstream to the confluence area. Similar fish migration patterns were manifested over the years for other periods with increased spills.

Upstream migration from the bypass channel to the fish ladder

After passing the rapid at Baggböle, the salmon migrated relatively quickly, of the order of 1–2 days, c. 6–7 km, upstream to the rapids immediately below the entrance to the fish ladder. At these rapids (N. Kungsmofallet, Ö. Kungsmofallet and Laxhoppet) the salmon had additional problems to pass, showing a slight delay and unsuccessful upstream passages; however, the migration behaviour of fish at these rapids was not so closely monitored. Nevertheless, the receiver at the fish ladder area indicated that both up- and downstream movements occurred at the uppermost

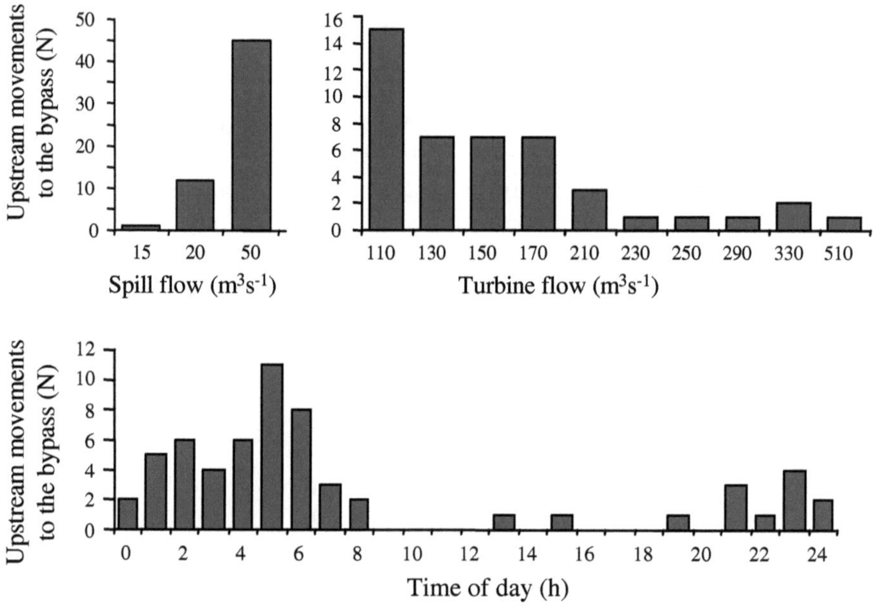

Fig. 6 Radio-tagged salmon at the confluence area ($n = 34$) entered the bypass channel more frequently at turbine flows below 200 $m^3 s^{-1}$, at higher dam spills and during the night hours. (Data from year 2003)

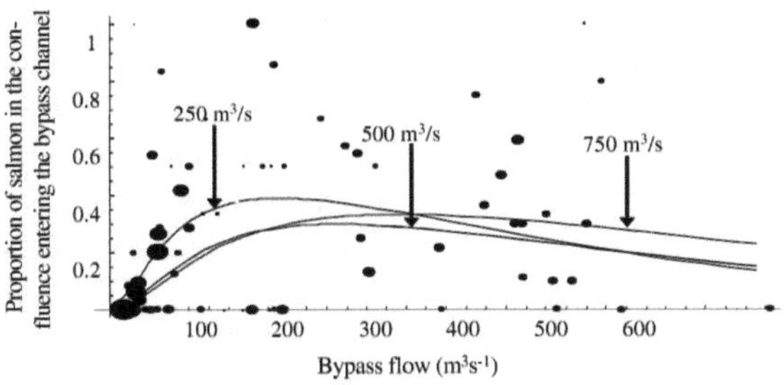

Fig. 7 Salmon in the confluence area mainly entered and passed further upstream in the bypass channel at spill flows below 150–200 $m^3\ s^{-1}$. Non-linear response curves for turbine flows of 250–750 $m^3\ s^{-1}$ shows the effects on the ascent of fish to the bypass. The size of each plot point is scaled from 1 to 597 to indicate the number of salmon registrations (from Leonardsson et al., 2005)

rapid below the fish ladder and that some salmon repeatedly entered the ladder without ascending. The receiver at the fish ladder showed that fish (n = 58 in 2001, 2002, 2004 and 2005) entered the ladder at spill flows that varied between 15 and 105 $m^3\ s^{-1}$, with an average of 32 $m^3\ s^{-1}$. Upstream movement of radio-tagged salmon through the ladder was highest when attraction water of 19 $m^3\ s^{-1}$was directed towards the ladder entrance; the proportion of ascending fish decreased at spills in excess of this. There was a tendency for salmon to enter the ladder more frequently in afternoons and early evenings than at night hours. Ladder flows were normally held constant at c. 1 $m^3\ s^{-1}$, yet the swim through time among radio-tagged individuals in the ladder showed large disparity, varying from 3 to 133 h with an average of 35 h. Fish size (P = 0.62, t-ratio = 0.502) or day at tagging (P = 0.94, t-ratio = 0.074) was not related to the travel time through the fish ladder (n = 58, d.f. = 2, Cox regression). However, the average duration in the ladder for male salmon (mean = 25.8 h, S.D. ± 16.8) was significantly faster ($P < 0.05$, $\chi^2 = 5.438$, Cox regression) than for females (mean = 41.0 h, S.D. ± 28.5). During these periods, the river temperature ranged from 9.2 to 20.8°C (average 17.3°C), yet no influence of river temperature on salmon travel time through the ladder was detected.

Salmon reaching the top of the fish ladder had spent an average of 44 days from the river mouth to the top of the ladder (Table 3). Individual migration time to the ladder differed greatly within years, but not between years, according to the data obtained from all tagged fish (n = 565) in 1997–2005 (P = 0.195, d.f. = 6, durations $\log_{10}$-transformed, Tukey's Post Hoc). Similarly, travel time to the ladder was independent of the sex of the fish (P = 0.377, d.f. = 1, $\chi^2 = 0.781$, Cox regression) or size (P = 0.628, t-ratio = −0.485), yet travel time was related to the day of tagging; fish tagged early had longer travel time before passing the fish ladder ($P < 0.01$, t-ratio = 4.337).

Overall migration success, cumulative losses and population modelling

The overall results of taggings from 1995 to 2005 (n = 2650) revealed that most of the salmon that

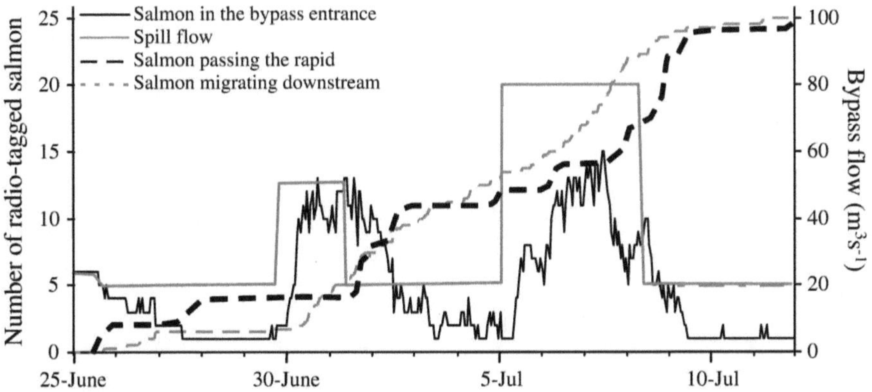

Fig. 8 Radio-tagged salmon from the confluence area rapidly entered the bypass (black line, left axis) when spill flow was increased from 20 $m^3\ s^{-1}$ to 50 or 80 $m^3\ s^{-1}$ for two and four days, respectively (grey line, right axis). About half of the fish passed upstream (cumulative number of salmon passing the rapid, black dashed line) both at increased and decreased flows, while the other half migrated down to the confluence area (cumulative number of salmon moving downstream to confluence area, grey dashed line)

initiated their migration to the spawning grounds of the Vindelälven were unable to migrate from the river mouth to the top of the fish ladder. On average only 30% (a weighted average that accounts for the numbers of tagged fish per year) of all tagged fish were able to pass the regulated river stretch (Fig. 9). The proportion of wild salmon passing the fish ladder varied from 14% to 47% from 1996 to 2005, and none of the radio-tagged fish succeeded in 1995. The correlation between the percentage of tagged fish that passed the ladder and the mean annual spill was statistically significant ($P < 0.05$, $r = -0.73$, $r^2 = 0.53$, d.f. = 1, 7, Linear regression), but this result was heavily influenced by the results of 1995 when the spring flood was high (up to 2000 $m^3 s^{-1}$) in most of June. If this outlier was omitted from the analysis, the correlation between average spill and migration success was not statistically significant ($P = 0.247$, $r = -0.46$, $r^2 = 0.21$, d.f. = 1, 6, Linear regression), although we note that the second highest mean annual spill (2004) concurred with the second lowest migration success at the ladder. The highest overall success rates were observed in 2002 and 2005 (Table 3), years with relatively low and stable spills, perhaps increased by the artificial freshets provided by the regulations to ensure minimum spill volumes. In these years flows of 80 and 90 $m^3 s^{-1}$ were spilled, while maximum spills rarely exceeded these amounts. On the whole, over the years, radio-tagged salmon showed prolonged migration time at the confluence area, that explained the long average travel time of 44 days from the river mouth to the fish ladder. Salmon were partially hindered at rapids in the bypass channel that resulted in delays and reduced upstream passage success according to the flow regime. Of all observed losses of upstream migrants over the years (Fig. 9), a loss of c. 50% was observed at the confluence area and the first rapid (Baggböle) in the bypass channel. Additional losses of c. 20% occurred at the rapids in the upper part of the bypass channel (N. Kungsmofallet, Ö. Kungsmofallet and Laxhoppet), and the remaining c. 30% was related to problems for the salmon to find and pass the fish ladder. After the radio-tagged fish were released upstream of the fish ladder they reached their main spawning areas, 210–250 km upstream in Vindelälven in about 10–15 days after passing several major rapids and climbed an altitude of about 200 m (Lundqvist et al., 2006).

In 1997 data obtained from radio-tagged hatchery salmon indicated that a majority of these fish were unable to find and pass the bypass channel, and none of the hatchery salmon passed the ladder, perhaps because their release locations were immediately below the dam. Complementary data from 1996 demonstrated that a lower proportion of hatchery salmon (8%) than wild salmon (18%) passed that ladder. Additional data, analysed for the years 2002 and 2003, demonstrated that radio-tagged wild salmon had the same migration success from the tagging site to the fish ladder as the control group of PIT-tagged salmon (Rivinoja et al. 2006).

The population dynamics model suggests that if the passage problems in the regulated river sections

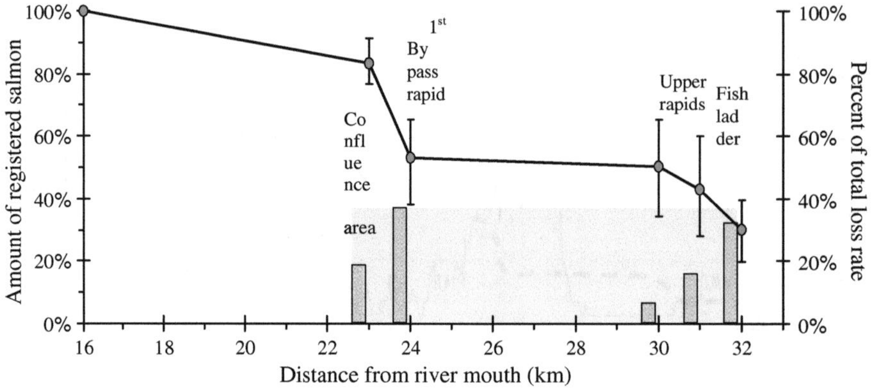

Fig. 9 The cumulative passage success of wild salmon past various areas upstream of the tagging-site (left axis) where the bars indicate the 75% percentiles. A total of 478 radio-tagged salmon that entered River Umeälven from 1995 to 2005 is included in the data. The proportion of the total losses of salmon (unsuccessful passages) at problematic areas is denoted by the grey boxes (right axis)

were alleviated, an increase of 500% in the escapement returns should be expected after 10 years if the number of fish reaching the spawning areas could be improved from the current 30% to 75% (Fig. 10), assuming that spawning and rearing habitats are not limiting the natural production of juveniles in the Vindelälven. The corresponding increase for a migration success of 50% gives a yearly population abundance increase of 9%; the total increase after 10 years would be 160% over current levels.

Discussion

The probability of wild salmon successfully migrating through the regulated part of the River Umeälven from the estuary to the fish ladder is low, with average losses of c. 70% of potential salmon spawners. Gowans et al. (2003) reported large cumulative negative effects on fish migration in rivers with numerous obstacles and that this affected the escapement returns to the spawning areas. The large variation in migration success among years (min–max: 0–47%) is not easily explained. It may have been the result of extra spill water with stable discharges in the bypass, as in 2002 and 2005. Years of generally unsuccessful upstream migration had large differences in bypass flows and high spills, which can explain the low success rates in 1995 and 2004.

In River Umeälven, migration problems occurred in different parts of the flow-controlled areas, but the greatest losses (50%) of salmon took place where the turbine discharge water joins the bypass flow. Here the complex flow patterns and large turbine flows directed salmon away from the upstream routes. Areas with these characteristics can be major

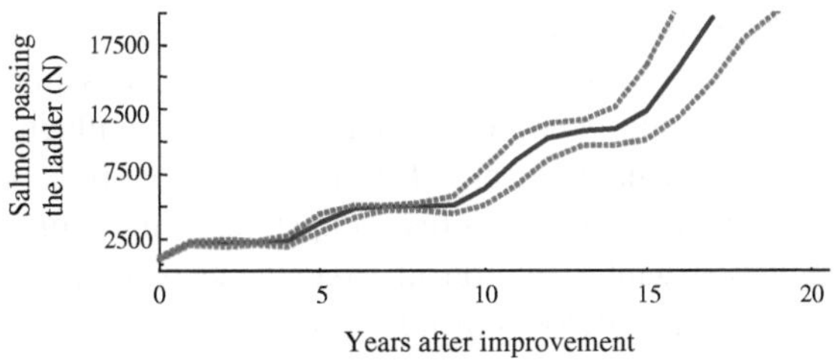

Fig. 10 Predicted number of adult salmon females passing the fish ladder during a 20-year period after improved upstream migration from 30% to 75% at the regulated part of the River Umeälven. An estimated yearly population increase of 18% is expected, and after 10 years the population has increased about 500%. Stroked line shows the 95% CI

hindrances for upstream migrating fish (Arnekleiv & Kraabøl, 1996; Karppinen et al., 2002; Thorstad et al., 2003). Ferguson et al. (2002) explained the discharge-seeking behaviour as an evolved mechanism that maximises spawning success since fish attracted to the highest discharge normally follow the main branches of rivers on their way to the spawning grounds. As shown here and also by Arnekleiv & Kraabøl (1996) in studies on brown trout, successful upstream migration of fish to bypasses was positively related to spillway flow, and fish could stop their migration if they were guided towards turbines. In addition, Ferguson et al. (2002) pointed out that during situations of low spill, fish might lose the attraction cues from bypasses, which may prevent or impede adult fish from migrating upstream.

Another important finding in this study was that salmon that reached the turbine outlet and confluence area moved several kilometres downstream, mainly when turbine discharge was lowered. Arnekleiv & Kraabøl (1996) also observed this restless behaviour of fish and noted that up- and downstream movements of several kilometres occurred. These "yo-yo migrations" delayed migration and caused increased swimming behaviour with associated energetic costs for the fish. These costs cannot be recovered because maturing anadromous salmon do not feed while in freshwater. Lower fat reserves will potentially lower the fitness of individuals during competition for mates and may lead to lower overwinter survival, which would amplify the negative effects on the population.

Salmon positioned at various sites in the confluence area responded strongly to increased spill in the bypass channel in combination with lower flows from the turbine outlet. At these events fish generally migrated quickly into the bypass channel, but occasionally without passing the first rapid c.1 km upstream of the confluence area. During periods with high spill flows (e.g. spring flood) a subsequent reduction in spill volume could facilitate passage of the rapid. Likewise, a reduction in spill flow may also cause downstream migration from the rapid. It is commonly known that salmon tend to aggregate in areas with partial barriers and the rapids in the bypass could act as such barriers. In addition, these delays prolong the upstream migration time which can also cause failure to pass further upstream (Power & McCleave, 1980; Webb, 1990; Rivinoja et al., 2001). Migrating salmon in natural flows have also been shown to exhibit upstream migration at both decreasing and increasing flows (Trépanier et al., 1996). It might be that fish have locally adapted behaviours in relation to river-specific conditions and their physiological status, e.g. salmon might stop and wait at rapids for suitable conditions for passage, and as was shown in this study, high flows can hinder and delay the upstream migration. This behaviour indicates the complex nature of salmon upstream migration, since the location of a fish at any time probably influences its response to increasing or decreasing flows.

Regardless of the problems in the confluence area and the first rapid in the bypass channel, an equivalent loss was observed when one sums the losses at rapids further upriver (20%) and in the fish ladder area (30%). Ferguson et al. (2002) argued that the existing fish ladder is not designed properly to attract fish and secure their passage. They stressed that a successful upstream passage facility should pass more than 95% of the migrating adult fish. In regulated rivers where fish ladders are adjacent to the spillways, water is generally diverted into the lowest parts of the ladders to get better fish attraction at the entrances. Nevertheless, these areas might cause problems for migrating fish since they are influenced by the discharge from the dam combined with the attraction flow leading to fish ladder entrances (Quinn et al., 1997). In the present study, both the proportion of radio-tagged salmon that entered and the proportion that passed the fish ladder decreased when surplus water was spilled outside the ladder. This happened only when the total amount spilled was higher than the normal volume of attraction water supplied to the ladder. Consequently high dam spills could cause difficulties for the fish to locate the fishway entrance and delay the migrants, as has also been shown previously (Bjornn & Peery, 1992; Quinn et al., 1997). Other studies have described the searching behaviour of fish near fishway entrances (Williams, 1998; Gowans et al., 1999; Karppinen et al., 2002). Laine (1995) mentioned that fish may need to become familiar with the lower parts of the ladder before continuing upstream. Laine (1995) found an average delay of 14 days from the first approach of Atlantic salmon to a ladder until they finally entered, whilst Webb (1990) found delays of 0.6–43 days. Even if it is not possible to determine whether the large variation in passage time among individuals at the ladder in the River Umeälven is normal or not, the relatively long time for salmon to travel through the ladder (up to 133 h), independent of fish size, points out that this ladder is not optimally designed. On the other hand, data presented by Bjornn & Peery (1992) indicated that duration of passage through fish ladders can vary widely among Pacific adult salmonids (*Oncorhynchus* sp.). Although river temperature was not found to have an effect on the passage through the ladder in River Umeälven, Gowans et al. (1999) stressed that the ratio of salmon ascending a fish ladder can be positively correlated with temperature.

Our observation that it took on average 44 days for the salmon to migrate the relatively short distance of 32 km from the mouth of the river to the fish ladder is consistent over decades (McKinnell et al. 1994). They compared the timing of the migration based on numbers of salmon caught daily in the fishery in the lower part of River Umeälven in the early 1980s with the daily counts of salmon at the fish ladder and reported a travel time for multi sea-winter fish from the coast to the fish ladder of c. 40 days. Byström (1867) mentioned a migration time of about 4–6 weeks for salmon from the river mouth to the rapids where the current dam and fish ladder are situated. In that era, before the hydropower developments, other large man-made obstacles (e.g. fish traps) might also have affected or delayed the upstream migration. The initial slow migration process of salmon in River Umeälven might also be an evolved characteristic for the salmon population. The relative steepness of the river from coast to the ladder area (c. 75 m), combined with the seasonal high forest- and mountain floods, might have caused an adaptive response for the salmon to wait for decreasing flows in the lower part of the river.

Bjornn & Peery (1992) found that temperature and turbidity can delay fish migrations. We do not expect that the small daily temperature differences of c. 0.2°C between the colder water from the turbine outlet versus the bypass spill would cause the salmon to be directed in any particular way at the confluence area. McKinnell et al. (1994) found no effect of ambient river temperatures on upstream migration of multi-sea winter salmon in River Umeälven. Trépanier et al. (1996) showed only limited effects of temperature on salmon upstream migration. Still, Jensen et al. (1986) observed that Atlantic salmon passages upstream of rapids in a Norwegian river were correlated to increasing water temperature.

The low proportion of salmon migrating from the coast to the ladder might have been influenced by their physiological condition. In the last 24 years, outbreaks of the disease M74 have been observed among Baltic salmon stocks (Bengtsson et al., 1999). From 1994 to the beginning of 2000 it occurred in 16–78% of fish sampled, but thereafter, the incidence of M74 has decreased during the last five years from 45% in 2000 to 4% in 2004 (data from Vattenfall AB). No study has revealed a relationship between M74 and migration performance of Atlantic salmon. It is not known if this disease weakens the upstream migration of adults, and even if it did, M74 is not expected to be the one single factor that causes the observed upstream migration patterns of salmon in the regulated part of the river.

Conclusions and management implications for sustaining a salmon stock

In this study, the major passage problems for migrating wild salmon spawners at the regulated section in River Umeälven occurred at the confluence area. These disturbances caused most Vindelälven salmon to abandon their upstream spawning migration. These migration problems were caused by large variability in flow regimes in the confluence area. The upstream migration was enhanced by increased discharge in the bypass, but too much could hinder the upstream passages. If too low, the salmon entered the turbine outlet. Leonardsson et al. (2005) verified that certain combinations of spillway and turbine flows were found to be beneficial to guide and pass salmon in the correct upstream route, depending on season. Previous observations that enhanced directional cues could affect the fish migration positively (Mills, 1989) and findings that upstream migration rate could be increased by spills (Arnekleiv & Kraabøl, 1996) or bypass constructions close to turbine outlets (Calles & Greenberg, 2005) indicate that the upstream migration of salmonids can be managed. Consequently, the migration problems found for adult salmon in the regulated part of River Umeälven could be lowered by: (1) construction of a fishway in the turbine outlet, so salmon could easily find an upstream route, (2) regulation of spill flows to secure successful attraction and passage efficiency of the bypass, and (3) reconstruction of the current fish ladder at the dam to improve passage speed and success. These implementations could be highly favourable for the salmon stock in River Vindelälven since the population models suggested a five-fold increase in spawner abundance within 10 years if the losses at the regulated area could be lowered. Restoration programmes (Nilsson et al., 2005) now undertaken in the River Vindelälven system will increase the amount of spawning habitat, which can enhance future population growth if more spawners were added to the system. Furthermore, efforts taken by the power station owners in River Umeälven to establish a new fish ladder with a downstream guidance device for smolts and kelts suggest a promising scenario for how anadromous fish can be preserved and even enhanced in a regulated river. A variety of designs and techniques to improve migration conditions have been implemented in the USA and Canada for Pacific salmonids (Clay, 1995; Williams, 1998) and more recently in Europe for Atlantic salmon (Larinier, 2002a, b; Larinier et al., 2005).

In conclusion, our demonstration that a majority of the upstream migrating salmon in this river stock had problems to bypass the existing hydropower complex in their search for natural spawning areas upriver is in conflict with sustainable management of the anadromous fish resources. If these problems are not taken into account and solved, we will compromise the future of the salmon population for the generations to come.

Acknowledgments Thanks to Elforsk, the Swedish Energy Agency, the National Board of Fisheries, the Swedish Environmental Protection Agency and Vattenfall AB who financed this study as a part of a Swedish R&D programme. Bo-Sören Wiklund is thanked for his never-failing support and technical skills. Johan Östergren, Nils-Erik and Björn Sjöström, and the staff at the Norrfors hatchery are thanked for their co-operation. Without the kind help of Vattenfall AB, this study could not have been undertaken. The National Board of Fisheries in Luleå financed the Carlin taggings. This study was carried out under licence from the Swedish Animal Welfare Agency (CFN Dnr. A70-03 and A90-03).

References

Arnekleiv, J. O. & M. Kraabøl, 1996. Migratory behaviour of adult fast-growing brown trout (*Salmo trutta*, L.) in relation to water flow in a regulated Norwegian river. Regulated rivers: Research and Management 12: 39–49.

Balk, H. & T. Lindem, 2004. Sonar4 and Sonar 5-Pro, Post Processing System, Operator Manual v5.9.3. Lindem Data acquisition. Norway.

Banks, J. W., 1969. A review of the literature on the upstream migration of adult salmonids. Journal of Fish Biology 1: 85–136.

Bengtsson, B.-E., C. Hill, Å. Bergman, I. Brandt, N. Johansson, C. Magnhagen, A. Södergren & J. Thulin, 1999. Reproductive disturbances in Baltic fish: a synopsis of the FiRe Project. AMBIO. Special Issue 28(1): 2–8.

Bjornn, T. C. & C. A. Peery, 1992. A review of literature related to movements of adult salmon and steelhead past dams and through reservoirs in the lower Snake river. Moscow, Idaho, USA: US Army Corps of Engineers, Walla Walla District and Idaho Cooperative Fish and Wildlife Research Unit, Draft Technical Report 92-1, 107 pp.

Byström, C., 1867. Om fisket i Umeå elf och skärgård, Lantbruksakademiens Tidskrift, December 1867, Berättelse 20: 91–100 (In Swedish).

Calles, E. O. & L. A. Greenberg, 2005. Evaluation of natural-like fishways for re-establishing connectivity in fragmented salmonid populations in the River Emån. River Research and Applications 21: 951–960.

Clay, C. H., 1995. Design of Fishways and Other Fish Facilities, 2nd edn. Lewis Publishers, CRC Press, Inc., Boca Raton, Florida.

Coutant, C. C. & R. R. Whitney, 2000. Fish behavior in relation to passage through hydropower turbines: a review. Transactions of the American Fisheries Society 129: 351–380.

Eriksson, T. & L.-O. Eriksson, 1993. The status of wild and hatchery propagated Swedish salmon stocks after 40 years of hatchery releases in the Baltic rivers. Fisheries Research 18: 147–159.

Ferguson, J. W., J. G. Williams & E. Meyer, 2002. Recommendations for improving fish passage at the Stornorrfors Power Station on the Umeälven, Umeå, Sweden. U.S. Department of Commerce, National Marine Fisheries Service, Northwest Fisheries Science Center, Seattle, Washington, February 2002.

Gowans, A. R. D., J. D. Armstrong & I. G. Priede, 1999. Movements of Atlantic salmon in relation to a hydroelectric dam and fish ladder. Journal of Fish Biology 54: 713–726.

Gowans, A. R. D., J. D. Armstrong, I. G. Priede & S. Mckelvey, 2003. Movements of Atlantic salmon migrating upstream through a fish-pass complex in Scotland. Ecology of Freshwater Fish 12: 177–189.

ICES, 2001. Report of the Baltic salmon and trout assessment working group. Pärnu, Estonia 28 March–6 April 2001. International Council for the Exploration of the Sea, CM 2001/ACFM, 14.

ICES, 2005. Report of the Baltic salmon and trout working group (WGBAST). Helsinki, Finland, 5-14 April 2005. International Council for the Exploration of the Sea, CM 2005/ACFM, 18.

Jensen, A. J., T. G. Heggberget & B. O. Johnsen, 1986. Upstream migration of adult Atlantic salmon, *Salmo salar* L., in the River Vefsna, northern Norway. Journal of Fish Biology 29: 459–465.

Karppinen, P., T. S. Mäkinen, J. Erkinaro, V. V. Kostin, R. V. Sadkovskij, A. I. Lupadin & M. Kaukoranta, 2002. Migratory and route-seeking behaviour of ascending Atlantic salmon in the regulated River Tuloma. Hydrobiologia 483: 23–30.

Katopodis, C., 1990. Advancing the art of engineering fishways for upstream migrants. In Proceedings of the International Symposium on Fishways, 90 in Gifu, Japan, October 8–10, 1990: 19–28.

Laine, A., 1995. Fish swimming behaviour in Finnish fishways. In Komura, S. (ed.), Proceedings of the International Symposium on Fishways '95 in Gifu, Japan, October 24–26, 1995. Gifu: Organising Committee for International Symposium on Fishways '95 in Gifu, 323–328.

Larinier, M., 1998. Upstream and downstream fish passage experience in France. In: Jungwirth, M., S. Schmutz & S. Weiss (eds), Fish Migration and Fish Bypasses. Cambridge University Press, Fishing News Book, 127–145.

Larinier, M., 2002a. Biological factors to be taken into account in the design of fishways, the concept of obstruction to upstream migration. Bulletin Français de la Pêche et de la Pisciculture 364(suppl.): 28–38.

Larinier, M., 2002b. Pool fishways, pre-barrages and natural bypass channels. Bulletin Français de la Pêche et de la Pisciculture 364(suppl.): 54–82.

Larinier, M., M. Chanseau, F. Bau & O. Croze, 2005. The use of radiotelemetry for optimizing fish pass design. In Spedicato, M. T., G. Lembo & G. Marmulla (eds), Aquatic Telemetry: Advances and Applications. FAO/COISPA, Rome: 53–60.

Leonardsson, K., Y. Belyaev, P. Rivinoja & H. Lundqvist, 2005. Modelling upstream migration of Atlantic salmon as a function of environmental variables. In Rivinoja, P. (ed.), Migration Problems of Atlantic Salmon (*Salmo salar* L.) in Flow Regulated Rivers, Chapter III, Doctoral thesis 2005:114. ISSN: 1652-6880. ISBN: 91-576-6913-9. Department of Aquaculture, Swedish University of Agricultural Science, Umeå, Sweden.

Lilja, J., 2004. Assessment of fish migration in rivers by horizontal echo sounding: problems concerning side-aspect target strength, Doctoral thesis. ISBN 951-39-1725-8. University of Jyväskylä, Finland.

Lundqvist, H., S. McKinnell, H. Fängstam & I. Berglund, 1994. The effect of time, size and sex on recapture rates and yield after river release of *Salmo salar* smolts. Aquaculture 121: 245–257.

Lundqvist, H., J. Östergren & S. McKinnell, 2006. Interannual variation in distance of spawning migration of radio tagged wild Atlantic salmon (*Salmo salar*) females. In Östergren, J. (ed.), Migration and genetic structure of anadromous *Salmo salar* & *Salmo trutta* in northern Swedish rivers, Chapter 2, Doctoral thesis 2006:112. ISSN: 1652-6880. ISBN: 91-576-7261-X. Department of Aquaculture, Swedish University of Agricultural Science, Umeå, Sweden.

McKinnell, S., H. Lundqvist & H. Johansson, 1994. Biological characteristics of the upstream migration of naturally and hatchery-reared Baltic salmon, *Salmo salar* L. Aquaculture and Fisheries Management 25: 45–63.

Mills, D. H., 1989. Ecology and Management of Atlantic Salmon. Chapman & Hall, London, 351 pp.

Montén, E., 1985. Fish and Turbines; Fish Injuries During Passage Through Power Station Turbines. Vattenfall, AB, Stockholm, Sweden.

Nilsson, C., F. Lepori, B. Malmqvist, E. Törnlund, N. Hjerdt, J. M. Helfield, D. Palm, J. Östergren, R. Jansson, E. Brännäs & H. Lundqvist, 2005. Forecasting environmental responses to restoration of rivers used as log floatways: an interdisciplinary challenge. Ecosystems 8: 779–800.

Northcote, T. G., 1998. Migratory behaviour of fish and its significance to movement through riverine fish passage facilities. In: Jungwirth, M., S. Schmutz & S. Weiss (eds), Fish Migration and Fish Bypasses. Cambridge University Press, Fishing News Book, 3–18.

NRC, 1996. Upstream: salmon and society in the Pacific Northwest. Report of the committee on protection and management Pacific Northwest anadromous salmonids. National Academy Press, Washington D.C.

Power, J. H. & J. D. McCleave, 1980. Riverine movements of hatchery-reared Atlantic salmon (*Salmo salar*) upon return as adults. Environmental Biology of Fishes 5: 3–13.

Quinn, T. P., S. Hodgson & C. Peven, 1997. Temperature, flow, and the migration of adult sockeye salmon (*Oncorhynchus nerka*) in the Columbia River. Canadian Journal of Fisheries and Aquatic Sciences 54: 1349–1360.

Rivinoja, P., 2005. Migration Problems of Atlantic Salmon (*Salmo salar* L.) in Flow Regulated Rivers. Doctoral thesis 2005:114. ISSN: 1652-6880. ISBN: 91-576-6913-9. Department of Aquaculture, Swedish University of Agricultural Science, Umeå, Sweden.

Rivinoja, P., K. Leonardsson & H. Lundqvist, 2005. Size dependent power-station induced mortality of smolts (*Salmo sp.*) and the potential effects on the spawning stock. In Rivinoja, P. (ed.), Migration Problems of Atlantic Salmon (*Salmo salar* L.) in Flow Regulated Rivers, Chapter VI. Doctoral thesis 2005:114. ISSN: 1652-6880. ISBN: 91-576-6913-9. Department of Aquaculture, Swedish University of Agricultural Science, Umeå, Sweden.

Rivinoja, P., S. McKinnell & H. Lundqvist, 2001. Hindrances to upstream migration of Atlantic salmon (*Salmo salar*) in a northern Swedish river caused by a hydroelectric power-station. Regulated Rivers: Research and Management 17: 101–115.

Rivinoja, P., K. Leonardsson & H. Lundqvist, 2006. Migration success and migration time of gastrically radio-tagged v. PIT-tagged adult Atlantic salmon. Journal of Fish Biology 69: 304–311. DOI:10.1111/j.1095-8649.2006.01101.x

Scruton D. A., R. S. McKinley, N. Kouwen, W. Eddy & R. K. Booth, 2002. Use of telemetry and hydraulic modeling to evaluate and improve fish guidance efficiency at a louver and bypass system for downstream-migrating Atlantic salmon (*Salmo salar*) smolts and kelts. Hydrobiologia 483: 83–94.

Thorstad, E. B., F. Økland, F. Kroglund & N. Jepsen, 2003. Upstream migration of Atlantic salmon at a power station on the River Nidelva, Southern Norway. Fisheries Management and Ecology 10: 139–146.

Trépanier. S., M. A. Rodríguez & P. Magnan, 1996. Spawning migrations in landlocked Atlantic salmon: Time series modeling of river discharge and water temperature effects. Journal of Fish Biology 48: 925–936.

Vasemägi, A., R. Gross, T. Paaver, M. L. Koljonen & J. Nilsson, 2005. Extensive immigration from compensatory hatchery releases into wild Atlantic salmon population in the Baltic sea: spatio-temporal analysis over 18 years. Heredity 95: 76–83.

Webb, J., 1990. The movements of adult Atlantic salmon in the River Tay and Tummel to Pitlochry Dam. Scottish Fisheries Research Report 48: 27 pp.

Williams, J. G., 1998. Fish passage in the Columbia River, USA and its tributaries: problems and solutions. In Jungwirth, M., S. Schmutz & S. Weiss (eds), Fish Migration and Fish Bypasses. Fishing News Book, University Press, Cambridge: 180–191.

Hydrobiologia (2008) 602:129–143
DOI 10.1007/s10750-008-9279-2

FISH AND DIADROMY IN EUROPE

Poor water quality constrains the distribution and movements of twaite shad *Alosa fallax fallax* (Lacépède, 1803) in the watershed of river Scheldt

Joachim Maes · Maarten Stevens · Jan Breine

Abstract Worldwide, river fragmentation is primarily responsible for the decline of populations of migrating fish. In particular, anadromous fish species, which necessarily migrate to fresh water to reproduce, are endangered since many are no longer able to reach their natural spawning sites. In addition, pollution of rivers effectively prevents upstream or downstream movements and blocks access to spawning grounds. This article investigates how poor water quality interferes with the life history cycle of twaite shad *Alosa fallax fallax* (Lacépède, 1803), an anadromous clupeid fish, in the watershed of River Scheldt, a heavily impacted environment in West Europe. We used two models based on known ecological and environmental information to explain past and present twaite shad distribution within the watershed and to make inferences about a future population recovery and juvenile habitat value. We demonstrated that historical spawning areas satisfy water quality conditions necessary to support spawning and successful development of early life history stages of the twaite shad. However, poor water quality conditions just upstream the freshwater–saltwater boundary still act as an effective migration barrier for upstream movement. As a consequence, spawning grounds are inaccessible and the population is dominated by seasonal adults occurring in the lower estuarine part of the watershed. This article provides testable and diagnostic information to the watershed management in that it identifies habitat and water quality requirements needed to support the expected recovery of an endangered anadromous fish population.

Keywords Fish migration · Twaite shad · Logistic model · Bioenergetics model · Water pollution · Freshwater tidal reach · Anadromy · River Scheldt

Guest editors: S. Dufour, E. Prévost, E. Rochard & P. Williot
Fish and diadromy in Europe (ecology, management, conservation)

J. Maes
Flemish Institute for Technological Research, Integrated Environmental Studies, VITO, Boeretang 200, 2400 Mol, Belgium

M. Stevens
Laboratory of Aquatic Ecology, Katholieke Universiteit Leuven, Debériotstraat 32, 3000 Leuven, Belgium

M. Stevens · J. Breine (✉)
Research Institute for Nature and Forest, Duboislaan 14, 1560 Groenendaal, Belgium
e-mail: jan.breine@inbo.be

Introduction

Predicting changes in individual traits, population sizes, community interactions or ecosystem functioning is a challenging scientific task. Too often, ecological theory fails at application (Hansson, 2003). Large simulation models with many variables and parameters may come closer but are hard to develop, hard to communicate and hard to understand (Grimm, 1999; Hansson, 2003). In a local context, biological inference from simple observations may still offer superior advice (Hansson, 2003). Empirical models may be more robust in predicting possible changes, especially when they describe changes along ecological gradients.

In this article, we used the latter strategy to make predictions about the future recovery of a migrating fish population in a heavily impacted and modified watershed. We summarized the known ecological and environmental information into simple models in order to predict migration opportunities and juvenile habitat quality. In particular, we focussed on the distribution of the twaite shad *Alosa fallax fallax* (Lacépède, 1803) in the basin of River Scheldt, a medium-sized West European lowland river with origin in France, main drainage area in Belgium and river delta in the Netherlands. The recent history of River Scheldt is characterized by a century of serious pollution and habitat deterioration. The major cause of environmental degradation of River Scheldt and its major tributaries can be linked to the high population density within the watershed combined with a historical lack of proper land use planning, especially in Flanders, the northern part of Belgium. Fragmentation of the landscape by roads and infrastructure constrains river channel dynamics. Scattered housing and the large amount of impervious cover limit the efficiency of wastewater treatment facilities. Livestock wastes accumulate in soils or run off into surface waters. As a result, exceptionally high nutrient loads reach the main river, especially in periods with increased rainfall (Struyf et al., 2004). The direct consequence is a hypoxic zone in the freshwater tidal area of the Scheldt. Dissolved oxygen levels reached an absolute minimum in the mid-1970s when the freshwater area of the river was virtually anoxic. Since then, and due to efforts to better treat wastewaters, average dissolved oxygen in the river increased by about 1 mg l^{-1} per decade (Van Damme et al., 1995).

Although seriously impacted, the river ecosystem still has important nature values. The estuary has a complete salinity gradient and includes extensive freshwater, brackish and salt marshes to its ecosystem. Tides penetrate as far as 160 km land inward and influence some of the major contributories of river Scheldt. It follows that, due to the absence of flow-regulating constructions, there are unique opportunities for migratory fish populations in the basin, all of which are now under threat or locally extinct. Sturgeon (*Acipenser sturio* L.), Atlantic salmon (*Salmo salar* L.) and allis shad (*Alosa alosa* L.) all disappeared by the end of the 1930s, while populations of twaite shad, flounder (*Platichthys flesus* L.) and smelt (*Osmerus eperlanus* L.) are confined to the brackish and marine parts of the estuary. Only the river lamprey (*Lampetra fluviatilis* L.) seemed to have persisted, as well as populations of more tolerant species such as eel (*Anguilla anguilla* L.) and three-spined stickleback (*Gasterosteus aculeatus* L.) (Van Damme et al., 1994).

Here, we focused on the ecological status of the twaite shad in the tidal part of the watershed of River Scheldt. Twaite shad is an iteroparous, anadromous fish species (Aprahamian, 1988). At maturity, adult shad gather in estuaries in April and May and move upstream to the head of estuaries or above to spawn from May to July (Aprahamian & Aprahamian, 2001; Oesmann & Thiel, 2001; Maitland & Hatton-Ellis, 2003). Once the eggs hatch nearby the spawning sites, the young fish quickly move downstream in the current to the quieter waters of the upper estuary, where they start to feed and grow (Aprahamian & Aprahamian, 2001; Gerkens & Thiel, 2001; Maitland & Hatton-Ellis, 2003). The juveniles are present in the estuary from July until they migrate seaward in the autumn (Aprahamian & Aprahamian, 2001). In the Elbe estuary, adult shad spawn in the tidal freshwater region, so that the spawning area and nursery for the juveniles spatially overlap. Historical evidence (Vrielynck et al., 2003) suggests that this was also the case for the shad population of the Scheldt basin.

In this article, we addressed the question as to why the life cycle of twaite shad in the watershed of the Scheldt is interrupted. In particular, we statistically investigated how poor water quality in the middle part of the River Scheldt between the lower estuary and the historical spawning grounds hinders the spawning run of mature shad. In addition, we modelled the future opportunities for growth of early juveniles in the estuary provided that migration would end in successful spawning above historical spawning grounds in the freshwater tidal reaches of the watershed. We used field data of water quality and fish catches to construct an empirical model that described the presence or absence of adult fish as a function of environmental variation. Additionally, we modelled habitat quality for young-of-the-year fish using a spatially explicit model of growth rate potential. Using the models and assuming a further rise of, particularly, dissolved oxygen concentrations in the watershed, we illustrate how the models can be applied to provide future spatio-temporal distribution patterns. With this information, local watershed managers can anticipate to expected changes by identifying, delineating and protecting essential fish habitats.

Materials and methods

Study area

River Scheldt has its origin in the north of France and discharges into the North Sea near Vlissingen (The Netherlands). It is a lowland river with a total length of 355 km and a fall of 100 m at most. The catchment area is approximately 21,000 km^2 with a population of 10 million inhabitants (Van den Bergh et al., 2005). This study focuses on the tidal part of the watershed, which is presented in Fig. 1. The lower estuary (Westerschelde) is characterized by flood and ebb channels, separated by sandy or muddy intertidals. Due to the funnel shape of the lower estuary the maximum vertical tidal range is about 100 km upstream, in the freshwater zone (Van den Bergh et al., 2005). The tidal influence thus extends much further land inward than does the freshwater–saltwater boundary (Fig. 1). As a result, an extensive freshwater region under tidal influence is present. The tidal excursion goes as far as Gent, 160 km from the river mouth, where the tide is stopped by sluices (Fig. 1). Also the tributaries Durme, Rupel, Nete, Kleine Nete, Grote Nete, Dijle and Zenne are under tidal influence and are therefore considered as an integral part of the estuary (Fig. 1). The historical spawning sites, where twaite shad used to spawn until the 1930s above sandbanks (Vrielynck et al., 2003) are situated within the freshwater tidal reach of the watershed (Fig. 1).

Sampling

Between 1995 and 2004, we collected fish samples along the River Scheldt using pairs of fyke nets. Fish sampling sites are presented on a map in Fig. 1. A fyke net is essentially a fish trap consisting of a long bag net distended by hoops, into which fish can pass, without being able to return. At each sampling site a pair of two fyke nets was deployed on the low water mark and emptied after 24 h. The fishing gear consisted of two 7.7 m fykes between which an 11-m lead net was suspended. The first hoop of each fyke is horseshoe shaped with a basis of 120 cm and a diameter of 80 cm. Fish can be removed on the other end of the fyke where the mesh size is 8 mm. The fishing gear was placed parallel to the river border on the tidal mudflats. During rising or falling tides, fish that encounter the leader net are guided into the fykes. Hence, both fish movements as well as mesh size influence the selectivity of fyke nets.

All fieldwork was done by trained fish biologists using a standardized protocol. Between 1995 and 2004 fishing occurred for 48 h at each site between March and October, yielding a total of 114 samples. Fish captured were identified on site using a single field guide (Nijssen & De Groot, 1987) but quality assurance of the fish identifications was performed by occasional cross-examination in the laboratory, especially of small sized specimens. Fish data recorded included species-specific fish frequencies, individual total lengths (±1 mm) and wet weights (±1 g).

Environmental data that was included in the different models was obtained from different sources. Water temperature (°C) was measured coincident with the fish sampling. Dissolved oxygen (mg l^{-1}) data was provided by Van Damme et al. (2005) for the period 1995–2000 and downloaded from the Flemish Environmental Agency internet database for other months of the study period (www.vmm.be). Freshwater flow rate data (m^3 s^{-1}) was obtained from the Dutch Rijkswaterstaat database (www.waterbase.nl).

Models

We used two models to predict the basin wide distribution patterns of twaite shad and juvenile habitat value, respectively. Presence or absence of adult shad was modelled using logistic regression

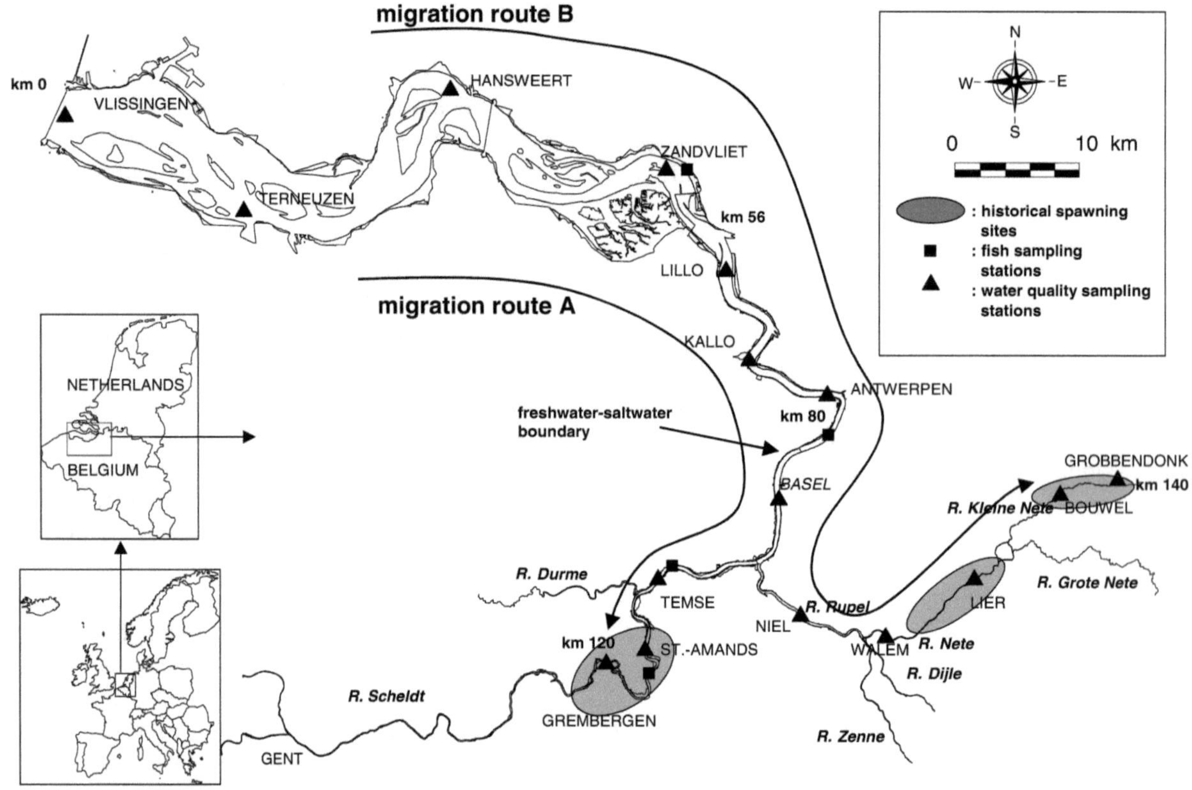

Fig. 1 Map of the tidal part of the Scheldt basin indicating the River Scheldt and tributary rivers. End point of the rivers represent the tidal limit. The map shows the fish sampling stations (squares) and the water quality sampling stations (triangles). Historical spawning sites of twaite shad are indicated as well based on a review of Vrielynck et al. (2003). Km indications measure the distance to the river mouth

based on field samples. In the model, a binary response variable (presence/absence) was expressed as a linear combination of freshwater flow, temperature and dissolved oxygen concentration through a logit link function. A priori, we expected that the response of twaite shad to oxygen concentration was increasing, following a logistic S-shaped function. In estuarine fish species, the avoidance of low oxygen concentration differs, as some species exhibit an avoidance threshold, while others exhibit a graded response (Wannamaker & Rice, 2000). A graded response is best modelled using linear regression, but this technique is less appropriate to model binary data such as the presence or absence of species. Stier & Crance (1985) constructed habitat suitability models for *Alosa sapidissima* (Wilson, 1811) and assumed a threshold response for dissolved oxygen following a study by Chittenden (1973). We expected an opposite relationship with freshwater flow, following observations that the upstream movement behaviour of American shad decreases at stream velocities above 1 m s^{-1} (Stier & Crance, 1985). Similarly, Acolas et al. (2006) observed that migration activity in allis shad increased while temperature increased and water flow decreased, and slowed when temperature decreased. The response to temperature was assumed to follow a bell shaped curve which can be approximated by a second order polynomial, which results in a temperature optimum. The model design is:

$$\text{logit}\ p = \log_e[p/(1-p)] = b_0 + b_1 \times [O_2] + b_2 \times Q + b_3 \times T + b_4 \times T^2 + \varepsilon \quad (1)$$

where p is the probability to capture adult twaite shad in a fish fyke over a 24 h period; $[O_2]$ represents the dissolved oxygen concentration (mg l^{-1}), Q represents the average monthly freshwater flow or debit

($m^3\ s^{-1}$) and T is the ambient temperature (°C). ε is the error term. The regression parameters b_i were estimated using the maximum likelihood procedure. Model goodness of fit was evaluated using the model deviance. It is defined as $-2 \times (L_M - L_S)$ where L_M denotes the maximized log-likelihood value for the model of interest, and L_S is the log-likelihood for the saturated model. Under the null hypothesis that the logistic model is true, the deviance is χ^2-distributed. The null hypothesis of $b_i = 0$ was tested using the Wald statistic. The fishing gear that was used captures pelagic species at low efficiency, hence we consider the predictions as lower limits.

Since adult shad is, at present, not able to spawn, we assessed the possible habitat value of the estuary for juveniles using a bioenergetic growth model. The assumption is that growth rate potential can be used as a criterion to evaluate habitat suitability (Brandt et al., 1992). In spatially explicit habitat models, the growth rate potential describes the amount of growth achievable by a size and species-specific predator when placed in a predetermined habitat characterised by a particular suite of environmental conditions (Roy et al., 2004). Several authors have used growth rates potential of habitats to describe the relative quality of habitats and environments for fish populations in lakes, rivers and estuaries (Brandt et al., 1992; Luo et al., 2001; Höök et al., 2003). We used the basic Wisconsin bioenergetics-modelling framework (described by Kitchell et al., 1977) to develop a model of juvenile twaite shad. Bioenergetics models balance the flow of energy through an individual fish over consumption, growth, and energy loss (respiration, specific dynamic action, egestion and excretion). We refer to Hanson et al. (1997) for a complete description of bioenergetic models for fish. We used parameters of the congeneric *Alosa pseudoharengus* (Wilson, 1811) (Stewart & Binkowski, 1986) and of herring *Clupea harengus* L. (Rudstam, 1988) to model the bioenergetics of twaite shad (Table 1). Following Luo et al. (2001), oxygen dependence of maximum consumption was modelled as a hypothetical function based on the general concept of fish physiological processes in response to stress (Bartell, 1990):

$$f(\mathrm{DO}) = \left[1 + \exp\left(4.77 - 1.59 \times [\mathrm{O_2}]\right)^{-1}\right] \qquad (2)$$

This function limits consumption by 50% at a DO of 3 mg l^{-1} (Stier & Crance, 1985).

Environmental scenarios

Models were used to predict spatially explicit probabilities of past and present distributions and to assess juvenile habitat value. Past conditions referred to the year 1990. Present conditions refer to the situation as observed in 2003. We also made predictions based for 2010 assuming that observed historical trends in dissolved oxygen concentration between 1989 and the present will continue into the future. In each case, we focussed on two possible migration routes, which adult shad should follow in order to reach historical spawning sites (Fig. 1). The first route (route A) leads shad to spawning grounds situated in the tidal, freshwater part of river Scheldt at 120 km of the river mouth. The second route (route B) leads shad to an historical spawning area in a smaller tributary river (Kleine Nete), situated 140 km upstream. We estimated future environmental conditions using a statistical model, assuming that temperature and river flow in 2010 can be approximated using monthly averages for the period 1993–2004 and that DO in 2010 can be approximated using monthly averages for the period 1989–2004 superimposed on a long-term linear trend. Predicted oxygen concentrations in 2010 satisfied the following equation:

$$[\mathrm{O_2}] = \alpha \times \mathrm{Trend} + \beta_j \times \mathrm{Month}_j + \varepsilon \qquad (3)$$

Equation 3 is a general linear model with the oxygen concentration $[O_2]$ as dependent variable, with slope α and intercept $\beta_j \times \mathrm{Month}_j$. Trend as continuous predictor variable, which measures the number of days that passed since 1 January 1989 divided by 30. The categorical variable Month_j takes 12 different codes (January, ..., December) and was encoded into 12 binary variables. These variables assume a value of one for samples taken in the designated month, zero otherwise. As a result, categorical predictors either increase or decrease the model intercept but the slope α remains unaffected. It follows that, under the null hypothesis of a non-significant slope α, oxygen concentrations are estimated as monthly averages. ε is the error term, which is normally distributed around zero. Models were constructed for 16 fixed-sampling stations situated in the watershed (Fig. 1).

Table 1 Parameters used in the twaite shad bioenergetics model

Symbol	Description	Value	Unit
P-value	Proportion of maximum consumption	0.27	Dimensionless
Ca	Intercept for maximum consumption	0.8464	$g\ g^{-1}\ d^{-1}$
Cb	Exponent for maximum consumption	−0.3	Dimensionless
f(T)	Temperature dependence of maximum consumption *f*(T) = *f*(K1, K2, K3, K4, T1, T2, T3, T4)		Dimensionless
f(O_2)	Oxygen dependence of maximum consumption see text for description		Dimensionless
K1	Proportion of maximum consumption at T1	0.17	Dimensionless
K2	Proportion of maximum consumption at T2	0.98	Dimensionless
K3	Proportion of maximum consumption at T3	0.98	Dimensionless
K4	Proportion of maximum consumption at T4	0.01	Dimensionless
T1	Temperature for K1 in *f*(T)	3	°C
T2	Temperature for K2 in *f*(T)	27	°C
T3	Temperature for K3 in *f*(T)	27	°C
T4	Temperature for K4 in *f*(T)	32	°C
Ra	Intercept for maximum standard respiration	0.00367	$g\ g^{-1}\ d^{-1}$
Rb	Exponent for maximum standard respiration	−0.2152	Dimensionless
RQ	Slope for temperature dependence of standard respiration	0.0548	Dimensionless
RTO	Coefficient for swimming speed dependence on metabolism	0.03	$s\ cm^{-1}$
RTL	Cut-off temperature at which the activity relationship changes	9	°C
RK1	Intercept for swimming speed above the cut-off temperature	15	$cm\ s^{-1}$
RK4	Mass dependence coefficient for swimming speed at all water temperatures	0.13	Dimensionless
ACT	Intercept for a 1 g fish at 0°C of the relationship for swimming speed versus mass at water temperatures less than RTL	3.9	$cm\ s^{-1}$
BACT	Temperature dependence coefficient of swimming speed at temperatures below RTL	0.149	$°C^{-1}$
SDA	Specific dynamic action coefficient	0.175	Dimensionless
Fa	Proportion of consumed food egested	0.16	Dimensionless
Ua	Proportion of assimilated food excreted	0.1	Dimensionless

Model symbols and conventions correspond with the Wisconsin Bioenergetics Model (Hanson et al., 1997). The model describes growth during the first year for all weights of twaite shad between 0.25 g and 5 g. Prey (zooplankton) and predator energy density were set at 2,580 J g^{-1} (Hanson et al., 1997) and 5,233 J g^{-1} (Stewart and Binkowski, 1986), respectively

Results

Distribution of twaite shad in the Scheldt basin

Regular sampling of fish populations in the estuary of River Scheldt using fyke nets started in 1995. Adult twaite shad first appeared in samples in 1996. Since then, annual landings, initially rose until 1999 but fluctuated strongly during the remainder of the study period (Fig. 2A). Pooling monthly data shows that shad abundance generally peaked in June and August. Twaite shad left the estuary before November to winter in the North Sea (Fig. 2B). Total shad body length averaged 38 cm (Fig. 2C). The 0-group shad (<10 cm) were never caught. The spatial distribution of twaite shad appeared to be limited to the saline and brackish parts of the estuary. So far, no shad have been captured in the part of the Scheldt basin upstream the city of Antwerp. Apparently, adult shad were unable to reach upstream historical spawning

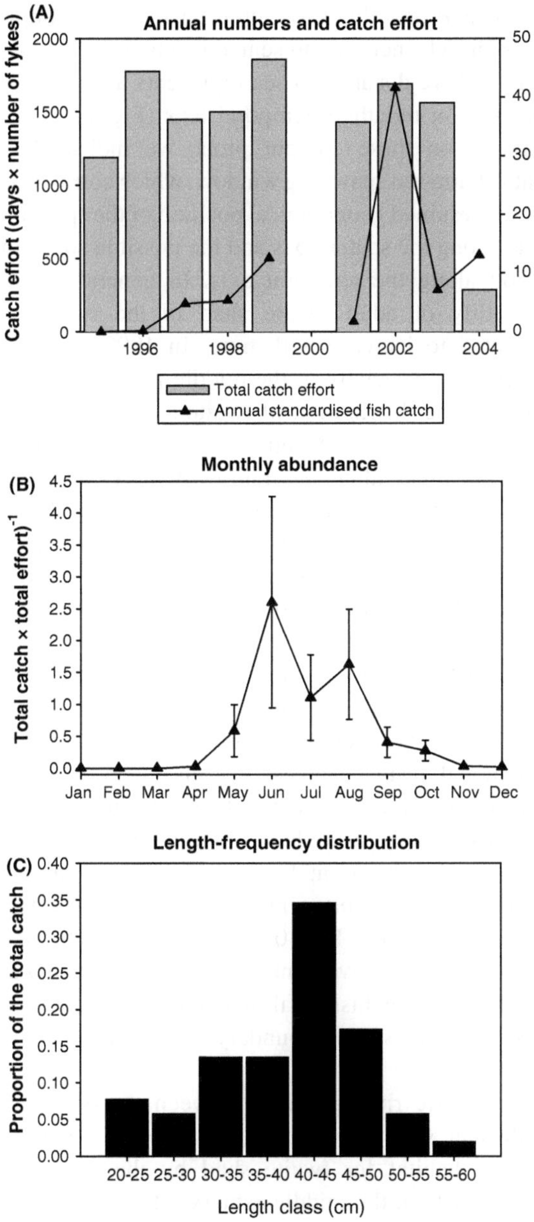

Fig. 2 Abundance patterns and population structure of twaite shad in the Scheldt estuary based on fyke net data taken between 1995 and 2004 at a permanent fish monitoring station in the Scheldt estuary (km 58, Fig. 1). (**A**) Total annual catch effort (number of fishing days × number of fyke nets) and standardised annual catch (numbers of shad per 100 fishing days using 1 fyke net). (**B**) Average monthly distribution of adult shad (error bars are standard deviations). (**C**) Length frequency distribution of shad over 5 cm length classes ($N = 1260$)

sites which are situated in the tidal freshwater part of the River Scheldt and in the tributaries of the Scheldt (River Nete and River Kleine Nete, Fig. 1).

Logistic model

Logistic regression with the presence of twaite shad in fyke nets as dependent variable and ambient oxygen concentration, freshwater flow and temperature as independent predictor variables yielded a statistically significant model. The model deviance was 68.7, with 108 degrees of freedom. Since $\chi^2_{p=0.05}$ is 85.7, we accept the logistic regression with equation logit $(p) = -10.85 + 0.77 \times [O_2] - 0.014 \times Q + 0.67 \times T - 0.014 \times T^2 + \varepsilon$. Increasing dissolved oxygen significantly affected the probability to capture adult shad in fyke nets, given temperature and flow (Wald test: df = 108; $W = 16.5$; $P < 0.001$, Fig. 3). Shad were negatively affected by high river flow rates, given temperature and DO (Wald test: df = 108; $W = 4.01$; $P < 0.05$, Fig. 3). Essentially, this variable does not affect the probability of capturing shad in fyke nets since freshwater flow rates rarely exceed 400 $m^3\ s^{-1}$. The likelihood for shad to occur in fyke nets increased with temperature, reflecting their seasonal occurrence between April and October (Fig. 3).

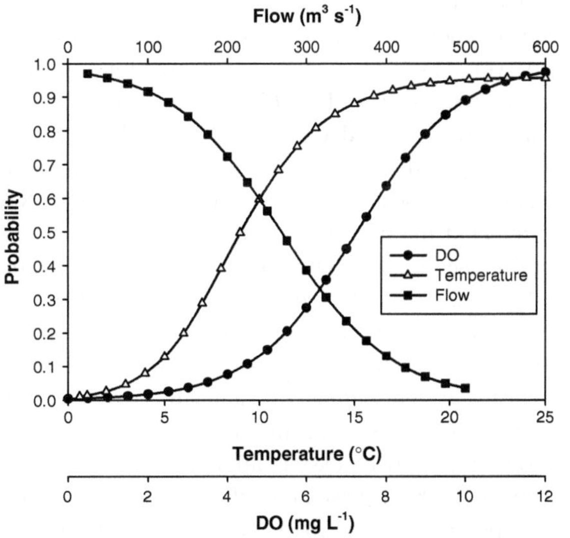

Fig. 3 Logistic response of twaite shad occurrence as a function of temperature, dissolved oxygen (DO) and monthly river flow. Probabilities represent the statistical likelihood to capture twaite shad in 24 h fyke net samples. Probabilities are calculated using Eq. 1 and are presented along three univariate axes, given constant values for the two other environmental variables (DO = 10 mg l^{-1}; temperature = 15.4°C, river flow 121 $m^3\ s^{-1}$)

Spatio-temporal distribution patterns of twaite shad in the Scheldt basin

The logistic model was used to plot basin-wide probability distributions of adult twaite shad occurrence under past, present and future environmental conditions. DO concentrations for 2010 were predicted using linear regression models for 16 stations situated in the tidal watershed of the Scheldt (Fig. 1). Model parameters and diagnostics of the regression models are presented in Table 2. All models were significant at $P < 0.001$ and the explained variance varied between 36% for station Lier to 75% for station Terneuzen. Five stations in the watershed, of which three in River Scheldt and two in River Kleine Nete, displayed a slope that was not significantly different from zero (Table 2). DO predictions for these stations calculated for the year 2010 are therefore approximated as monthly averages. The regression slopes for other stations were significantly positive (Table 2), evidencing an increasing DO trend in the watershed. As an example, we presented the historical and predicted time series of DO for two water quality stations. In Fig. 4, we regressed the DO time series against the number of days since 1 January 1989 using Eq. 3. At Basel, a station in the River Scheldt nearby the freshwater–saltwater boundary (Fig. 1), DO displays a significantly positive trend (see slope α, Table 2). Hence DO predictions for 2010 are based on historical monthly averages superimposed on an increasing long-term trend. At station Grobbendonk, situated in River Kleine Nete (Fig. 1) at the limit of the tidal range in the Scheldt basin, the trend in DO has a non-significant slope (slope α, Table 2), so predictions for 2010 are estimated using monthly means for the period 1989–2005. Figure 5 summarizes the spatial trends in DO by presenting the annually averaged DO concentrations as a function of distance to the river mouth for the two possible migration routes. Data for 1990 and 2003 is based on field measurements, while DO predictions for 2010 are based on the general linear model. Between 1990 and 2003, the annual average DO concentration clearly increased in the river part upstream km 40 and DO conditions are expected to improve by 2010.

Environmental variables (temperature, DO and river flow) were used as input to make inferences about the spatial and temporal distribution of shad in the basin. Probabilities to capture shad in 24-h samples based on fyke nets are presented in a two-dimensional plane, where distance to sea represents a spatial axis and time in months a temporal axis (Fig. 6). Superimposed on these contour plots, we indicated the spatio-temporal spawning window, which corresponds to the presumed geographical position of the spawning areas along the spatial axis and the possible spawning period along the temporal axis. In general, habitat suitability of adult twaite shad in the basin was predicted to increase with time. In 1990, poor DO conditions, especially in the middle part of the river, confined the distribution of twaite shad to the lower estuary (km 0–60). Capture probabilities in more upstream areas along migration route A, were zero (Fig. 6). The spawning areas situated in River Kleine Nete (migration route B) already satisfied water quality requirements but again, a zone of hypoxia between km 60 and km 100 was unsuitable for twaite shad and possibly prevented movements between the upstream river parts upstream km 100 and the lower estuary downstream of km 60.

In 2003, the situation improved relative to 1990 and model results suggest that the spatio-temporal distribution of available twaite shad habitat in the basin expanded, particularly along migration route A (Fig. 6). Conditions at historical spawning grounds probably allowed spawning but poor water quality in the middle estuary (km 80–km 100) was predicted to block upstream movements. Again, twaite shad were unable to reach historical spawning sites, although these sites satisfy to boundary conditions for shad presence.

The entire river gradient between the ocean and freshwater spawning sites is predicted to become a suitable habitat for twaite shad by the year 2010 (Fig. 6). DO in the middle estuary (at about km 100) is expected to increase (Fig. 5), creating a time window during which the probability to pass this part of the river increases.

Juvenile habitat value

We used the observed and predicted environmental data for the three years considered in order to assess growth rate potential for young-of-the-year twaite shad. These simulations are based on a bioenergetic model that was parameterized for juvenile shad. We

Table 2 Regression model coefficients and diagnostics of a general linear regression model which predicts the concentration of dissolved oxygen for 16 water quality sampling stations in the Scheldt basin according to Eq. 3

Sampling station	Distance to the river mouth (km)	F	DF	R^2	α	β_1 Jan	β_2 Feb	β_3 Mar	β_4 Apr	β_5 May	β_6 Jun	β_7 Jul	β_8 Aug	β_9 Sep	β_{10} Oct	β_{11} Nov	β_{12} Dec
Vlissingen	0	30.2	186	0.64	0	10.37	10.24	10.17	10.39	9.95	8.39	7.71	7.48	7.14	7.78	8.80	9.03
Terneuzen	21.5	36.8	129	0.75	0	10.52	10.22	10.03	9.85	9.41	7.99	7.20	7.08	7.04	7.75	8.76	8.78
Hansweert	37.3	37.0	180	0.69	0	10.89	10.43	9.96	9.88	9.49	8.45	7.67	7.42	7.37	8.05	9.02	9.07
Zandvliet	57.5	7.4	100	0.41	0.010	7.10	6.49	5.63	4.32	3.52	3.30	4.03	3.38	4.33	4.64	4.61	4.70
Lillo	64.8	10.0	104	0.48	0.014	5.76	6.12	4.39	3.00	1.92	2.32	2.80	2.88	3.37	3.12	3.14	3.57
Kallo	70.3	10.9	103	0.51	0.015	5.21	4.55	3.61	2.12	1.17	1.41	1.69	1.88	1.77	2.06	2.20	2.53
Antwerpen	76.1	17.0	152	0.54	0.018	3.30	3.12	2.39	0.99	0.28	0.68	0.54	0.51	0.62	0.68	0.68	1.27
Basel	90.5	22.6	177	0.58	0.019	3.56	3.20	2.35	0.65	0.37	−0.16	0.32	−0.51	−0.46	−0.34	0.56	0.91
Temse	99.8	10.4	89	0.53	0.008	6.32	6.47	5.06	3.06	1.78	1.36	1.69	1.94	1.91	1.77	1.83	3.05
St.-Amands	110.7	11.5	133	0.46	0.016	4.19	4.82	4.43	1.87	0.97	0.51	0.68	0.71	0.56	0.41	1.03	1.86
Grembergen	121.6	26.3	154	0.65	0.026	3.93	3.67	3.34	2.76	0.75	0.50	0.96	1.00	0.88	0.86	1.30	1.89
Niel	96.0	17.0	148	0.55	0.017	2.68	3.80	2.39	0.71	-0.02	−0.17	−0.53	0.29	−0.22	−0.39	0.70	0.91
Walem	106.9	10.9	170	0.39	0.008	7.54	6.39	5.87	5.23	3.62	2.59	2.46	2.76	2.61	2.62	4.83	4.30
Lier	115.6	5.0	73	0.36	0.001	8.82	7.33	8.57	6.83	6.70	6.43	5.69	5.29	5.09	6.31	7.83	6.94
Bouwel	129.7	5.9	91	0.37	0	12.27	10.73	11.10	8.70	8.37	7.80	7.84	7.88	8.19	9.44	10.19	9.37
Grobbendonk	140.0	13.3	226	0.38	0	10.16	9.44	9.35	8.77	7.87	7.31	6.75	7.01	7.01	8.16	8.94	8.37

Sampling stations refer to the nearest city and are situated at different distances (km) to the mouth of the river (Fig. 1). All models were significant ($P < 0.001$) as evaluated using an F test. DF refers to the degrees of freedom; R^2 presents the explained variance by each model. Regression slopes (α) different from zero were significant ($P < 0.05$), slopes equal to zero were not significant. Model intercepts (β_j) differ for each month and can be interpreted as the concentration of dissolved oxygen if the variable 'trend' in Eq. 3 is equal to zero

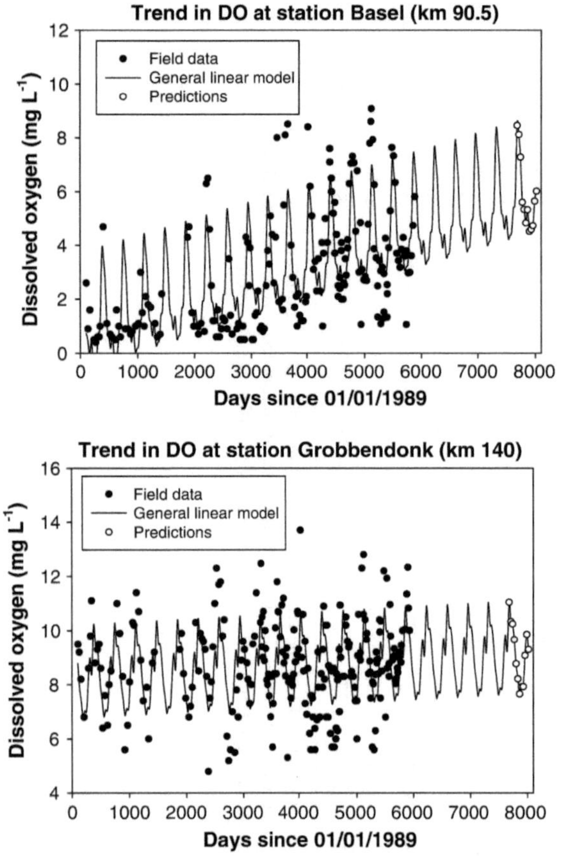

Fig. 4 Observed field data (black dots) and modelled trend of dissolved oxygen (DO) at two water quality stations in the Scheldt watershed (Basel and Grobbendonk) since 01/01/1989. The trendline is based on a general linear regression model according to Eq. 3. At station Basel, DO increased significantly with time (see α, Table 2) while at station Grobbendonk, the trend in DO has a non-significant slope (α, Table 2). The model was used to project DO concentrations for the year 2010 (white dots)

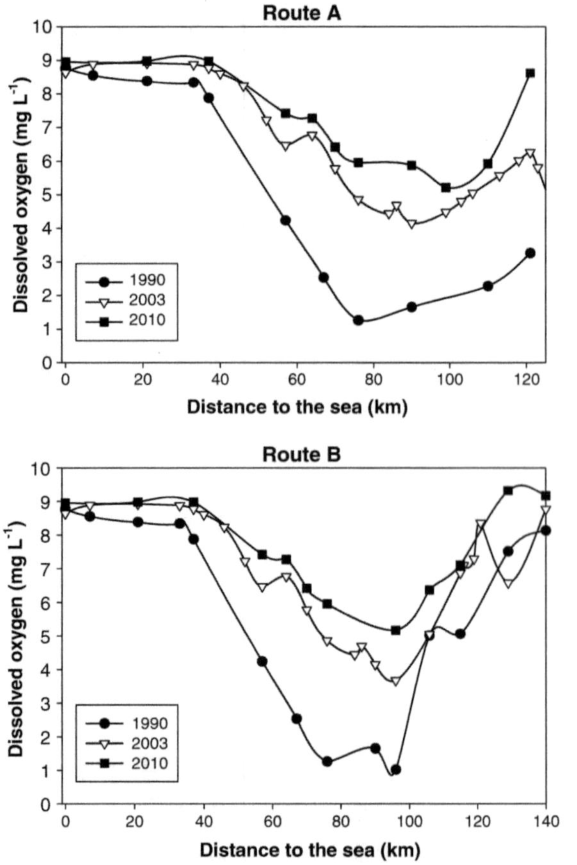

Fig. 5 Past (1990), present (2003) and predicted (2010) annual mean concentration of dissolved oxygen (DO) along two possible migration routes (route A, route B, see Fig. 1) as a function of distance to the sea. The data for 1990 and 2003 is based on DO water samples at 16 stations in the Scheldt basin (Fig. 1). Predicted values for the year 2010 were calculated using a general linear model (GLM) given in Eq. 3

assumed young shad with total body weights of 0.5 and 1.5 g, respectively. Spatial changes of juvenile growth rates are illustrated in Fig. 7. The calculations were made for the entire estuary including the tidal freshwater part which extents to Gent along route A and to Grobbendonk along route B (Fig. 1).

Similar spatial patterns for the two considered body weights emerged but the scales differed since both consumption and resting metabolic rate scale with body weight. The spatial distribution of habitat offering optimal growth rates was predicted to increase over the years. In the lower estuary (<40 km) where water is fully saturated with oxygen, differences in mean growth rates reflected differences in mean annual temperatures. Slower growth rates in 2003 predicted for this part of the estuary were produced by above-average summer temperatures following the heat wave in Europe.

The opportunity for growth of young shad gradually decreased in upstream direction as a result of deteriorated DO conditions relative to the downstream part of the watershed. Nearby the historical spawning sites in the upstream parts in the upstream tributary Kleine Nete along migration route B, environmental conditions permitted positive growth rates for the three years that were included in the simulation. Habitat quality nearby spawning sites in the tidal freshwater reach of river Scheldt (Fig. 7, route A) was low in 1990 but is expected to increase by 2010 as a result of the expected increase in DO concentration.

Discussion

Two important conclusions derived from this study. First, historical spawning sites in the Scheldt basin already satisfy water quality conditions necessary for spawning and successful development of early life history stages of twaite shad. However, available habitat still showed a bimodal distribution along the river gradient. Poor water quality conditions just upstream the freshwater–saltwater boundary still acted as an effective migration barrier for both upstream and downstream movements. A second conclusion is that there is a reasonable possibility that the twaite shad population recovers by the end of this decade. These conclusions are based on a number of assumptions with respect to twaite shad life history, habitat utilization and expected environmental recovery. We consider these assumptions before discussing model applications.

Model assumptions

In this article, we used water temperature, river flow and DO as spatio-temporal input in order to make predictions of juvenile and adult-habitat distribution within the watershed. These variables represent key ecological resources for the twaite shad population. Temperature directly or indirectly controls physiological rates, hatching success (Maitland & Hatton-Ellis, 2003), juvenile growth rate (Aprahamian, 1988), year class strength (Aprahamian & Aprahamian, 2001), juvenile seaward migration (Limburg, 1996), adult upstream movements (Aprahamian, 1988). More generally, temperature also relates to natural mortality of teleost fishes (Pauly, 1982) as well as to prey abundance. DO influences the behaviour of twaite shad through physiological tolerances. Twaite shad is reported as sensitive to pollution, but few data appear to be available (Maitland & Hatton-Ellis, 2003). The modelled logistic response to dissolved oxygen reported in this article suggests that adult shad become seriously stressed at DO less than 3 mg l^{-1}, a concentration at which the capture probability is less than 10%. This parallels the situation in the Elbe estuary (Germany) where twaite shad were absent in fish samples taken at sites where DO felt below 3 mg l^{-1} (Möller & Scholz, 1991). Stier & Crance (1985) used 5 mg l^{-1} DO as threshold level for adult American shad *Alosa sapidissima*. It is unclear whether the relationship between low DO and twaite shad occurrence is causal. Eggs and larvae do not tolerate poor oxygen conditions, so there is a clear fitness consequence when adults take the risk to continue upstream migration and face low DO concentrations. A similar conclusion may be drawn to explain the negative relationship between increasing river flow and shad presence. As appears to be the case with other migratory species (Maitland, 2003), high flows are likely to be detrimental to populations of twaite shad by limiting the access spawning grounds and by lowering recruitment after spawning, probably due to downstream advection of eggs and larvae.

To model estuarine habitat suitability of young-of-the-year twaite shad using a fresh body weight of either 0.5 g or 1.5 g, we assumed that the consumption rate was constrained by DO. Although we did not know the exact nature of this relationship, there is evidence low DO directly restricts consumption in fish (Secor & Niklitschek, 2001; Poon et al., 2001). Under hypoxia, fish have difficulties in supporting any activity that requires aerobic respiration and they shift to anaerobic pathways for energy expenditure (Poon et al., 2001). Feeding and, in particular, digestion of food are energetically demanding processes which need to be stopped during periods of hypoxia. Hence, it seemed appropriate to include a sigmoid function in the bioenergetic model to restrict food consumption.

In order to translate model results into predictions for a nearby future, we accepted that a linear path in the concentration of dissolved oxygen is a realistic environmental scenario to approximate future DO conditions. An analysis of monthly DO measurements taken between 1989 and 2005 at 16 sampling stations in the watershed corroborates the hypothesis of linearly increasing oxygen. The increasing trend is the result of increased efforts to clean wastewaters. Although global warming may affect DO in rivers on the long term, we do not expect sudden near-future declines in DO. At present, DO simulation models are under construction for the estuary of River Scheldt but so far, they are unavailable for predictions. Until then, a statistical model may be the only effective alternative to produce extrapolations.

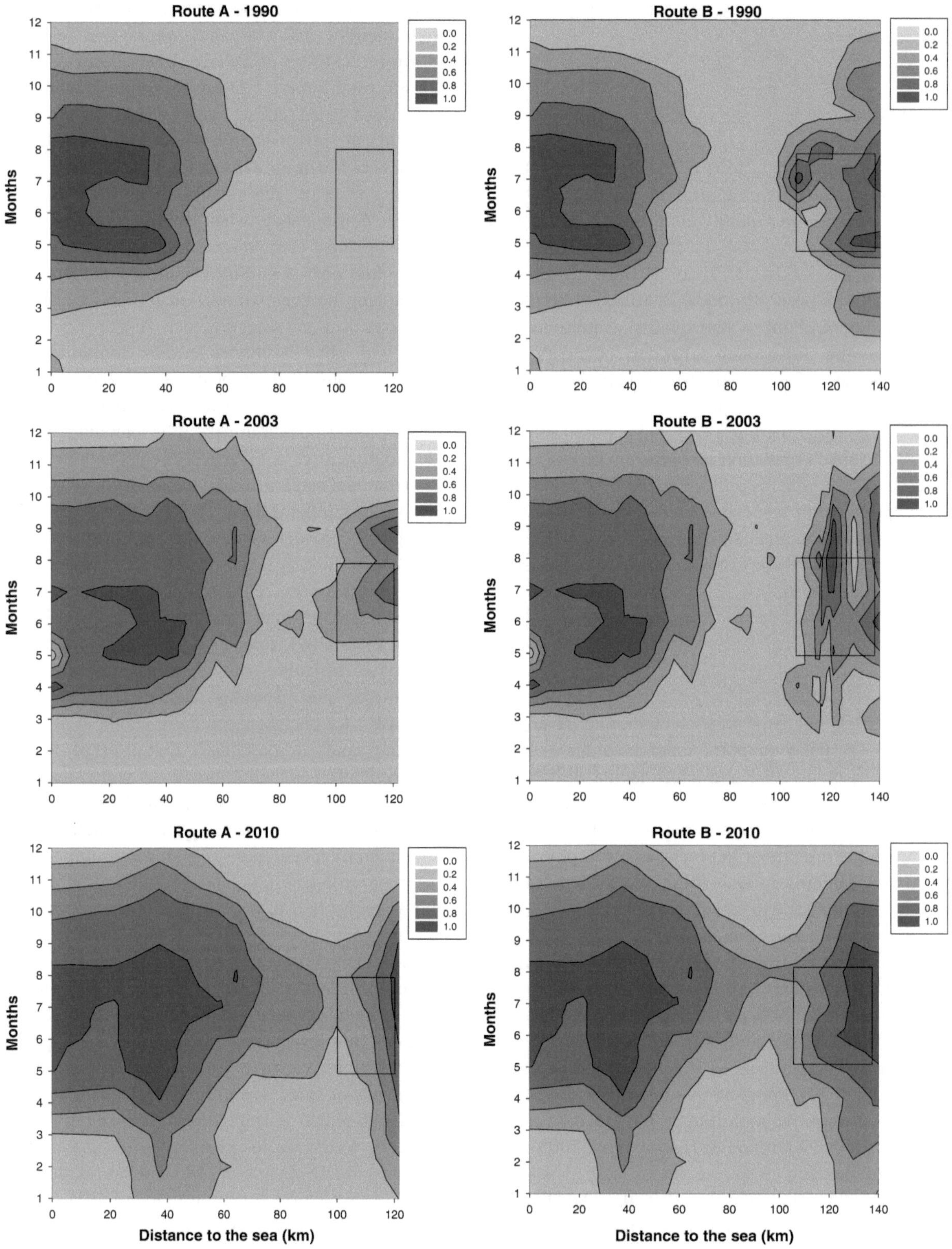
Route A - 1990
Route B - 1990
Route A - 2003
Route B - 2003
Route A - 2010
Route B - 2010
Months
Distance to the sea (km)
0.0
0.2
0.4
0.6
0.8
1.0
0 20 40 60 80 100 120
0 20 40 60 80 100 120 140
1 2 3 4 5 6 7 8 9 10 11 12

◀**Fig. 6** Spatio-temporal capture probabilities of adult twaite shad along two river gradients (migration route A and B). Contours define space-time areas with similar catch probability. The spatial axis represents the distance to the sea (river mouth at 0 km and most upstream area at 140 km). The temporal axis represents the months of the year (Jan = 1, ..., Dec = 12). Probabilities vary between 0 and 1. Three environmental scenarios simulate conditions for 1990, 2003 and 2010. Rectangles delineate historical spawning areas along the spatial axis and time available for reproduction along the time axis

Finally, we assumed that, once shad will reach historical spawning sites, they will successfully spawn. Hence, twaite shad were assumed to show sufficient plasticity in spawning behaviour and habitat requirements. Historical evidence shows that in the basin of River Scheldt twaite shad used to spawn in or just above the tidal reach (Vrielynck et al., 2003). Sand banks appeared to act as preferred spawning substrate, but probably any zone in the freshwater reach where egg retention is sufficiently high may provide suitable spawning sites. Retention zones may comprise areas with slower flow in shallow fast flowing waters as well as deep areas in the tidal part of the river where net particle displacement is limited due to periodic upstream flood currents. This would explain the variety of spawning habitats and substrates that are reported for European estuaries and rivers (Caswell & Aprahamian, 2001; Gerkens & Thiel, 2001).

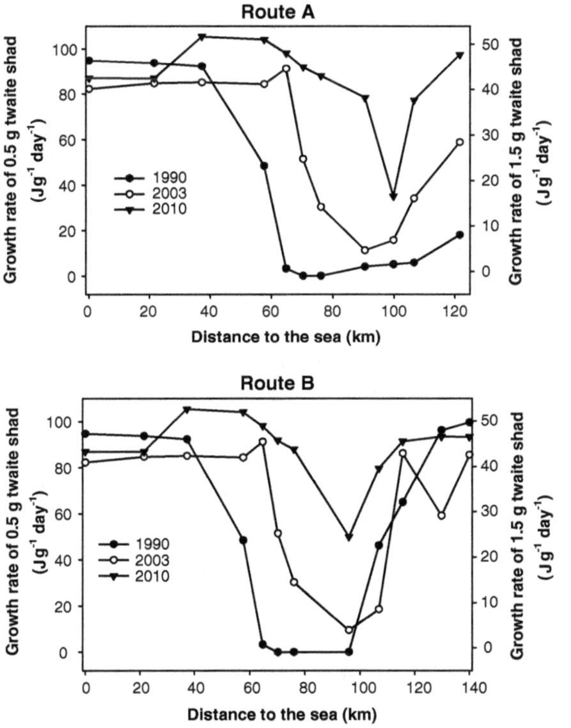

Fig. 7 Spatially explicit annually averaged growth rates (J g^{-1} day^{-1}) of juvenile twaite shad (0.5 and 1.5 g body weight) along two river gradients (migration route A and B, see Fig. 1). Three environmental scenarios simulate conditions for 1990, 2003 and 2010. The spatial axis represents the distance to the sea

Applications for watershed management

Once conditions are restored and ecosystem stressors are removed, fish populations may recover and recolonise watersheds quickly (Eklöv et al., 1998). Here, we predicted that twaite shad, an endangered and protected fish species, will probably recover after an absence of nearly one century. We consider twaite shad as a indicator species showing the possibilities of ecosystem recovery to a society, which has no more collective memory of migrating fish species. The species was historically abundant, is sensitive to pollution and necessarily needs the whole river gradient for completion of the life history. Its predicted return to historical spawning sites would be an important milestone after decades of decline and an environmental success.

We demonstrated that it is possible to make acceptable predictions about a future population recovery with limited information. The models that were used yielded testable predictions and should convince river managers to take a number of necessary steps in order to support the expected recovery. First, it is clear that at present the mid part of the estuary is the most critical part in which hypoxia and high temperature conditions cause severe habitat fragmentation. Movements of adults are blocked and the total nursery area for young-of-the-year fish is reduced. This part of the river should receive prompt attention. Second, the bioenergetics model can be applied to identify and protect essential habitats as nursery areas for early life history stages. The approach to couple habitat quality to growth rate potential has proven its value in applied ecosystem management issues (Brandt et al., 1992; Luo et al., 2001; Höök et al., 2003; Niklitschek & Secor, 2005). Amongst other factors such as field measurements of fish density or an estimate of habitat related survival, the calculation of spatially explicit growth rates is a suitable proxy to identify nursery areas for young fish (Beck et al., 2001). Especially for habitats where juveniles are absent because of a lack in connectivity between adult habitats at sea and spawning and nursery sites in fresh water spatially explicit bioenergetic growth models may offer an valuable alternative to the more classic habitat quality estimates that are based on direct fish densities measured in the field. Finally, in anticipation of population recovery, a survey of suitable spawning habitats and substrate is clearly needed to support successful restoration of the twaite shad population in the Scheldt basin.

The predictor variables used in the models represent true ecological recourses and data of dissolved oxygen, temperature and river flow is commonly, and often freely, available in databanks. The empirical models are simple and straightforward and can be used directly by watershed managers. Clearly, species distributions are affected by biotic interactions, and in particular, by the distribution and abundance of their prey. This interaction was not included in this study since predator prey relations require considerable modelling and assumptions about the distribution and behaviour of prey and predators. Therefore, practical use of models that include biotic interaction may be limited.

Acknowledgements This article benefited from several fish monitoring projects funded by VIBNA (Association of Industrial Companies of North Antwerp), AMINAL Afdeling Bos en Groen (Flemish Environmental Administration) and the Institute of Forestry and Game Management. This paper is a contribution to the European Interreg IIIb North Sea project Harbasins, aimed at harmonising European river basin management.

References

Acolas, M. L., V. Veron, H. Jourdan, M. L. Begout, M. A. R. Sabatie & A. L. Acolas, 2006. Upstream migration and reproductive patterns of a population of allis shad in a small river (L'Aulne, Brittany, France). ICES Journal of Marine Science 63: 476–484.

Aprahamian, M. W., 1988. The biology of the twaite shad *Alosa fallax fallax* (Lacépède) in the Severn Estuary. Journal of Fish Biology 33A: 141–152.

Aprahamian, M. W. & C. D. Aprahamian, 2001. The influence of water temperature and flow on year-class strength of twaite shad (*Alosa fallax*) from the River Severn, England. Bulletin Français de la Pêche et de la Pisciculture 362/363: 953–972.

Bartell, S. M., 1990. Ecosystem context for estimating stress induced reductions in fish populations. In Adams, S. M. (ed.), Biological Indicators of Stress in Fish. American Fisheries Society, 167–182.

Beck, M. W., K. L. Heck, K. W. Able, D. L. Childers, D. B. Eggleston, B. M. Gillanders, B. Halpern, C. G. Hays, K. Hoshino, T. J. Minello, R. J. Orth, P. F. Sheridan & M. R. Weinstein, 2001. The identification, conservation, and management of estuarine and marine nurseries for fish and invertebrates. Bioscience 51: 633–641.

Brandt, S. B., D. M. Mason & E. V. Patrick, 1992. Spatially-explicit models of fish growth rate. Fisheries 17: 23–33.

Caswell, P. A. & M. W. Aprahamian, 2001. Use of River Habitat Survey to determine the spawning habitat characteristics of twaite shad (*Alosa fallax fallax*). Bulletin Français de la Pêche et de la Pisciculture 362/363: 919–929.

Chittenden, M. E., 1973. Effects of handling on oxygen requirements of American shad (*Alosa sapidissima*). Journal of the Fisheries Research Board of Canada 30: 105–110.

Eklöv, A. G., L. A. Greenberg, C. Bronmark, P. Larsson & O. Berglund, 1998. Response of stream fish to improved water quality: a comparison between the 1960s and 1990s. Freshwater Biology 40: 771–782.

Gerkens M. & R. Thiel, 2001. A comparison of different habitats as nursery areas for twaite shad (*Alosa fallax* Lacépède) in the tidal freshwater region of the Elbe River Germany. Bulletin Français de la Pêche et de la Pisciculture 362/363: 773–784.

Grimm, V., 1999. Ten years of individual-based modelling in ecology: what have we learned and what could we learn in the future? Ecological Modelling 115: 129–148.

Hansson, L., 2003. Why ecology fails at application: should we consider variability more than regularity. Oikos 100: 624–627.

Hanson, P. C., Johnson T. B., Schindler D. E. & J. F. Kitchell, 1997. Fish Bioenergetics 3.0 for Windows. University of Wisconsin Sea Grant Institute.

Höök, T. O., E. S. Rutherford, S. J. Brines, D. M. Mason, D. J. Schwab, M. J. McCormick, G. W. Fleischer & T. DeSorcie, 2003. Spatially explicit measures of production of young alewives in Lake Michigan: linkage between essential fish habitat and recruitment. Estuaries 26: 21–29.

Kitchell, J. F., D. J. Stewart & D. Weiniger, 1977. Applications of a bioenergetics model to yellow perch (*Perca flavescens*) and walleye (*Stizostedion vitreum vitreum*). Journal of the Fisheries Research Board of Canada 34: 1922–1935.

Limburg, K., 1996. Modelling the ecological constraints on growth and movement of juvenile American shad (*Alosa sapidissima*) in the Hudson river estuary. Estuaries 19: 794–813.

Luo, J. G., K. J. Hartman, S. B. Brandt, C. F. Cerco & T. H. Rippetoe, 2001. A spatially-explicit approach for estimating carrying capacity: an application for the Atlantic menhaden (*Brevoortia tyrannus*) in Chesapeake Bay. Estuaries 24: 545–556.

Maitland, P. S., 2003. Ecology of the River, Brook and Sea Lamprey. Conserving Natura 2000 Rivers Ecology Series No. 5. English Nature, Peterborough.

Maitland, P. S. & T. W. Hatton-Ellis, 2003. Ecology of the Allis and Twaite Shad. Conserving Natura 2000 Rivers Ecology Series No. 3. English Nature, Peterborough.

Möller, H. & U. Scholz, 1991. Avoidance of oxygen-poor zones by fish in the Elbe River. Journal of Applied Ichthyology 7: 176–182.

Nijssen, H. & S. J. De Groot, 1987. De Vissen van Nederland. Stichting KNNV-Uitgeverij, Utrecht.

Niklitschek, E. J. & D. H. Secor, 2005. Modeling spatial and temporal variation of suitable nursery habitats for Atlantic sturgeon in the Chesapeake Bay. Estuarine, Coastal and Shelf Science 64: 135–148.

Oesmann, S. & R. Thiel, 2001. Feeding of juvenile twaite shad (*Alosa fallax* Lacepede, 1803) in the Elbe estuary. Bulletin Français de la Pêche et de la Pisciculture 362/363: 785–800.

Pauly, D., 1982. On the interrelationships between natural mortality, growth parameters, and mean environmental temperature in 175 fish stocks. Journal du Conseil international d'Exploration de la Mer 39: 175–192.

Poon, W. L., C. Y. Hung & D. J. Randall, 2001. The effect of aquatic hypoxia on fish. In Thurston, R.V. (eds), Fish Physiology, Toxicology and Water Quality. Environmental Protection Agency, USA, 31–50.

Roy, D., G. D. Haffner & S. B. Brandt, 2004. Estimating fish production potentials using a temporally explicit model. Ecological Modelling 173: 241–257.

Rudstam, L. G., 1988. Exploring the dynamics of herring consumption in the Baltic: applications of an energetic model of fish growth. Kieler Meeresforschungen Sonderheft 6: 312–322.

Secor, D. H. & E. J. Niklitschek, 2001. Sensitivity of sturgeons to environmental hypoxia: a review of physiological and ecological evidence. In Thurston, R. V. (ed.), Fish Physiology, Toxicology and Water Quality. Environmental Protection Agency, USA, 61–78.

Stewart, D. J. & F. P. Binkowski, 1986. Dynamics of consumption and food conversion by Lake Michigan alewives – an energetics modelling synthesis. Transactions of the American Fisheries Society 115: 643–661.

Stier, D. J. & J. H. Crance, 1985. Habitat suitability index models and instream flow suitability curves: American shad. U.S. Fish and Wildlife Service Biological Report 82, 1–34.

Struyf, E., S. Van Damme & P. Meire, 2004. Possible effects of climate change on estuarine nutrient fluxes: a case study in the highly nutrified Schelde estuary (Belgium, The Netherlands). Estuarine, Coastal and Shelf Science 60: 649–661.

Van Damme, P. A., K. Hostens & F. Ollevier, 1994. Fish species of the lower Zeeschelde (Belgium) – a comparison with historical checklists. Belgian Journal of Zoology 124: 93–103.

Van Damme, S., P. Meire, H. Maeckelberghe, M. Verdievel, L. Bougoing, E. Taverniers, T. Ysebaert & G. Wattel, 1995. De waterkwaliteit van de Zeeschelde: evolutie in de voorbije dertig jaar. Water 85: 244–256.

Van Damme, S., E. Struyf, T. Maris, T. Ysebaert, F. Dehairs, M. Tackx, C. Heip & P. Meire, 2005. Spatial and temporal patterns of water quality along the estuarine gradient of the Scheldt estuary (Belgium and the Netherlands): results of an integrated monitoring approach. Hydrobiologia 540: 29–45.

Van den Bergh, E., S. Van Damme, J. Graveland, D. de Jong, I. Baten & P. Meire, 2005. Ecological rehabilitation of the Schelde Estuary (The Netherlands-Belgium; Northwest Europe): Linking ecology, safety against floods, and accessibility for port development Source. Restoration Ecology 13: 204–214.

Vrielynck, S., C. Belpaire, A. Stabel, J. Breine & P. Quataert, 2003. De visbestanden in Vlaanderen anno 1840–1950. Een historische schets van de referentietoestand van onze waterlopen aan de hand van de visstand, ingevoerd in een databank en vergeleken met de actuele toestand. Instituut voor Bosbouw en Wildbeheer en Afdeling Water (AMINAL), Groenendaal, 271 pp.

Wannamaker C. M. & J. A. Rice, 2000. Effects of hypoxia on movements and behavior of selected estuarine organisms from the southeastern United States. Journal of Experimental Marine Biology and Ecology 249: 145–163.

Hydrobiologia (2008) 602:145–154
DOI 10.1007/s10750-008-9280-9

FISH AND DIADROMY IN EUROPE

Aspects of anadromous Allis shad (*Alosa alosa* Linnaeus) and Twaite shad (*Alosa fallax* Lacépède) biology in four Irish Special Areas of Conservation (SACs): status, spawning indications and implications for conservation designation

James J. King · William K. Roche

Abstract Four Special Areas of Conservation (SACs) have been designated in the Republic of Ireland for Twaite shad, *Alosa fallax* (Lacépède), based on recent historical information on spawning activity—the estuarine reaches of the Rivers Munster Blackwater, Suir, Barrow–Nore and Slaney. The spawning status of some of these populations is considered to have declined substantially in recent years, and no known spawning areas of Allis shad, *Alosa alosa* (L.), are known for Ireland. Data are presented confirming the presence of both Allis shad and Twaite shad in all four SACs. The majority of the material has come as by-catch from estuarine commercial salmon netsmen. Angling is the principal source of material from the R. Barrow. Examination of scale and gonad material of Allis shad indicates some degree of readiness for spawning as well as spawning having taken place. Gonadosomatic index (GSI) data from the R. Barrow for Twaite shad in the May spawning period (2000–2002) had mean GSIs of 13.7–15% with maxima in the range of 21–28%. GSI values of 18% and 22% have been recorded in the July period for individual Allis shad from the upper estuaries of the R. Slaney and R. Suir, respectively, with both fish displaying gonad maturity stage V or VI (after Pina, 2000). Low GSI values have been recorded from other Allis shad in the June–July period displaying gonad condition varying from stage III (immature) to stage VIII (spent). These findings point to a possibility that Allis shad do engage in spawning activity, whether mono-specifically or with Twaite shad. The possibility of the latter is supported by gill raker count data presented. These show clearly defined value ranges for Twaite and for Allis as well as a series of fish with intermediate counts. The apparently low population levels of Twaite shad within the SACs identify a need to examine what measures might be undertaken to ensure the favourable conservation status of this species.

Keywords Allis shad · Twaite shad · Irish SACs · Status · Gonadosomatic index · Gill raker count · Hybridization

Submitted for oral presentation to Fish Diadromy in Europe conference, 27.1.2005.

Guest editors: S. Dufour, E. Prévost, E. Rochard & P. Williot
Fish and diadromy in Europe (ecology, management, conservation)

J. J. King (✉)
Central Fisheries Board, Swords Business Campus, Balheary Road, Swords, Co., Dublin, Republic of Ireland
e-mail: jimmy.king@cfb.ie

W. K. Roche
Central Fisheries Board, Unit 4, Irish Specimen Fish Committee, Swords Business Campus, Balheary Road, Swords, Co., Dublin, Republic of Ireland

Introduction

Studies on shad in a number of European countries (Rameye et al., 1976; Claridge & Gardiner, 1978;

Aprahamian, 1981; Taverny, 1988; Assis, 1990; Thiel et al., 1996; Alexandrino, 1996) have elucidated aspects of the ecology and evolutionary history of these species. However, they have also identified the decline in shad populations and the factors contributing to this decline. Within this context, both the Allis (*Alosa alosa* L.) and the Twaite shad (*Alosa fallax* Lacépède) were included as Annex II species in the European Union's Habitats Directive of 1992. Such designation has obliged the Republic of Ireland, as a member state, to list Special Areas of Conservation (SACs) in the jurisdiction within which the conservation status of the target species is to be ensured at a favourable level. The National Parks and Wildlife Service (NPWS) implements the Habitats Directive in Ireland and designated four river corridor segments for the Twaite shad—Rivers Slaney, Barrow–Nore, Suir and Munster Blackwater (Fig. 1). As is the case in the U.K. (Hillman, 2002) no designations have been made in regard to the Allis shad as there are no known spawning sites for this species in the state.

Went (1953) reviewed the distribution of anadromous shad in Ireland, while Bracken & Kennedy (1967) provided additional records and information. Unpublished recent records compiled by Flannery and Quigley (pers. comm.) and the present authors indicate the presence of both Twaite and Allis shad from around the Irish coast, with some concentration of Twaite records in the south-east and of Allis along the western seaboard. However, numbers of records are small. The Irish SAC list for shad was compiled on the basis of former known spawning areas for Twaite shad as reported by Went (1953) and Bracken & Kennedy (1967). This absence of up-to-date scientifically based knowledge has also been identified by Maitland & Lyle (1991). The Irish approach to shad data collection as presented here parallels one currently being developed in the UK (Hillman, 2002; Hillman et al., 2003) and involves data collection from commercial sources at sea and in estuaries, from angling sources and from scientifically based surveys. Some of the findings have been presented in respect of the Slaney and Munster Blackwater SACs (King & Linnane, 2004). This study confines itself to the authors' recent findings on the ecology and reproductive status of anadromous Allis and Twaite shad in the four Irish SACs and to adjacent marine-caught material to assess the species' status, the appropriateness of the current SAC designations and the appropriateness, or otherwise, of SAC designation for Allis shad.

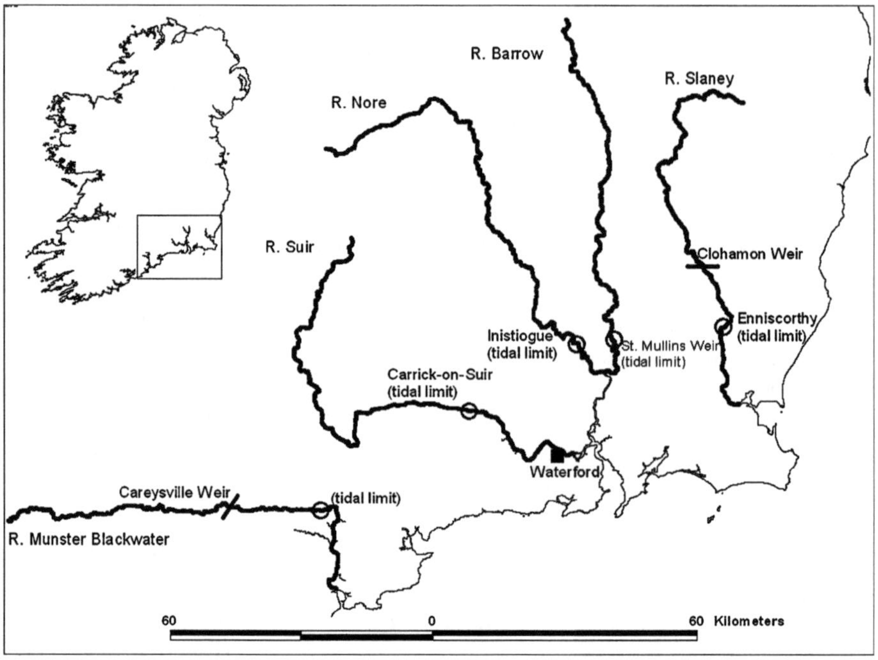

Fig. 1 Map to show the rivers studied, locations referred to in text and location of sites in Ireland

Materials and methods

Sources of material

The commercial salmon netsmen operating in the estuaries of the four SACs retained shad mortalities, taken as by-catch, and passed these to officers of the Eastern or Southern Regional Fisheries Boards (RFBs), together with information on date and location of capture, to the authors. The principal salmon-netting practises are drift, draft and snap netting (Table 1). In all cases mesh size ranges between 45 and 55 mm knot-to-knot (180–210 mm in-the-round). Drift nets function as gill nets with long sheets of netting hanging vertically in the water column suspended from a floating line at the surface. The nets are marked by buoys at either end and are allowed to travel with the tidal flow. Draft nets are set from shore, generally by small boat, in a ring to form a large bag or trap with both ends being drawn together to the shore. Snap netting is an indigenous netting technique (Wilkins, 1998) in which a net is suspended between two small canoe-like boats ('cots') and fishing is done with the tide. Guide ropes enable the fishermen to feel when fish have contacted the net, and manipulation of these ropes enables the net to be 'snapped' shut, trapping the fish in the bag thus formed. Sea fisheries staff based in commercial ports around the Irish coast were requested to be watchful for shad by-catch and to retain this material for the authors.

Drift net sampling (55-mm mesh knot-to-knot) surveys for adult shad were undertaken in the Barrow–Nore, Suir and Blackwater estuaries during April–May 2005. Drifts were of short duration (5 min) due to frequency of submerged obstructions and to capture of fish. Shad were found to be very susceptible to mortality due to stress in net entanglement and handling, and the short period of detention in nets enhanced chances of live return to the water. Snap net trials were undertaken for the present study by experienced local netsmen on the R. Suir over 5 days in the period 2nd May–2nd June 2000 at the top of the tidal limit. Rod-caught shad came from the St. Mullins area of the R. Barrow (Fig. 1), where Twaite shad ascend to spawn in May annually. Fish above a specified weight, currently set at 1.1 kg, qualify as 'specimens' by the Irish Specimen Fish Committee (ISFC, 2004). Bodies must be submitted for ratification, and all material submitted to ISFC was made available to the authors for scientific study.

Autumn sampling for juvenile shad was conducted on two occasions in each of the estuaries in the 2000–2004 period using a beach seining technique. The seine was 30 m in length, a maximum of 3 m depth in its centre and had a mesh size of 3 mm knot-to-knot. Trial sampling using paired-trawling was undertaken in autumn 2005 in the lower reaches of the Suir and Barrow estuaries and the shared waters of Waterford Harbour. The trawl had a width of 20 m when fully extended, a height of 2 m and mesh size at the cod end was 18 mm (knot-to-knot). Trawls were of 30-min duration.

Data collection from individual fish

Fish were speciated on the basis of initial examination of the gill rakers of the first gill arch. This arch was subsequently removed and gill rakers counted under binocular microscope. Counts included the

Table 1 Numbers of Twaite, Allis and presumed hybrid shad collected in the present study 2000–2005

Location	Twaite shad	Allis shad	Presumed hybrids	Principal method	Length of estuary (km)
Slaney SAC[a]	8	2	2	Draft	22
Barrow–Nore SAC	195	1	3	Angling	32
Suir SAC	15	5	1	Snap net	40
Waterford Harbour	70	23	2	Drift net	10
Munster Blackwater SAC	43	9	0	Drift net	29
Coastal Waters due S. of Waterford Harbour (5-6/2001)	0	78	5	Trawling	
	331	118	13		

[a] SAC = Special Area of Conservation, under EU Habitats Directive of 1992

small number of rakers, normally present at each extremity, not fully developed or extended. All fish were measured for fork length (with a subset measured for total length) in mm and weighed (g). Length data are presented below as fork length unless otherwise stated. Scale sampling, annotation of age and occurrence and frequency of spawning marks, in the form of erosion and non-uniform deposition of cerculi at the outer edge of scales, followed Bagliniere et al. (2000). Fish were dissected to ascertain gender, degree of gonad maturity (Pina, 2000 as in Aprahamian et al., 2003) and gonadosomatic index (gonad weight × 100/total weight). The Roman numeric I–VIII scale of Pina (2000) (as reproduced in Aprahamian et al., 2003) was used to allocate a maturity stage to gonad material examined.

Results

Species determination

Gill raker counts indicated a clear distinction between the Twaite and Allis material (Fig. 2) with a small number of specimens with intermediate counts. The Twaite material had a narrow modal peak centred on 42–43 rakers, whereas the Allis material was more widely dispersed, ranging from 99 to 134 rakers. Values lying between 50 and 90 were considered to belong to presumed-hybrid material.

Status within the Special Areas of Conservation

Both Allis and Twaite shad were collected and examined from the four Rivers Slaney, Barrow–Nore,

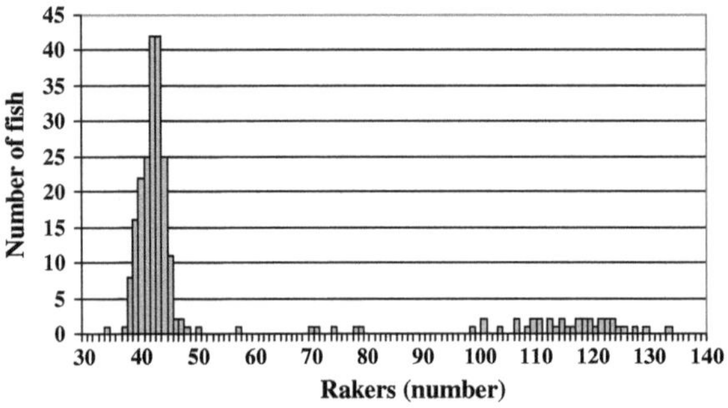

Fig. 2 Gill raker count from adult shad (>300 mm fork length, $n = 241$) taken in the R. Slaney, R. Barrow–Nore, R. Suir and R. Munster Blackwater in the period 2000–2005

Suir and Munster Blackwater in the period 2000–2005 and from Waterford Harbour, the estuarine area formed by the convergence of the R. Barrow and R. Suir estuaries. In addition, a marine sample of Allis shad, taken as by-catch due south of Waterford Harbour, was examined. An overview of numbers, by species, waterbody and sampling source (Table 1) indicated the predominance of angling-caught fish on the Barrow and of commercially taken fish on the R. Slaney, R. Suir and R. Blackwater.

Twaite shad records were most numerous from the R. Barrow, due to angler catches taken in May at the spawning area at St. Mullins, 32 km up the estuary from the confluence with the R. Suir estuary. Drift net sampling by the authors in April–May 2005 identified the presence of Twaite shad along the full length of the estuarine R. Barrow channel, above the R. Nore confluence (Fig. 2). By comparison, such sampling at the top of the tidal limit in the R. Suir and R. Munster Blackwater in 2005 yielded very low numbers. The majority of the Twaite shad sample on the R. Suir and R. Munster Blackwater was taken by commercial salmon netsmen. Numbers taken as by-catch depended to some extent on the status of the Atlantic salmon run and, as shad were frequently taken alive in snap nets, the netsmen operated a policy of releasing these fish. Snap net trials on the R. Suir, undertaken by experienced local netsmen, yielded a total of four shad in 36 h of netting effort spread over 5 days in the period 2nd May–2nd June 2000 at the top of the tidal limit, suggesting low numbers of adult shad. The lowest returns of shad came from the R. Slaney, where draft netting, exclusively, is used by commercial salmon netsmen (Table 1).

Allis shad were taken in all four SAC estuaries in very low numbers. Almost all material was taken as commercial salmon by-catch. The majority of these were taken in the shared tidal waters downstream of the Barrow–Suir confluence, either by salmon drift nets in June–July or by bottom-fishing gill nets in autumn–winter, set for marine species. A single fish, in prime spawning condition, was taken by angling in the R. Barrow in May 1998. Two fish were taken by angling, in 2003 and 2005, on the R. Munster Blackwater 25 km into freshwater above the tidal limit (Fig. 1). Samples of Allis shad were confined to the lower reaches of the Slaney estuary.

Indications of spawning success, however small, were provided by juvenile Twaite shad taken in the present studies. Autumn beach seine sampling yielded a single 0-year-old Twaite shad, only, on the R. Suir in 2000. A greater degree of success was achieved in paired, bottom-trawling operations in October 2005, when two 0-group fish were taken in the R. Suir and 20 in the R. Barrow from a total of four 30-min trawls. A sample of immature Twaite shad, 200–240 mm, was taken in one area of the R. Munster Blackwater during scientific netting in May 2003 (King & Linnane, 2004).

The trawler-caught samples from May–June 2001 consisted of 78 fish, with 55 fish taken by one trawler on one day, and were taken due south of Waterford Harbour in marine waters. These samples consisted of Allis shad with a small amount of presumed hybrids and were landed at the commercial fishing port of Dunmore East, at the mouth of Waterford Harbour. Similar large samples were not subsequently reported from here.

Biometry

The length frequency distribution of fish reflected the various sources of available data (Fig. 3)—with all fish less than 300 mm being the product of specific scientific samplings. Adult Twaite shad ranged from 320 to 480 mm. Both male and female fish were found across the full size range, although female fish were predominant over 400 mm. This was due to the large size of the rod-caught sample from St. Mullins on the R. Barrow, with ripe female fish being in the majority of material submitted to the Irish Specimen Fish Committee. No Twaite shad longer than 500 mm was captured. Allis shad ranged from 360 to 580 mm with both genders represented across the full size range. The small number of presumed-hybrid shad examined spanned a length range of 360–470 mm. The 2001 marine-caught sample of 78 fish, consisting almost entirely of Allis shad, ranged in size from 270 mm to 410 mm (Fig. 4). The pronounced modal peak at 320 mm may have indicated a degree of net selectivity, with smaller fish less vulnerable to capture. There was a marked difference in length distribution between this marine-caught sample and the combined samples of Allis shad from the four rivers studied.

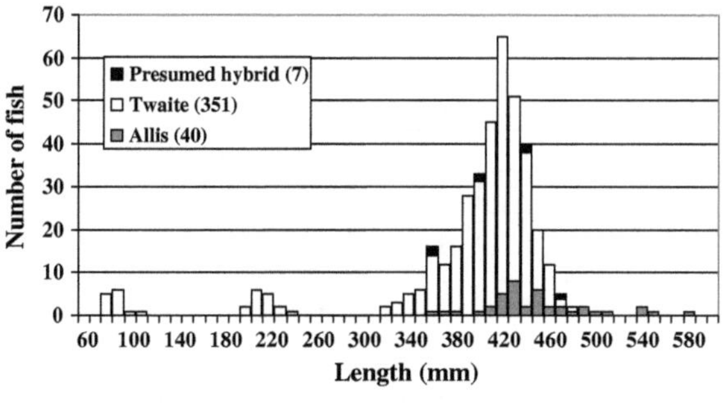

Fig. 3 Length frequency of Allis, Twaite and presumed-hybrid shad examined in the R. Slaney, R. Barrow–Nore, R. Suir and R. Munster Blackwater, 2000–2005

Age

The Twaite shad sample analysed consisted of fish from the R. Barrow SAC, only, whereas the Allis shad sample consisted of material from the other three waters. The Twaite shad sample showed a marked sexual difference in age distribution (Fig. 5a) with male fish ranging in age from 4 to 7 years and females ranging from 4 to 10. The male sample had a modal peak at 5 years whereas the female sample had a modal peak at 7 years. A large proportion of the female fish were 'specimen' shad, representing larger fish as opposed to the full size and age range present in the spawning waters on the R. Barrow. In contrast, the Allis shad sample had almost equal numbers of male and female fish, and both genders spanned a similar age range (Fig. 5b). The majority of male fish were 4 years of age, while the female modal peak was at 5 years of age.

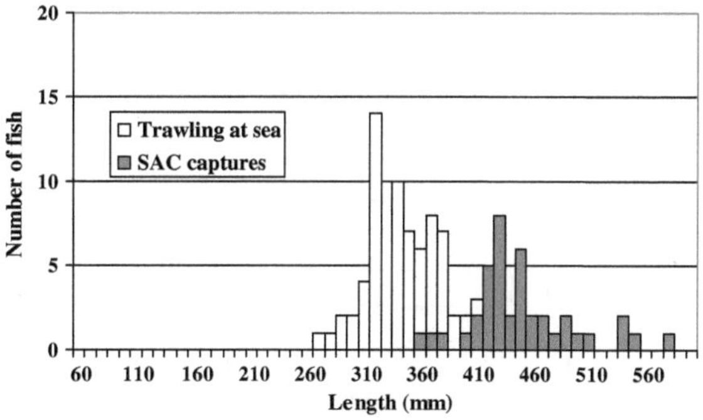

Fig. 4 Comparison of length frequency of Allis shad from the R. Slaney, R. Barrow–Nore, R. Suir and R. Munster Blackwater (2000–2005), with captures from marine trawling in early summer 2001. (SAC = Special Area of Conservation)

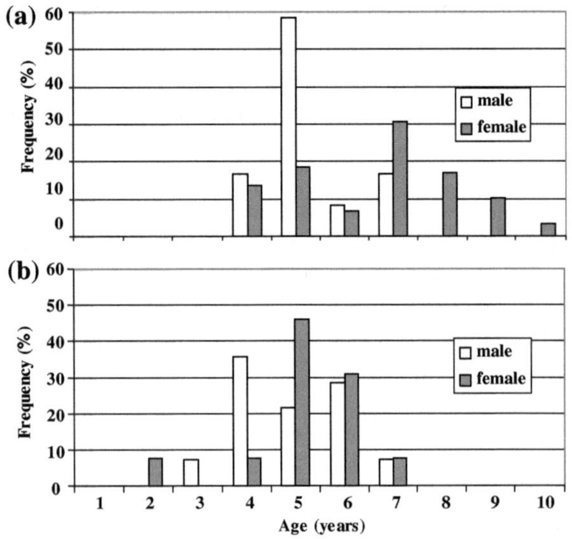

Fig. 5 Age distribution of (**a**) Twaite shad (12 male, 59 female) and (**b**) Allis shad (13 males, 14 females). Twaite shad captured in R. Barrow; Allis shad captured in R. Slaney, R. Nore, R. Suir and R. Munster Blackwater (2000–2005)

The Twaite shad sample showed clear evidence of repeat spawning with approximately 70% of male and female fish having at least two spawning marks (Fig. 6a). A small number, only, of virgin fish was recorded. The pattern of frequency of prior spawning differed between the sexes. The male fish had a modal peak at two previous spawning events whereas the female sample contained almost equal numbers of fish with 2-, 3- and 4 previous spawning marks. In excess of 70% of both male and female Allis sample had not spawned previously. However, among a small proportion of Allis shad examined, of both genders (Fig. 6b), there was clear evidence of prior spawning, in the form of erosion and non-uniform deposition of cerculi at the outer edge of scales.

A small sample, only, of scales from presumed-hybrid shad was available, and examination indicated evidence of previous spawning.

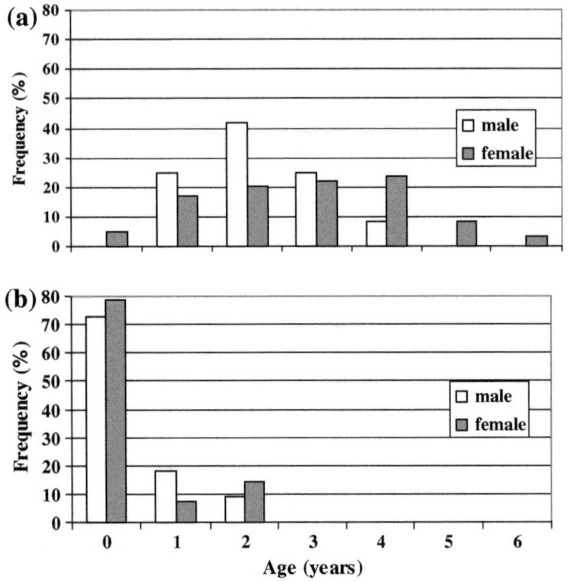

Fig. 6 Frequency of spawning of (**a**) Twaite shad (12 male, 59 female) and and (**b**) Allis shad (13 males, 14 females). Twaite shad captured in R. Barrow; Allis shad captured in R. Slaney, R. Nore, R. Suir and R. Munster Blackwater (2000–2005)

Gonadosomatic Index (GSI) in female fish

In the April–June period the majority of Twaite shad was taken in the upper reaches of the estuarine SACs, at or close to traditional spawning grounds. Mean GSI values ranged from 11 to 14%, with maxima exceeding 18% in April and reaching 25% in both May and June (Fig. 7). Mean GSI declined from June, with the majority of fish taken in the lower reaches of the estuarine SACs. The female Allis shad sample size was small ($n = 16$). Mean GSI values

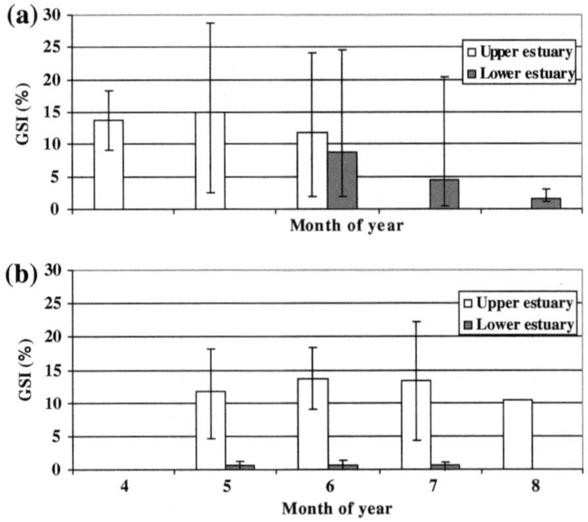

Fig. 7 Gonadosomatic Index (GSI) (Mean and max–min) for (**a**) female Twaite shad ($n = 165$) and (**b**) female Allis shad ($n = 16$) for the April–August period (2000–2005). Twaite shad captured in R. Barrow; Allis shad captured in R. Slaney, R. Nore, R. Suir and R. Munster Blackwater

exceeded 10% each month from May to August in the upper reaches of the estuaries, while GSI values less than 1% were recorded simultaneously in the lower reaches of the same waters (Fig. 7). Maximum values in excess of 18% were recorded in each month from May to July.

Gonad maturity stage in female fish

This complemented the GSI data collected as, while samples with high GSI values would clearly indicate a level of gonad development, low GSI values could indicate either immature or spent gonad. Twaite shad of maturity stages IV–V were present during most of the April–August period (Table 2). These fish were initially present in upper estuarine reaches but were found in lower reaches from June onwards. These latter data may indicate a degree of preparedness for late spawning or may indicate that not all ripe/ripening fish do actually spawn. Condition values of VII, indicative of spent/recovering fish, were recorded from June to August. The small sample of female Allis shad indicated the presence of mature fish from May to August in upper estuarine reaches (Table 2). The maximum value in July was indicative of a spent fish (stage VIII).

Table 2 Gonad maturity stage (after Pina, 2000) for female Twaite shad ($n = 135$) from R. Barrow and for Allis shad ($n = 12$) from R. Nore, R. Suir and R. Munster Blackwater taken in the April–August period (2000–2005)

	Gonad maturity Stage	Month of year				
		April	May	June	July	August
Twaite shad	I				4	1
	II			1	11	
	III		**1**	4	8	
	IV	**1**	**29**	***5***	5	4
	V	**1**	**39**	***7***	5	
	VI		**3**			
	VII			***2***		2
	VIII				2	
	Total number	2	72	19	35	7
Allis shad	I			2	2	
	II		1	1		
	III					
	IV					
	V		**1**	**2**	**1**	**1**
	VI					
	VII					
	VIII				1	
	Total number	0	2	5	4	1

Samples from lower part of estuaries in normal case; upper parts in bold; samples with fish from both parts are in bold italics

Discussion

The four Irish waters designated for Twaite shad conservation have been considered, historically, as spawning areas for this species (Went, 1953; Bracken & Kennedy, 1967). Fahy (1982) reported the small-scale commercial exploitation of shad (presumed to be Twaite) from the R. Slaney in the period 1965–76. Annual evidence from angling-caught fish in May is indicative of the ongoing presence of a spawning Twaite shad population on the R. Barrow (ISFC, 2004). The capture of young-of-year Twaite shad on the R. Barrow and R. Suir and of 2+ year old Twaite shad in the Blackwater (King & Linnane, 2004) is indicative of some successful spawning occurring here. However, anecdotal reports from anglers and netsmen from the R. Suir and R. Slaney indicate a very significant decline in shad numbers over the last 20 years. Previous references to Allis shad in the four SACs appear confined to those found by Doherty & Mc Carthy (2002) in Waterford Harbour—the lower reaches of the shared Barrow–Nore–Suir estuary.

Evidence from scales, from GSI values and gonad maturity stage as well as the recording of presumed hybrids indicate that shad of both species were in spawning condition in all four Irish SACs. Given the high degree of fidelity to natal waters exhibited by shad species on the American east coast (Hendricks et al., 2002) and the genetic indicators pointing to natal fidelity (Alexandrino, 1996), it is suggested that Twaite shad spawning, either mono-specifically or with Allis shad, in the four Irish SAC waters examined here are the progeny of Irish parentals. This may also be true of Allis shad taken, contrasting with the view of these fish as vagrants, moving around coastal waters and wandering into estuaries (Doherty & Mc Carthy, 2002). Support for these allis shad being of Irish origin lies with the ripe and spent condition of some fish, the occurrence of presumed hybrid fish in spawning condition and the presence of presumed hybrids with scale evidence of prior spawning. Bracken & Kennedy (1967) reported trawler catches of adult Allis shad (260–473 mm), in Dingle Bay in the Irish southwest, in the February–April period ($n = 36$) and again in July–August ($n = 39$) whereas two fish, only, were taken in May and June. This prompted Bracken & Kennedy (1967) to postulate that spawning might take place in one of the rivers discharging to Dingle Bay. The recording of individual allis shad in two separate years 25 km into freshwater on the Munster Blackwater may also be more than a mere coincidence of vagrancy. Maitland & Lyle (2001), based on their evidence from the Solway in Scotland, proposed that either the Allis shad move into estuaries in search of spawning grounds or that they follow Twaite shad inshore to common spawning grounds where they may spawn, either mono-specifically or with Twaite shad. Those authors found hybrids in their studies also. Viable hybrids between the two taxa have been reported previously (Boisneau et al., 1992; Alexandrino et al., 1996), and the presence of spawning marks on scales of some presumed hybrids from the present study points to the ability of hybrids to spawn successfully.

The occurrence of hybrids may be indicative of a pressure, either in the form of low population levels or of restrictions, natural or man-made, preventing access to spawning (Rameye et al., 1976; Manyukas, 1989; Menesson-Boisneau et al., 1993). Carstairs (2000) concluded that such restriction may have led to interbreeding with a consequent introgression or loss of genetic uniqueness of the less well-represented species and that this may have led to the decline of, and in some cases demise of, Allis shad in some UK channels. Allis shad have a pronounced capacity to travel up estuaries and deep into freshwater to spawning grounds (Aprahamian, 1981; Manyukas, 1989). Such capacity is obviously facilitated, in part, in the R. Munster Blackwater but is impeded in both the R. Barrow and R. Suir by navigation weirs and locks and by an area of steep bed slope with stony, shallow bed type, respectively. Improvement in fish passage facilities in major Irish channels could permit a spatial, and hence genetic, separation of Allis and Twaite shad in the same catchment. It would be imperative that the upstream channel provide suitable spawning habitat including extensive areas of fast-flowing shallows over cobble and gravel as well as pool areas and backwaters (Maitland & Hatton-Ellis, 2003). Such terrain is present in the R. Nore and R. Suir, whereas dredging and navigation weirs on the riverine R. Barrow render it less suitable as spawning habitat. The addressing of such obstructions may be required under both Water Framework Directive and Habitats Directive. Such measures would provide ‘infrastructural support’ for the shad species but could only be beneficial to shad conservation if shad were available to penetrate these new areas in sufficient numbers to find mates and engage in successful spawning.

Shad by-catch has been the principal source of information for this study, as with Maitland & Lyle (2001) and Hillman (2002), indicating susceptibility to capture during the licensed salmon-netting season (May–July) in Irish estuaries. Doherty et al. (2004) considered that levels of shad by-catch in the estuarine waters examined here were low, and this contention is supported by the snap net trials in the present study. Shoaling species such as shad may be particularly susceptible to marine capture. The marine sample reported here may reflect this vulnerability, with 55 fish taken by one trawler on one day. This sample was taken due south of the entry to the Barrow–Nore and Suir SACs and due west of the Bristol Channel, into which major English and Welsh SACs for Twaite shad discharge. The pronounced difference in length frequency distribution between this marine sample and those from within the Irish SACs at the same time of year may indicate a temporal separation of age groups, with larger fish moving into estuarine waters. Such a separation might also indicate a deliberate spawning movement of larger Allis shad into estuarine waters in the May–June period, as previously postulated by Bracken & Kennedy (1967). Major loss of pre-spawning or virgin adult fish would be particularly damaging in the case of Allis shad, shown to be predominantly semelparous. Marine interception may have even more widespread implications if mixing of shoals from different geographic areas occurs at sea, as could be the case off the Irish south-east coast. While management of commercial netting at sea with a view to the prevention of by-catch, such as shad, may be difficult it is conceivable that effort could be implemented to reduce by-catch in estuaries. The Scientific Committee of the Irish National Salmon Commission considered that the precautionary method to meet national and international objectives was to operate salmon fisheries on individual river stocks shown to be exceeding Conservation Limits (Anon, 2006). This would shift exploitation from open sea to individual rivers. In many cases, commercial exploitation would not be feasible in these rivers due to their not exceeding their Conservation Limits (Anon, 2006). Such restrictions on commercial netting would have, potentially, positive implications for shad survival in spawning migrations.

While the Irish Environmental Protection Agency has classified most of the estuarine segments of the current shad SACs as being eutrophic (EPA, 2001), both species of shad continue to be taken in these waters, albeit in small numbers, as well as smelt (*Osmerus eperlanus* L.) and Atlantic salmon (*Salmo salar* L.). Implementation of the EU Water Framework Directive, the Nitrates Directive and the Urban Wastewater Directives should be conducive to the improvement of water quality in Irish rivers and estuaries, which could, in turn, be conducive to shad conservation in the four Irish SACs. Improved water quality has led to the upstream displacement of the major shad spawning areas in the R. Elbe (Thiel et al., 1996) and contributed to the re-appearance of shad in the Thames (Colclough et al., 2002) and the Clyde (Maitland & Lyle, 2001). While none of the four current Irish SACs is designated for Allis shad, it is likely that, if Twaite shad reach a satisfactory conservation status within these waters then the water quality and habitat conditions would be conducive for Allis shad to attain a similar status.

The present study has confirmed the presence of shad of both species in reproductive condition in all four Special Areas of Conservation designated for Ireland, but it is evident that the population size of both species is low. Improvement of fish passage facilities and enhanced water quality conditions could contribute to the well-being of both species. However, the low level of presence recorded may indicate a more fundamental problem of diminished spawning population, in the case of the Twaite shad, and of extremely low numbers, in the case of Allis shad. In such circumstances it may be necessary to explore such measures as reduction of exploitation or by-catch, both at sea and in estuaries, and to assess the feasibility of a husbandry programme to augment the impoverished stocks to the level of favourable conservation status—a level required under the Habitats Directive.

Acknowledgements Financial support for some of the investigations reported here was provided by National Parks and Wildlife Service, Dept. of Environment, Heritage and Local Government and by the Office of Public Works. Thanks are due to management and staff of the Southern and Eastern Regional Fisheries Boards for assistance in surveys and in liaison with netsmen. The commercial salmon netsmen on the four SACs are thanked for their support in providing samples of moribund by-catch, which formed the basis of the sampling regime. The Irish Specimen Fish Committee and the committee of the St. Mullins Angling Club made specimens of R. Barrow fish available for examination. Miran Aprahamian is thanked for his advice and support and provision of access to review literature. Colleagues and summer students in Central Fisheries Board assisted in data collation.

References

Alexandrino, P., 1996. Genetic and morphological differentiation among some Portuguese populations of Allis shad *Alosa alosa* (L., 1758) and Twaite shad *Alosa fallax* (Lacépède, 1803). Publicacion Especial Instituto Espanol de Oceanografica 21: 15–24.

Alexandrino, P., N. Ferrand & J. Rocha, 1996. Genetic polymorphism of a haemoglobin chain and adenosine deaminase in European shads: evidence for the existence of distinct genetic entities with natural hybridisation. Journal of Fish Biology 48: 447–456.

Anon, 2006. The status of Irish salmon stocks in 2005 and Precautionary Catch Advice for 2006. Report of the Standing Scientific Committee of the National Salmon Commission, Dublin, 34 pp.

Aprahamian, M. W., 1981. Aspects of the biology of the Twaite shad (*Alosa fallax*) in the Rivers Severn and Wye. Proceedings of the 2nd British Freshwater Fish Conference, 373–381.

Aprahamian, M. W., C. D. Aprahamian, J. L. Bagliniere, R. Sabatie, & P. Alexandrino, 2003. *Alosa alosa* and *Alosa fallax* spp.: literature review and bibliography. R&D Technical Report W1-014/TR. Environment Agency, Bristol, UK, 349 pp.

Assis, C. A., 1990. Threats to the survival of anadromous fishes in the R. Tagus in Portugal. Journal of Fish Biology 37(suppl A): 225–226.

Bagliniere, J. L., M. R. Sabatie, M. W. Aprahamian, P. Alexandrino, C. D. Aprahamian, C. A. Assis, J. J. Cassou-Leins, M. Le Corre, C. Mennesson-Boisneau, D. Martin-Vandembulcke, E. Rochard & C. Teixeira, 2000. A guide to scale interpretation and age estimation for the east-Atlantic and west-Mediterranean Shads (*Alosa* spp.). Bulletin Francais de la Peche et de la Pisciculture 359: 545–587.

Boisneau, P., C. Menesson-Boisneau & R. Guyomard, 1992. Electrophoretic identity between Allis shad, *Alosa alosa* (L.), and Twaite shad, *Alosa fallax* (Lacépède). Journal of Fish Biology 40: 731–738.

Bracken, J & M. Kennedy, 1967. Notes on some Irish estuarine and inshore fishes. Irish Fisheries Investigations, Series B (Marine), No. 3. 28 pp.

Carstairs, M., 2000. The ecology and conservation of Twaite and Allis Shad. British Wildlife 11: 159–166.

Claridge, P. N. & Gardner D. C., 1978. Growth and movements of the Twaite shad, *Alosa fallax* (Lacépède), in the Severn estuary. Journal of Fish Biology 12: 203–211.

Colclough, S., G. Gray, A. Bark & B. Knights, 2002. Fish and fisheries of the tidal Thames: management of the modern resource, research aims and future pressures. Journal of Fish Biology 61(A): 64–73.

Doherty, D. & T. K. McCarthy, 2002. Aspects of the ecology, parasites and future conservation of Twaite shad (*Alosa fallax*) and Allis shad (*Alosa alosa*) in south-eastern Ireland. In Collares-Pereira, M. J., I. G. Cowx & M. M. Coelho (eds), Conservation of Freshwater Fishes: Options for the Future. Fishing News Books, Oxford, 98–112.

Doherty, D., N. O Maoleidigh & T. K. McCarthy, 2004. The biology, ecology and future conservation of Twaite shad (*Alosa fallax* Lacépède), Allis shad (*Alosa alosa* L.) and Killarney shad (*Alosa fallax* Killarnensis Tate Regan) in Ireland. Biology and Environment: Proceedings of the Royal Irish Academy 104 B, 93–102.

EPA, 2001. An Assessment of the Trophic Status of Estuaries and Bays in Ireland. EPA Wexford, 12 pp.

Fahy, E., 1982. A commercial net fishery taking Twaite shad *Alosa fallax* (Lacépède) in the estuary of the river Slaney. The Irish Naturalist's Journal 20: 498–500.

Hendricks, M. L., R. L. Hopes, D. A. Arnold & M. L. Kaufmann, 2002. Homing of hatchery-reared American shad to the Lehigh River, a tributary to the Delaware River. North American Journal of Fisheries Management 22: 243–248.

Hillman, R., 2002. The distribution, biology and ecology of shad in South-West England. R&D Technical Report W1-047. Environment Agency, Bristol, UK, 148 pp.

Hillman, R. J., I. G. Cowx & J. Harvey, 2003. Monitoring the Allis and Twaite shad. Conserving Natura 2000 Rivers Monitoring Series No. 3. English Nature, Peterborough, 23 pp.

ISFC, 2004. Annual Report of the Irish Specimen Fish Committee. Dublin, 64 pp.

King, J. J. & S. M. Linnane, 2004. The status and distribution of lamprey and shad in the Slaney and Munster Blackwater SACs. Irish Wildlife Manuals, No. 14. National Parks and Wildlife Service, Dept. of Environment, Heritage and Local Government, Dublin Ireland, 74 pp.

Maitland, P. S. & T. W. Hatton-Ellis, 2003. Ecology of the Allis and Twaite shad. Conserving Natura 2000 Rivers Ecology Series No. 3. English Nature, Peterborough, 28 pp.

Maitland, P. S. & A. A. Lyle, 1991. Conservation of freshwater fish in the British Isles: the current status and biology of threatened species. Aquatic Conservation: Marine and Freshwater Ecosystems 1: 25–54.

Maitland, P. S. & A. A. Lyle, 2001. Shad and smelt in the Cree Estuary, South West Scotland. Scottish Natural Heritage Research, Survey and Monitoring Report no. 6, 92 pp.

Manyukas, Y. L., 1989. Biology of the Atlantic shad, *Alosa fallax fallax*, in Kurshiskiy Bay. Journal of Ichthyology 29: 125–128.

Mennesson-Boisneau, C., P. Boisneau & J. L. Bagliniere, 1993. Migration de reproduction des aloses dans le bassin de la Loire. Incidents des obstacles. Bulletin scientifique et technique. Departement d'Hhydrobiologie, Institut de la Recherche Agronomique (France). No. 27, 23 pp.

Pina, T. A. S., 2000. Aspectos da reproducao de savelha, *Alosa fallax fallax* (Lacépède 1803), nos rios Mira e Guadiana. Teste de Mestre. Universidade de Coimbra, Coimbra, 85 pp + anexos.

Rameye, L., A. Kiener, C. P. Spillmann & J. Biousse, 1976. Aspects de la biologie de l'alose du Rhone. Peche et difficultes croissantes de ses migrations. Bulletin Francais de Pisciculture 263: 50–76.

Taverny, C., 1988. An attempt to estimate *Alosa alosa* and *Alosa fallax* juvenile mortality caused by three types of human activity in the Gionde estuary, 1985-1986. In van Densen, W. L. T., B. Steinmetz & R. H. Hughes (eds), Management of Freshwater Fisheries. Pudoc, Wageningen, 215–229.

Thiel, R., A. Sepulveda & S. Oesmann, 1996. Occurrence and distribution of Twaite shad (*Alosa fallax* Lacépède) in the lower Elbe River, Germany. Conservation of endangered fish species in Europe, Birkhaeuser Verlag, Basel, 157–170.

Went, A. E. J., 1953. The status of the shads, *Alosa finta* and *A. alosa* Cuvier, in Irish waters. The Irish Naturalists Journal 11: 8–11.

Wilkins, N. P., 1998. Men, Times and Tides: Snap-netting on the Barrow, Nore and Suir. Southern Regional Fisheries Board, Clonmel, 64 pp.

Hydrobiologia (2008) 602:155–159
DOI 10.1007/s10750-008-9292-5

Potential re-establishment of diadromous fish species in the River Scheldt (Belgium)

D. Buysse · J. Coeck · J. Maes

Abstract This study (2002) documents on the inland penetration off diadromous fish species into the tidal and non-tidal part of the River Scheldt and on the impact of two lock-weir complexes. Long-term trends in oxic conditions show the river is undergoing a natural recovery process, nevertheless five of ten diadromous species recorded were still restricted to the brackish part of the estuary. Despite poor conditions, five species reached the upper freshwater part of the estuary. Erratic free-flowing conditions at the tidal weir offered limited opportunities for some diadromous species to migrate into the non-tidal part of the river. Upstream migration over the second barrier is almost completely blocked. Rehabilitation schemes to restore self-sustaining populations of diadromous fish in the cross-border River Scheldt need to include the building of fish bypasses and improved wastewater treatment and habitat restoration programmes.

Guest editors: S. Dufour, E. Prévost, E. Rochard & P. Williot
Fish and diadromy in Europe (ecology, management, conservation)

D. Buysse (✉) · J. Coeck
Research Institute for Nature and Forest, Kliniekstraat 25, Brussels 1070, Belgium
e-mail: david.buysse@inbo.be

J. Maes
Laboratory of Aquatic Ecology, Katholieke Universiteit Leuven, Ch. De Bériotstraat 32, Leuven 3000, Belgium

Keywords Diadromous fish · Upstream migration · Pollution · Dissolved oxygen · Migration barriers · Habitat fragmentation

The status of the 14 diadromous fish species in Flanders (northern part of Belgium) is critical according to the Red List of Vandelannoote & Coeck (1998), which is based upon the IUCN (1994) criteria. Eight species are extinct (European Atlantic sturgeon *Acipenser sturio* L., allis shad *Alosa alosa* L., twaite shad *Alosa fallax* Lacepède, European whitefish *Coregonus lavaretus* L., houting *Coregonus oxyrhynchus* L., sea lamprey *Petromyzon marinus* L., Atlantic salmon *Salmo salar* L., sea trout *Salmo trutta* L.), four are rare (i.e. river lamprey *Lampetra fluviatilis* L., smelt *Osmerus eperlanus* L., thin-lipped grey mullet *Liza ramada* Risso and flounder *Platichthys flesus* L.) and two species are currently considered not to be threatened (i.e. three-spined stickleback *Gasterosteus aculeatus* L. trachurus form and eel *Anguilla anguilla* L.). However the latter species is considered as "outside safe biological limits" (ICES, 1999) and it is planned by the European Union to decide on emergency steps to protect the eel in the autumn of 2006.

Estuaries provide a vital migratory route for different species. Blockage of this route through migration barriers or other environmental impacts, such as pollution, can prevent species to reach crucial habitats and therefore disrupt the completion of their life cycles. In several European countries all diadromous species, except for the European eel, have become extinct as a result of habitat fragmentation (Lucas & Baras, 2001).

In the late 1970s symptoms such as temporal anoxia and organism death were common features in the upstream part of the River Scheldt (Soetaert et al., 2006). The start of water treatment programmes around this period, however, resulted in an improvement of the water quality (Soetaert & Herman, 1995) and as a consequence fisheries researchers gained interest in the brackish region. Maes et al. (1998a) recently caught six diadromous species in the brackish water part of the Scheldt estuary (i.e. eel, flounder, river lamprey, thin-lipped grey mullet, smelt and twaite shad, which was believed to be extinct).

Up till recently the upper freshwater tidal zone has faced a lack of interest by fisheries researchers and is therefore poorly documented. In this article, we report on a year-round (2002) study and we illustrate how far upstream the diadromous species penetrate into the River Scheldt. Water quality and opportunities for diadromous species to negotiate the first two migration barriers are discussed.

The estuary of the river Scheldt extends from the mouth at Vlissingen (km 0—The Netherlands) to Ghent (km 160—Belgium), where the first river obstruction, a lock-weir complex, impairs the tidal wave in the Upper Scheldt (Fig. 1). It is one of the few remaining estuaries in Europe with an exceptional tidal freshwater area (Meire et al., 2005). The non-tidal part of the River Scheldt between the French–Belgian border and Ghent is strongly fragmented by weirs equipped with shipping locks (Asper—km 170, Oudenaarde—km 180,...).

Fish were sampled with fyke nets (i.e. double traps). Each trap had a length of 5 m with a mesh size of 8 mm in the latter part. Traps were emptied after approximately 24 h.

Detailed information on sampling is given in Fig. 1. One sampling point was situated in the brackish part while the other sampling points covered the entire freshwater part of the estuary. Since migrating fish accumulated in front of the weirs, two traps were set in the main channel right under the weirs of Ghent, Asper and Oudenaarde to assess the barrier effect of the first two complexes. Handling of the fish was restricted to a minimum. All fish were identified, measured, weighted and released. The three-spined stickleback trachurus forms were distinguished from other three-spined stickleback forms and river lampreys were sexed if possible.

The River Scheldt is a typical rain-fed lowland river with maximum discharge during winter months (maximum daily discharge recorded in the Upper Sea

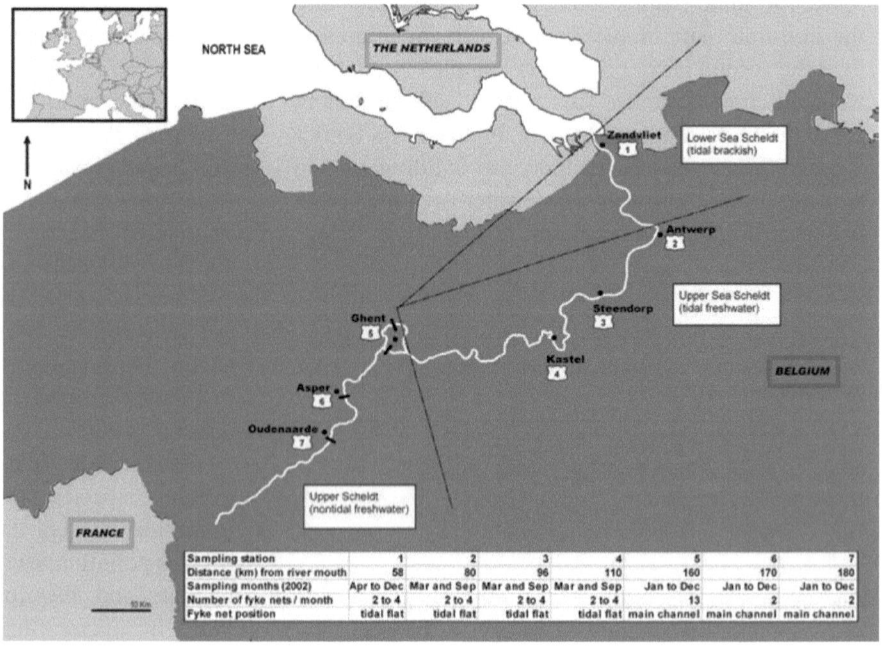

Sampling station	1	2	3	4	5	6	7
Distance (km) from river mouth	58	80	96	110	160	170	180
Sampling months (2002)	Apr to Dec	Mar and Sep	Mar and Sep	Mar and Sep	Jan to Dec	Jan to Dec	Jan to Dec
Number of fyke nets / month	2 to 4	2 to 4	2 to 4	2 to 4	13	2	2
Fyke net position	tidal flat	tidal flat	tidal flat	tidal flat	main channel	main channel	main channel

Fig. 1 Sampled section of the River Scheldt (Belgium). From upstream to down-stream there are: the Upper Scheldt, the Upper Sea Scheldt (under the influence of dynamic tide) called tidal freshwater, and so on. Lock-weir complexes are located by ▌ and sampling sites by •. Map showing the estuary, non-tidal part and nomenclature of the River Scheldt, the location of the tidal lock-weir complex at km 160, consecutive lock-weir complexes at km 170 and 180 and all sampling points in the brackish (km 58), freshwater (km 80, 96, 110 and 160) and non-tidal part (km 170, 180) of the river

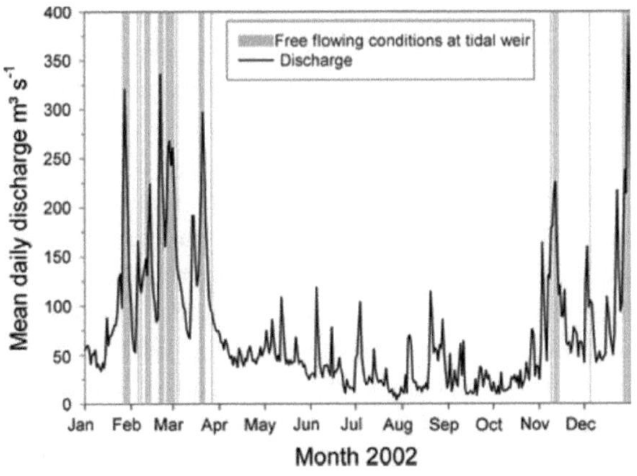

Fig. 2 Mean daily discharge (m^3 s^{-1}) off the River Scheldt at Ghent in 2002 and days when free-flowing conditions occurred for at least 4 h a day at the tidal weir of Ghent indicated by |

Scheldt at Ghent: 395.2 m^3 s^{-1} in December) (Fig. 2). During events of heavy rainfall, when water from the Upper Scheldt Basin needs to be evacuated rapidly along the tidal weir (km 160), free-flowing conditions occur by lifting both weir-valves. In 2002, free-flowing conditions at the tidal weir occurred for at least 4 h a day during 33 days (Fig. 2). This suggests that in only 9% of the days during the year 2002 free-flowing conditions occurred, which is even an overestimation since the event of free-flowing conditions did not always last for the entire 24 h. These free-flowing conditions were only observed during November up to March. Free-flowing conditions never occurred at weirs at km 170 and 180 because of different management: a permanent water level difference (>2 m) up- and down-stream of these weirs is maintained in function of navigation. Fish passage is possible only via the shipping locks, but due to a lack of attraction flow fish passage through the shipping lock is accidental.

A total of 48 fish species was recorded. The number of diadromous species per 24-h per fyke net (N/24 h) recorded at up- and down-stream sampling points is given in Table 1. Most diadromous species (n = 10) were recorded at Zandvliet (km 58), five species, i.e. twaite shad, thin-lipped grey mullet, Atlantic salmon, sea trout and Siberian sturgeon were

Table 1 List of diadromous species caught at sampling points in the brackish (km 58) and freshwater (km 80, 96, 110 and 160) part of the Scheldt estuary and in the fragmented part of the River Scheldt at sampling points down-stream the second (km 170) and third (km 180) lock-weir complex between January and December 2002

Scientific name	Common name	Life cycle category	Zandvliet (km 58)	Antwerp (km 80)	Steendorp (km 96)	Kastel (km 110)	Ghent (km 160)	Asper (km 170)	Oudenaarde (km 180)
Acipenser baerii	Siberian sturgeon	A	<0.1						
Alosa fallax	Twaite shad	A	0.4						
Anguilla anguilla	Eel	C	0.1	1.3		1.5	2.4	4.5	2.4
Gasterosteus aculeatus[a]	Three-spined stickleback[a]	A[a] and F[a]	<0.1	0.5		0.3	1.1	9.2	6.0
Lampetra fluviatilis	River lamprey	A	<0.1				0.9	7.3	0.7
Liza ramada	Thin-lipped grey mullet	C	<0.1	0	0	0	0	0	0
Osmerus eperlanus	Smelt	A	0.2					<0.1	
Platichtys flesus	Flounder	FC	6.3	16.5	5.5		17.5		
Salmo salar	Atlantic salmon	A	<0.1						
Salmo trutta	Sea trout	A	<0.1						

Life cycle categories according to Thiel & Potter (2001) (F = freshwater, S = saltwater or marine straggler, C = catadromous, A = anadromous, O = marine estuarine opportunist) except for flounder (FC = facultative catadromous) and thin-lipped grey mullet (C = catadromous (Béguer et al. 2007)), and mean number per 24 h (N/24 h) based on the total catch are presented

[a] Mean number per 24 h for the three-spined sticklebacks trachurus (A), semi-armatus and leiurus (F) form are summed

limited to this down-stream sampling point. Exotic Siberian sturgeons are probably escaped fish farm individuals. At Antwerp (km 80), Steendorp (km 96) and Kastel (km 110) the recorded number of diadromous species was lower (respectively $n = 3$, $n = 1$ and $n = 2$) due to limited sampling efforts.

Investigation of the oxygen conditions between the mid-1960s and now show that the oxygen concentrations were always highly undersaturated in the riverine part (6–35% saturation) (Soetaert et al., 2006) although a slow improvement of the average oxygen saturation can be noted from the 1980s onward. In summer, oxygen concentrations often reach very low levels in the most upstream part (Herman & Heip, 1999). As observed during our sampling in September 2002 temporarily anoxic conditions at the upstream freshwater sampling points cause fish mortality. The absence of salmon, thin-lipped grey mullet, sea trout and twaite shad at upstream freshwater sampling points can be explained by the fact that until now the estuary is only frequented by straying individuals. Hypoxic events are probably the most important factor limiting opportunities for diadromous species to reach the upper estuary (Maes et al., 2005). Although Maes et al. (1998b) reported that these migrating fish still were restricted to the brackish environment, our results prove the recovery process is expanding towards the upper freshwater part of the estuary. Five diadromous species reached the most inner part of the estuary namely river lamprey, eel, flounder, smelt and three-spined stickleback trachurus form (Table 1). Since pollution does no longer seem to form a permanent migration barrier for at least a part of the diadromous fish population, they still have to negotiate the tidal lock-weir complex in Ghent (km 160) to reach the non-tidal part of the river. During high discharge and high tides, the tidal weir is completely lifted creating opportunities for upstream migrating fish to pass the free-flowing tidal weir (Fig. 2). Similar results were reported by Winter & Van Densen (2001) for non-salmonid fish to pass weirs in the River Vecht (The Netherlands). Peak abundances of river lamprey and three-spined stickleback trachurus form were recorded in February under the second weir at Asper (km 170). A small fraction of them was recorded under the third weir at Oudenaarde (km 180) from February to April. Chances of meeting a free-flowing tidal weir were highest in February. On the other hand chances to pass the non-tidal complex at Asper (km 170) were very limited. Due to the different management of this weir, free-flowing conditions were never reached. Upstream migration via the shipping lock was the only possibility to negotiate the complex. The few smelts were recorded in March only under the second weir. High numbers of upstream migrating young of the year (0+) flounder were recorded in June at sampling points down-stream of the tidal weir. The 0+ flounder resided in the inner part of the estuary until September. Chances for flounder to meet a free-flowing tidal weir from June until September were absent. Mainly from April to October eel appeared under the three consecutive weirs with peak abundances in June and August (mean size 337 mm) under the second weir. The higher catches of eel in front off the weirs during summer months are an indication of upstream dispersal attempts of eel. A study in the River Fremur (France) showed that high dams inhibited upstream eel migration and reduced recruitment by yellow eels. After eel passes were installed, fish became more abundant upstream. It was concluded that eel passes are important to conserve and/or to recover eel stocks (Laffaille et al., 2005).

The Mersey estuary (England) can be set as an example of how wastewater treatment programmes can remedy the poor oxygen conditions in estuaries and rivers. Until relatively recently the estuary had long anoxic reaches in the upper estuary during summer. Dissolved oxygen in the Mersey estuary is now generally above 60% saturation, with salmon now being found at the tidal-limit (Jones, 2006). In Flanders, the households connected to a purification plant increased from almost none in 1970 to 30% in 1990, then to 60% in 2002. It was estimated that the discharge of biochemical and chemical oxygen demand (BOD, COD), phosphorus (P) and nitrogen (N) were reduced with 28%, 51%, 60% and 60%, respectively, from 1994 to 2002 (Maeckelberghe, 2002). Due to the overall improved water quality the upper estuary and the non-tidal river have now re-established some of their importance as migratory route (flounder, eel, river lamprey, smelt, three-spined stickleback trachurus form) and/or as nursery area (flounder). Through wastewater treatment of Brussels from 2006 onwards (water treatment plants are under construction) a further and consistent improvement of the water quality is expected (Van Damme et al., 2005). Next to pollution it is clear that the presence of consecutive lock-weir complexes on the River Scheldt also has a negative impact on the upstream migrating diadromous fish populations. Wastewater treatment programmes have to be extended over the whole catchment area of this cross-border river, fish passes need to be built at the tidal weir and in the non-tidal river part and present and future fish assemblages need to be monitored.

Acknowledgement We thank Raf Baeyens and Seth Martens for their assistance in the field.

References

Béguer, M., L. Beaulaton & E. Rochard, 2007. Distribution and richness of diadromous fish assemblages in Western Europe: large-scale explanatory factors. Ecology of Freshwater Fish 16: 221–237.

Herman, P .M. J. & C. H. R. Heip, 1999. Biogeochemistry of the MAximum TURbidity Zone of Estuaries (MATURE): some conclusions. Journal of Marine Systems 22: 89–104.

ICES, 1999. ICES cooperative research report no. 229. Report of the ICES Advisory Committee on Fisheries Management, 1998. Part 2, 446 pp.

IUCN, 1994. IUCN red list categories. Approved by the 40th Meeting of the IUCN Council, 30 November 1994. Gland, Switzerland.

Jones, P. D., 2006. Water quality and fisheries in the Mersey estuary, England: a historical perspective. Marine Pollution Bulletin 53: 144–154.

Laffaille, P., A. Acou, J. Guillouet & A. Legault, 2005. Temporal changes in European eel, *Anguilla anguilla*, stocks in a small catchment after installation of fish passes. Fisheries Management and Ecology 12: 123–129.

Lucas, M. C. & E. Baras, 2001. Applied aspects of freshwater fish migration. In Lucas, M. C. & E. Baras (eds), Migration of freshwater fishes, Chap. 7. Blackwell Science Ltd., Oxford, 271–310.

Maeckelberghe, H., 2002. De kwaliteit van het Vlaamse oppervlaktewater en de afval-waterlozingen in 2002. Een toelichting bij de resultaten van de meetnetten van de Vlaamse Milieumaatschappij. VMM report (in Dutch). Report of the Vlaamse Milieumaatschappij (VMM), Erembodegem, Belgium, 167 pp.

Maes, J., A. Taillieu, P.A. Van Damme, K. Cottenie & F. Ollevier, 1998a. Seasonal patterns in the fish and crustacean community of a turbid temperate estuary (Zeeschelde Estuary, Belgium). Estuarine, Coastal and Shelf Science 47: 143–151.

Maes, J., P. A. Van Damme, A. Taillieu & F. Ollevier, 1998b. Fish communities along an oxygen-poor salinity gradient (Zeeschelde Estuary, Belgium). Journal of Fish Biology 52: 534–546.

Maes, J., M. Stevens & F. Ollevier, 2005. The composition and community structure of the ichthyofauna of the Upper Scheldt estuary: synthesis of a 10-year data collection (1991–2001). Journal of Applied Ichthyology 21: 86–93.

Meire, P., T. Ysebaert, S. Van Damme, E. Van den Bergh, T. Maris & E. Struyf, 2005. The Scheldt estuary: a description of a changing ecosystem. Hydrobiologia 540: 1–11.

Soetaert, K. & P. M. J. Herman, 1995. Nitrogen dynamics in the Westerschelde estuary (S.W. Netherlands) estimates by means of the ecosystem model MOSES. Hydrobiologia 311: 225–246.

Soetaert, K., J. J. Middelburg & C. Heip, 2006. Long-term change in dissolved inorganic nutrients in the heterotrophic Scheldt estuary (Belgium, The Netherlands). Limnology and Oceanography 51: 409–423.

Thiel, R. & I. C. Potter, 2001. The ichthyofaunal composition of the Elbe estuary: an analysis in space and time. Marine Biology 138: 603–616.

Van Damme, S., E. Struyf, T. Maris, T. Ysebaert, F. Dehairs, M. Tackx, C. Heip & P. Meire, 2005. Spatial and temporal patterns of water quality along the estuarine salinity gradient of the Scheldt estuary (Belgium and The Netherlands): results of an integrated monitoring approach. Hydrobiologia 540: 29–45.

Vandelannoote, A. & J. Coeck, 1998. Rode Lijst van de Inheemse en Ingeburgerde Zoet-en Brakwatervissen en van de Rondbekken in Vlaanderen, p. 259–264 (in Dutch). In: Vandelannoote, A., R. Yseboodt, B. Bruylants, R. Verheyen, J. Coeck, C. Belpaire, G. Van Thuyne, B. Denaeyer, J. Beyens, J. Maes & P. Vandenabeele (eds), Atlas van de Vlaamse beek-en riviervissen. WEL v.z.w., Antwerpen, 303 pp.

Winter, H. V. & W. L. T. Van Densen, 2001. Assessing the opportunities for upstream migration of non-salmonid fishes in the weir-regulated River Vecht. Fisheries Management and Ecology 8: 513–532.

Hydrobiologia (2008) 602:161–177
DOI 10.1007/s10750-008-9293-4

FISH AND DIADROMY IN EUROPE

Return of twaite shad *Alosa fallax* (Lacépède, 1803) to the Southern Baltic Sea and the transitional area between the Baltic and North Seas

Ralf Thiel · Philip Riel · Renate Neumann · Helmut M. Winkler · Uwe Böttcher · Tomas Gröhsler

Abstract The status of twaite shad was investigated within the Southern Baltic Sea (ICES subdivision 22–27) and transitional area between the Baltic and North Seas (division IIIa). The following sources of data were analysed: (i) commercial catch statistics and relevant publications, (ii) records from ichthyological museum collections, (iii) records from recent commercial and recreational fisheries, (iv) research fisheries with trawls. A total of 476 records of twaite shad including more than 16 million individuals were obtained for the time between the years 1836 and 2005. About 72.9% of all records originated from commercial catch statistics and publications, whereas 18.9% were received from ichthyological collections. Research fisheries provided 6.3%, and 1.9% of the records were obtained from recent commercial and recreational fisheries. Most records of twaite shad were estimated for subdivisions 24 (45.2%) and 26 (35.5%). From 1836 to 1959, 29.6% of the records date from the period until 1899. 70.4% of the records of twaite shad originate from the twentieth century until 1959. The mean annual catch of twaite shad between 1891 and 1959 amounted to 86,674 kg within subdivisions 24–26 of the Southern Baltic Sea. Catch data show an approximately 20-year-cyclicity of maximum yields and minimum catches, respectively. The maximum annual yield of twaite shad in subdivisions 24–26 (474,700 kg) was registered in 1940, the minimum annual yield was estimated in 1958 (10 kg). In the 1950s, the annual catches of twaite shad declined sharply. Until 1960 twaite shad catches and records originated mainly from the Pomeranian Bay/Pomeranian Coast and adjacent waters including the Szczecin Lagoon (subdivision 24), the Bay of Gdańsk, Vistula Lagoon and Vistula Spit (subdivision 26) and from the Curonian Lagoon and Curonian Spit (subdivision 26). The highest catches of twaite shad originated from the area of

Guest editors: S. Dufour, E. Prévost, E. Rochard & P. Williot
Fish and diadromy in Europe (ecology, management, conservation)

Electronic supplementary material The online version of this article (doi:10.1007/s10750-008-9293-4) contains supplementary material, which is available to authorized users.

R. Thiel (✉) · R. Neumann
Biocenter Grindel and Zoological Museum, University of Hamburg, Martin-Luther-King-Platz 3, 20146 Hamburg, Germany
e-mail: ralf.thiel@uni-hamburg.de

P. Riel
Zoological Institute and Museum, Ernst-Moritz-Arndt University Greifswald, Johann Sebastian Bach-Straße 11/12, 17487 Greifswald, Germany

H. M. Winkler
Department Biological Sciences, University of Rostock, Universitätsplatz 2, 18055 Rostock, Germany

U. Böttcher · T. Gröhsler
Institute of Baltic Sea Fisheries, Federal Research Institute for Rural Areas, Forestry and Fisheries, Alter Hafen Süd 2, 18069 Rostock, Germany

Curonian Lagoon/Curonian Spit from 1941 to 1960. Seasonal catches of twaite shad showed maximum values from May to July. From 1960 to 1989, only four records of *A. fallax* were registered in the Southern Baltic Sea which originated from German coastal waters in subdivisions 22 and 24. A total of 107 records of twaite shad was obtained in subdivisions 20–27 from 1990 to 2005. Most of these recent records originate from a twaite-shad stock in subdivisions 24, 25 and 26. There are indications which suggest a separate stock of twaite shad in subdivisions 20 and 21. Disappearance and recovery of twaite shad stocks were probably caused by the following factors: construction of barriers in rivers with spawning sites of twaite shad; habitat destruction in those rivers as consequence of gravel extraction and reengineering scheme to improve navigation and for flood defence purposes, water pollution in the lagoons of the Southern Baltic and in their tributaries, commercial fishery in the Southern Baltic and climatic variation in the Baltic Sea basin south of the latitude of 60° N.

Keywords Twaite shad · Baltic sea · Past distribution · Commercial catch · Recent records

Introduction

Like the other shad species, twaite shad *Alosa fallax* (Lacépède, 1803) belongs to the subdivision Clupeomorpha and is a member of the family Clupeidae. The determination of the six subspecies of *Alosa fallax* which have been recognized by Quignard & Douchement (1991a) based mainly on the number of gill rakers on the first gill arch and the geographical location of the different subspecies. Recently, Kottelat (1997) concluded that some of those previously recognized subspecies should be reclassified as species. However, the correct nomenclature for the twaite shad population(s) in the Atlantic seabord including the Baltic Sea has not been finally determined. Aprahamian et al. (2003a) proposed to refer to the anadromous species of *Alosa fallax* inhabiting the Baltic Sea as *Alosa fallax fallax*. This subspecies is distributed in the Baltic Sea and in the whole Atlantic seabord including the North Sea (e.g. Saemundsson, 1949; Kartas, 1981; Taverny, 1991; Sabatié, 1993). The general characteristics of *Alosa fallax fallax* have been described by Svetovidov (1952), Wheeler (1969), O'Maoileidigh et al. (1988), Quignard & Douchement (1991b), Sabatié (1993) and Alexandrino (1996).

Winkler et al. (2000) evaluated the distribution status of twaite shad within the framework of their checklist of fish species in the Baltic Sea. The authors described the distribution status of *A. fallax* in the Baltic Sea as "present" for Denmark, as "common" for Poland and Lithuania and as "very rare" for Russia, Finland, Latvia and Germany. According to Quignard & Douchment (1991a) the twaite shad, which was very common in a number of Baltic and other European waters about 100 years ago, was very rare or completely extinct in many waters of its former distribution area during the 1980s. There were just a few rivers left with higher population size like the Garonne–Dordogne river system in France or the Elbe River in Germany (Quignard & Douchment, 1991a).

Due to its decrease in number and distribution, the twaite shad has been included in Appendix III of the Bern Convention and Annexes II and V of the EC Habitats Directive. The incorporation of this species into the Habitats Directive places an obligation of members of the European Union to assess numbers and exploitation of the populations and to designate special areas for conservation (SACs) to safeguard populations (Aprahamian et al., 2003b). In conclusion from recently published reviews on twaite shad (Aprahamian et al., 2003a, b), there is a great need to get detailed information about the status of *A. fallax* in most parts of its distribution area. Especially for the western and southern Baltic Sea, a complete description of the past and present distribution and a detailed analysis of the temporal and spatial development of the commercial catches are completely lacking. However, the results of those studies are the basis to define measures for the conservation of Baltic stocks of twaite shad and for sustainable fisheries. In this regard, the aims of this article are: (1) to re-assess the commercial catch of twaite shad in ICES subdivisions 24–26 of the Southern Baltic between 1891 and 1959; (2) to describe the temporal trends and spatial characteristics of the commercial catch of twaite shad; (3) to analyse the past and present distribution of twaite shad in subdivisions 20–27 of the Southern Baltic Sea (Fig. 1) and in the transitional area between the Baltic and North Seas (division IIIa: subdivisions 20–21).

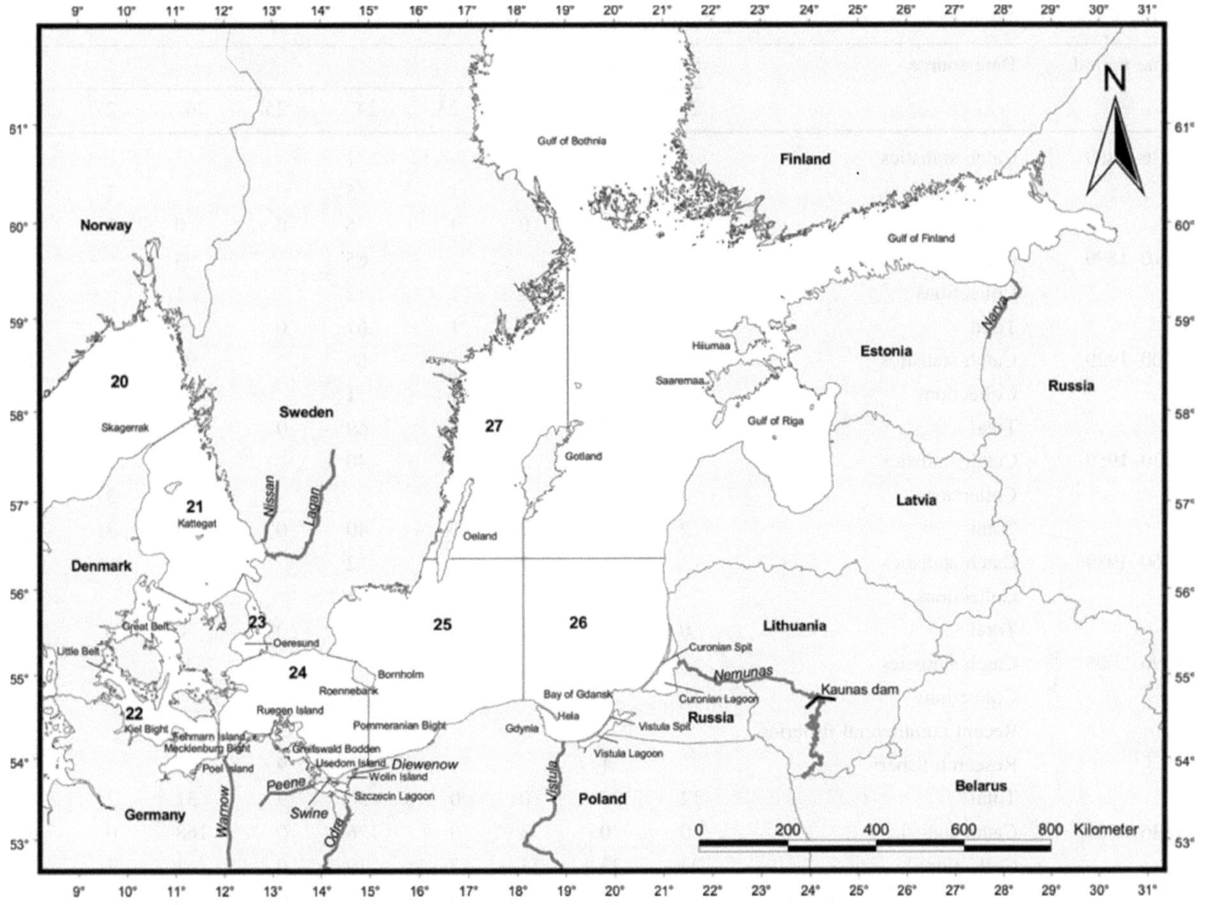

Fig. 1 Study area with geographical names, main tributaries and ICES subdivisions

Materials and methods

The status of twaite shad in the Baltic Sea was analysed on the basis of the following sources of data: (i) commercial catch statistics and relevant publications; (ii) records from ichthyological museum collections; (iii) recent records from commercial and recreational fisheries and (iv) research fisheries with trawls.

The origin of the data from different sources and for different time periods and subdivisions is presented in Table 1. The analysis of data is restricted to subdivisions 20–27 of the area of Division IIIa (subdivisions 20–21) and the major part of the Baltic Sea (subdivisions 22–27) (see Fig. 1). Within subdivision 20 only records of twaite shad from the Swedish Westcoast were considered.

Occurrence and distribution of twaite shad records in subdivisions 20–27 were investigated for the time interval between 1836 and 2005, whereas temporal trends and spatial characteristics of the commercial catch of twaite shad in the southern Baltic Sea were analysed in subdivisions 24–26 for the time period 1887–1959.

All collected records of twaite shad including information about locality and date of their catch are summarized in Tables 1–3. Records representing the same locality and caught during the same time period are summarized in Figs. 5–7.

The geographic positions of records of twaite shad were either obtained from available point positions of records or as midpoints of performed hauls, or were estimated for given locations using georeferenced maps. All geographic presentations of records of twaite shad were conducted with a geographic information system (GIS) using the software ArcView GIS (version 3.3, ESRI).

Table 1 Origin of records of twaite shad with regard to different time periods, data sources and subdivisions

Time period	Data source	Subdivisions								
		20	21	22	23	24	25	26	27	Total
1836–1869	Catch statistics					1				1
	Collections	1	2		1	5			1	10
	Total	1	2	0	1	6	0	0	1	11
1870–1899	Catch statistics					65		3		68
	Collections	7	5	13	1	2		1		29
	Total	7	5	13	1	67	0	4	0	97
1900–1929	Catch statistics					68		96		164
	Collections	1	2		1	1			1	6
	Total	1	2	0	1	69	0	96	1	170
1930–1959	Catch statistics			2		40		38		80
	Collections	2	2						3	7
	Total	2	2	2	0	40	0	38	3	87
1960–1989	Catch statistics			1		2				3
	Collections					1				1
	Total	0	0	1	0	3	0	0	0	4
1990–2005	Catch statistics							31		31
	Collections	12	22			1			2	37
	Recent commercial fisheries					9				9
	Research fishery		1			20	9			30
	Total	12	23	0	0	30	9	31	2	107
1836–2005	Catch statistics	0	0	3	0	176	0	168	0	347
	Collections	23	33	13	3	10	0	1	7	90
	Recent commercial fisheries	0	0	0	0	9	0	0	0	9
	Research fishery	0	1	0	0	20	9	0	0	30
	Total	23	34	16	3	215	9	169	7	476

Table 2 Historical records of twaite shad in subdivisions 20–27 of the Baltic Sea in the time period 1960–1989

No.	Year	Sub-division	Country	Locality	Number of individuals	Data source
1	1967	22	Germany	Wismar Bight	50	Schaarschmidt & Lemcke (2004)
2	1970	24	Germany	Strelasund	1000	Schaarschmidt & Lemcke (2004)
3	1974	24	Germany	Warnemuende	1	NHMB
4	1985	24	Germany	Greifswald Bodden	2	Schaarschmidt & Lemcke (2004)
Total	1960–1989	20–27	Baltic Sea		1053	This paper

For abbrevations see Chapter "Materials and methods"

Synopsis of data from commercial catch statistics and publications

Data from papers containing original information about catch statistics and records of twaite shad in subdivisions 20–27 were assessed, selected and than compiled. Such data were used from Anonymous (1887–1920), Ehrenbaum (1921), Eichelbaum (1926–1940), Manyukas (1989), Repečka (2003), Skóra (2003), Schaarschmidt & Lemcke (2004) and from Thiel et al. (2004, 2005a, 2006). Additional records of twaite shad from catch statistics of the Curonian lagoon (subdivision 26) were communicated by Repečka (pers. comm.).

Table 3 Recent records of twaite shad in subdivisions 20–27 of the Baltic Sea since 1990

No.	Year	Sub-division	Country	Locality	Number of individuals	Data source
1	1990	24	Denmark	Mon Island	1	ZMUC
2	1993	20	Sweden	West of Maseskaer Island	2	NRMS
3	1996	21	Sweden	Kattegat	1	NRMS
4	1996	26	Lithuania	Curonian Lagoon	39	Repečka (2003)
5	1997	20	Sweden	Skagerrak	1	NRMS
6	1997	26	Lithuania	Curonian Lagoon	480	Repečka (2003)
7	1997	26	Lithuania	Curonian Spit	1	Repečka (2003)
8	1998	26	Lithuania	Curonian Lagoon	171	Repečka (2003)
9	1998	26	Lithuania	Curonian Spit	8	Repečka (2003)
10	1999	20	Sweden	West of Maseskaer Island	6	NRMS
11	1999	20	Sweden	West of Maseskaer Island	1	NRMS
12	1999	21	Sweden	Kattegat	1	NRMS
13	1999	24	Germany	Adlerground	1	FRCF
14	1999	24	Germany	West of Bornholm Island	1	FRCF
15	1999	26	Poland	Bay of Gdansk	100	Skora (2003)
16	1999	26	Lithuania	Curonian Lagoon	654	Repečka (2003)
17	1999	26	Lithuania	Curonian Spit	64	Repečka (2003)
18	1999	26	Poland	Oksywie	2	Skora (2003)
18	2000	20	Sweden	Skagerrak	1	NRMS
19	2000	20	Sweden	West of Maseskaer Island	1	NRMS
20	2000	20	Sweden	West of Maseskaer Island	1	NRMS
21	2000	21	Denmark	Albaek Bugt	1	ZMUC
22	2000	21	Sweden	Goeteborg	1	NRMS
23	2000	21	Sweden	Kattegat	1	NRMS
24	2000	21	Sweden	Vaeroebacka	3	NRMS
25	2000	21	Sweden	Varberg	1	NRMS
26	2000	24	Germany	Baltic Sea	1	FRCF
27	2000	24	Germany	Baltic Sea	1	FRCF
28	2000	24	Germany	Northwest of Adlerground	2	Thiel et al. (2005a, b)
29	2000	25	Germany	Baltic Sea	14	FRCF
30	2000	25	Germany	Baltic Sea	9	FRCF
31	2000	25	Germany	Baltic Sea	8	FRCF
32	2000	25	Germany	Baltic Sea	1	FRCF
33	2000	25	Germany	Baltic Sea	8	FRCF
34	2000	25	Germany	Baltic Sea	2	FRCF
35	2000	25	Germany	Baltic Sea	1	FRCF
36	2000	26	Lithuania	Curonian Lagoon	1356	Repečka (2003)
37	2000	26	Poland	Jantar	3	Skora (2003)
38	2000	26	Poland	Jantar	17	Skora (2003)
39	2001	20	Sweden	West of Maseskaer Island	1	NRMS
40	2001	20	Sweden	West of Maseskaer Island	2	NRMS
41	2001	21	Sweden	Bohuslaen	1	NRMS
42	2001	21	Sweden	Falkenberg	2	NRMS
43	2001	21	Sweden	Halland	2	NRMS

Table 3 continued

No.	Year	Sub-division	Country	Locality	Number of individuals	Data source
44	2001	21	Sweden	Kattegat	1	NRMS
45	2001	21	Sweden	Kattegat	1	NRMS
46	2001	21	Sweden	Kattegat	1	NRMS
47	2001	21	Sweden	Kattegat	4	NRMS
48	2001	21	Sweden	Kattegat	2	NRMS
49	2001	24	Germany	West of Adlerground	1	Thiel et al. (2005a, b)
50	2001	26	Poland	Bay of Gdansk	73	Skora (2003)
51	2001	26	Lithuania	Curonian Lagoon	171	Repečka (2003)
52	2001	26	Poland	Jantar	3	Skora (2003)
53	2001	26	Poland	Jantar	4	Skora (2003)
54	2001	26	Poland	Jantar	1	Skora (2003)
55	2001	26	Poland	Jantar	3	Skora (2003)
56	2001	26	Poland	Jantar	5	Skora (2003)
57	2001	26	Poland	Oksywie	1	Skora (2003)
58	2001	26	Poland	Oksywie	3	Skora (2003)
59	2001	26	Poland	Oksywie	4	Skora (2003)
60	2001	26	Poland	Oksywie	2	Skora (2003)
61	2001	26	Poland	Oksywie	5	Skora (2003)
62	2001	26	Poland	Swibno	1	Skora (2003)
63	2001	26	Poland	Swibno	1	Skora (2003)
64	2001	27	Sweden	Oeland 13	1	NRMS
65	2002	20	Sweden	West of Maseskaer Island	1	NRMS
66	2002	20	Sweden	West of Maseskaer Island	1	NRMS
67	2002	24	Germany	Baltic Sea	1	FRCF
68	2002	25	Germany	Baltic Sea	2	FRCF
69	2002	26	Lithuania	Curonian Lagoon	14028	Repečka (pers.comm.)
70	2002	26	Poland	Hela Island	8	Skora (2003)
71	2002	26	Poland	Hela Island	2	Skora (2003)
72	2003	20	Sweden	Skagerrak	3	NRMS
73	2003	21	Sweden	Halland	2	NRMS
74	2003	21	Sweden	Halland	2	NRMS
75	2003	24	Germany	Oderbank, Stat. 10	2	Thiel et al. (2004)
76	2003	24	Germany	Oderbank, Stat. 4	1	Thiel et al. (2004)
77	2003	24	Germany	Oderbank, Stat. 6	2	Thiel et al. (2004)
78	2003	24	Germany	Peenestrom, Usedomer Hard	1	Thiel et al. (2004)
79	2003	24	Germany	Szczecin Lagoon east of Ueckermuende	2	Thiel et al. (2004)
80	2003	24	Germany	Szczecin Lagoon near Altwarp	6	Thiel et al. (2004)
81	2003	26	Lithuania	Curonian Lagoon	3702	Repečka (pers.comm.)
82	2004	21	Sweden	Aelvsborgsfjorden mouth	1	NRMS
83	2004	21	Sweden	Bohuslaen	1	NRMS
84	2004	21	Sweden	Goeteborg	1	NRMS
85	2004	21	Sweden	Halland	2	NRMS
86	2004	21	Denmark	Kattegat	1	FRCF
87	2004	21	Sweden	Varberg	1	NRMS

Table 3 continued

No.	Year	Sub-division	Country	Locality	Number of individuals	Data source
88	2004	24	Germany	Achterwasser	1	Thiel et al. (2005a, b)
89	2004	24	Germany	Achterwasser	1	Thiel et al. (2005a, b)
90	2004	24	Germany	Coastal Waters of Usedom Island, Stat. 20	2	Thiel et al. (2005a, b)
91	2004	24	Germany	Northwest of Adlerground	1	Thiel et al. (2005a, b)
92	2004	24	Germany	Northwest of Adlerground	4	Thiel et al. (2005a, b)
93	2004	24	Germany	Szczecin Lagoon near Ueckermuende	9	Thiel et al. (2005a, b)
94	2004	24	Germany	West of Adlerground	6	Thiel et al. (2005a, b)
95	2004	24	Germany	West of Adlerground	1	Thiel et al. (2005a, b)
96	2004	24	Germany	West of Adlerground	2	Thiel et al. (2005a, b)
97	2004	26	Lithuania	Curonian Lagoon	130,605	Repečka (pers.comm.)
98	2004	27	Sweden	East of Oeland	1	NRMS
99	2005	24	Germany	East of Adlerground	1	FRCF
100	2005	24	Germany	Goehren	4	Thiel et al. (2006)
101	2005	24	Germany	Northwest of Adlerground	1	FRCF
102	2005	24	Germany	Northwest of Kap Arkona	1	FRCF
103	2005	24	Germany	Roennebank	2	Thiel et al. (2006)
104	2005	24	Germany	Sassnitz	1	Thiel et al. (2006)
105	2005	24	Germany	West of Hiddensee	1	Thiel et al. (2006)
106	2005	25	Germany	Sea off Bornholm	1	FRCF
Total	1990–2005	20–27	Baltic Sea		151,682	This paper

For abbrevations see Section "Materials and methods"

For the period 1887–1900, data for a few locations were only available in an old German unit of measure (Schock) which corresponds to 60 individuals. Some of the annual yields from 1901 to 1915 were only available in currency (Mark). Conversion of those values into weight units was performed assuming that one kilogram of catch is equivalent to 0.7 German Mark. The number of individuals was obtained by using the assumption that 1 kg of catch contains three individuals of twaite shad. Conversion values were obtained as mean values from those annual yields from the relevant period which were given in units of Mark, kilogram and individuals as well. Converted values were rounded to the nearest hundred individuals.

Analysis of ichthyological collections

Large- and medium-sized ichthyological collections/databanks of museums and universities which were known by the authors to contain individuals of fish species from subdivisions 20 to 27 of the Baltic Sea in their collections were analysed with regard to records of twaite shad. In unclear cases of scientific determination, the shad specimens were re-determined. Ichthyological collections and/or databanks of 16 institutions were analysed. However, records of twaite shad were obtained from the following 7 institutions which are sorted according to the number of available records of twaite shad:

- Swedish Museum of Natural History, Stockholm, Sweden (NRMS, 57 records);
- Zoological Museum of the University of Copenhagen, Denmark (ZMUC, 13 records);
- Zoological Museum of the Christian-Albrechts-University Kiel, Germany (ZMUK, 9 records);
- Zoological Institute and Museum of the Ernst-Moritz-Arndt-University Greifswald, Germany (ZMUG, 6 records);
- Museum of Natural History of the Humboldt University in Berlin, Germany (NHMB, 3 records);
- Biocenter Grindel and Zoological Museum of the University of Hamburg, Germany (ZMUH, 1 record);
- Department Biology of the University of Rostock, Germany (DBUR, 1 record).

Compilation of recent catch records from commercial and recreational fisheries

In order to ensure that commercial and recreational fishermen communicate actual catch records of twaite shad, an information sheet containing drawings with the most important determination characteristics of the species was developed in 2003. This information sheet was then distributed along the entire Baltic coast of Germany. Furthermore, a catch award was announced to those fishermen who communicate verified and accurate catches of twaite shad. Overall, nine fishermen responded with catch records of twaite shad.

Evaluation of hauls from research fisheries

The Federal Research Centre for Fisheries (FRCF) contributed catch records of twaite shad from subdivisions 22 to 25 since 1991 based on data resulting from about 3,000 hauls, taken with bottom trawls. Some additional data originating from studies with bottom trawls in the region of Pommeranian Bay (subdivision 24) and from hydro-acoustic surveys (pelagic trawl) within the area of subdivisions 20–25 were available since 1978. More details regarding the sampling regime used by the Federal Research Centre for Fisheries are presented in the paper by Kloppmann et al. (2003).

The German Oceanographic Museum and the University of Rostock performed research fisheries with bottom trawls, semi-pelagic trawls and shrimp trawls in the Exclusive Economic Zone (EEZ) and in the coastal waters of Germany in the Baltic Sea region among the islands of Ruegen, Usedom and Bornholm during autumn 2003, in spring, summer and autumn 2004 and 2005. Fish sampling was performed at 22 stations. Overall, 189 hauls were carried out during that study. Trawls were towed during daylight and lasted for 30 min. After collection, adult and larger juvenile individuals were identified to species, counted, weighted, measured and released back into the water immediately. Fish larvae and small juveniles were investigated in the laboratory. More details of the construction of the trawls, the sampling procedures, catch efficiencies of the gears and details of the recorded twaite shads are given in Thiel et al. (2004, 2005a, 2006).

Results

Origin and quality of available records of twaite shad

A total of 476 records of twaite shad accounting for 16 million individuals were obtained based on the analysis of all available data sources in the time period between 1836 and 2005 (Tables 1–3). Most records (347 records, 72.9%) originated from commercial catch statistics and publications (Table 1). 90 records (18.9%) were received from the analysis of ichthyological collections. Research fishery supplied 6.3% (30 records) of the total number of records. 1.9% (9 records) of the records were provided by recent commercial and recreational fisheries (Table 1).

Records originated from commercial catch statistics, publications and ichthyological collections from the years 1836 until 1989. During that period, 85.6% of all records (316 records) were obtained from commercial catch statistics and publications, whereas 14.4% of the records (53 records) originated from the analysis of ichthyological collections (Table 1).

From 1990 to 2005, most records of twaite shad (37 records, 34.6%) originated from ichthyological collections, whereas catch statistics provided 30.0% (31 records). About 28.0% of the records (30 records) were obtained from research fishery. Between 1990 and 2005, 8.4% of the records (9 records) were attributed to recent commercial and recreational fisheries (Table 1).

Mean annual value, temporal trends and spatial characteristics of the commercial catches of twaite shad in subdivisions 24–26 of the Southern Baltic

High variations of annual catches are evident within subdivisions 24–26 of the Southern Baltic Sea. No catch statistics were available during the time periods 1920–1925 (crisis years of the Weimar Republic) and 1941–1946 (World War II, Fig. 2). However, it must be concluded from catch statistics that twaite shad was an important commercial species of clupeid fishes in the Baltic Sea during the last quarter of the nineteenth and the first half of the twentieth century.

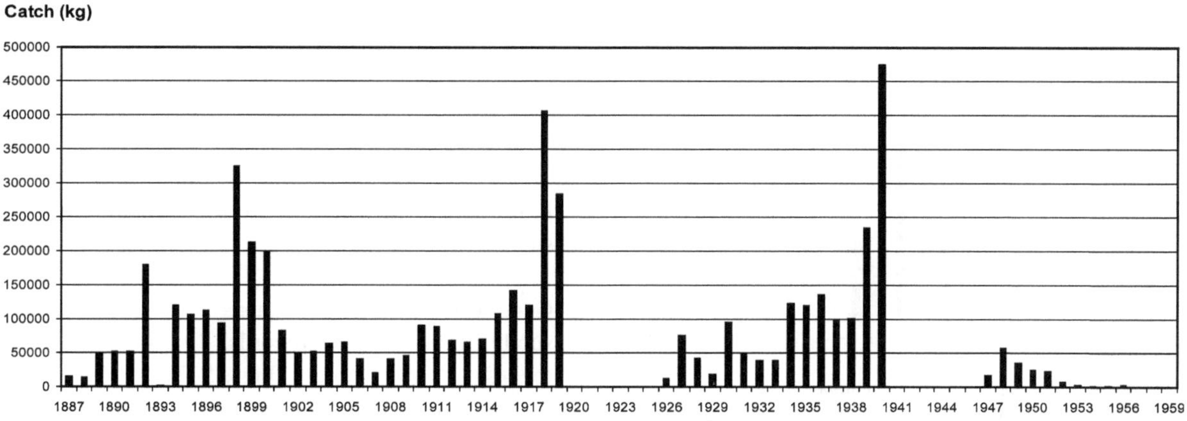

Fig. 2 Total landings of twaite shad in the Southern Baltic Sea (subdivisions 24–26) from 1887 to 1959

High annual catches of twaite shad of 324,500 and 406,425 kg were estimated in 1898 and 1918, respectively (Fig. 2). The maximum annual yield of twaite shad in the Southern Baltic (474,700 kg) was registered in 1940, the minimum annual yield was estimated in 1958 (10 kg). Catch data show an approximately 20-year-cyclicity of maximum yields (e.g. 1898–1900, 1918–1919, 1939–1940) and of minimum catches (e.g. 1887–1888, 1906–1908, 1926 and 1947), respectively.

However, the mean annual catch of twaite shad amounted to 86,674 kg in the years between 1891 and 1959. In the 1950s the annual catches of twaite shad declined sharply in subdivisions 24–26. Since 1960 no continuous catch statistics of twaite shad is recorded within that area.

From 1887 to 1960, twaite shad catches in subdivision 24–26 of the southern Baltic Sea originated mainly from the Pommeranian Bay/Pommeranian Coast (Germany, Poland), the Szczecin Lagoon and adjacent waters (Germany, Poland), the Bay of Gdansk (Poland, Russia), the Vistula Lagoon/Vistula Spit (Poland, Russia) and from the Curonian Lagoon/Curonian Spit (Russia, Lithuania; Fig. 3). From 1887 to 1900 the areas of Pommeranian Bay/Pommeranian coast and Szczecin Lagoon reached their highest importance, contributing together almost 100% to the twaite shad fishery in subdivisions 24–26 (Fig. 3). The contribution of Pommeranian Bay/Pommeranian Coast and Szczecin Lagoon decreased remarkably from 1901 to 1940. Generally, commercial catches of twaite shad were more or less equally distributed over the different fishery areas during the period 1901–1940, but did mainly concentrate in subdivision 26. After World War II regional catch values were only available for the Pommeranian Bay/Pommeranian Coast and the Curonian Lagoon/Curonian Spit (Fig. 3). However, the highest catches of twaite shad

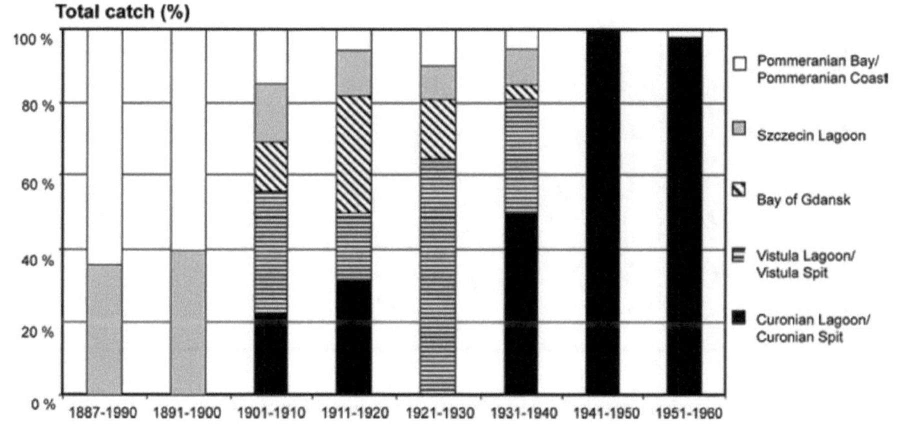

Fig. 3 Contribution (%) of the main fishery areas in the Southern Baltic Sea (subdivisions 24–26) to the total annual yield of twaite shad for different time periods from 1891 to 1960

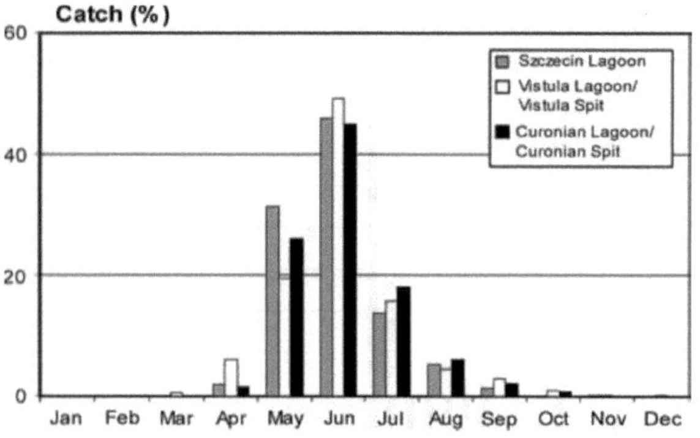

Fig. 4 Seasonal variation of total catches of twaite shad in the areas of Szczecin Lagoon, Vistula Lagoon/Vistula Spit and Curonian Lagoon/Curonian Spit from 1901 to 1919

originated from the area of Curonian Lagoon/Curonian Spit from 1941 to 1960.

Seasonal values of commercial catches of twaite shad were available for Szczecin Lagoon, Vistula Lagoon/Vistula Spit and Curonian Lagoon/Curonian Spit. The seasonal catches of twaite shad from all of those areas showed maximum values from May to July and peaked in June (Fig. 4). Additionally, all three areas are characterised by a sharp increase of catches from April to May, whereas the catches decreased more slowly between July and September (Fig. 4).

Past and present distribution of twaite shad in subdivisions 20–27

Generally, most twaite shads were recorded from subdivisions 24 (215 records, 45.2%) and 26 (169 records, 35.5%; Table 1, Figs. 5–7).

The oldest record originates from the year 1836 from the Greifswald Bodden (Pommeranian Coast, South of Ruegen Island) in subdivision 24 (see Electronic supplementary material). The longest time series of historical records of *A. fallax* was obtained for the Curonian Lagoon (subdivision 26) with the first record in 1875 (see Electronic supplementary material) and the last one in 2004 (Table 3).

The analysis of relevant records in commercial catch statistics, publications and ichthyological collections resulted in 365 records of twaite shad for the

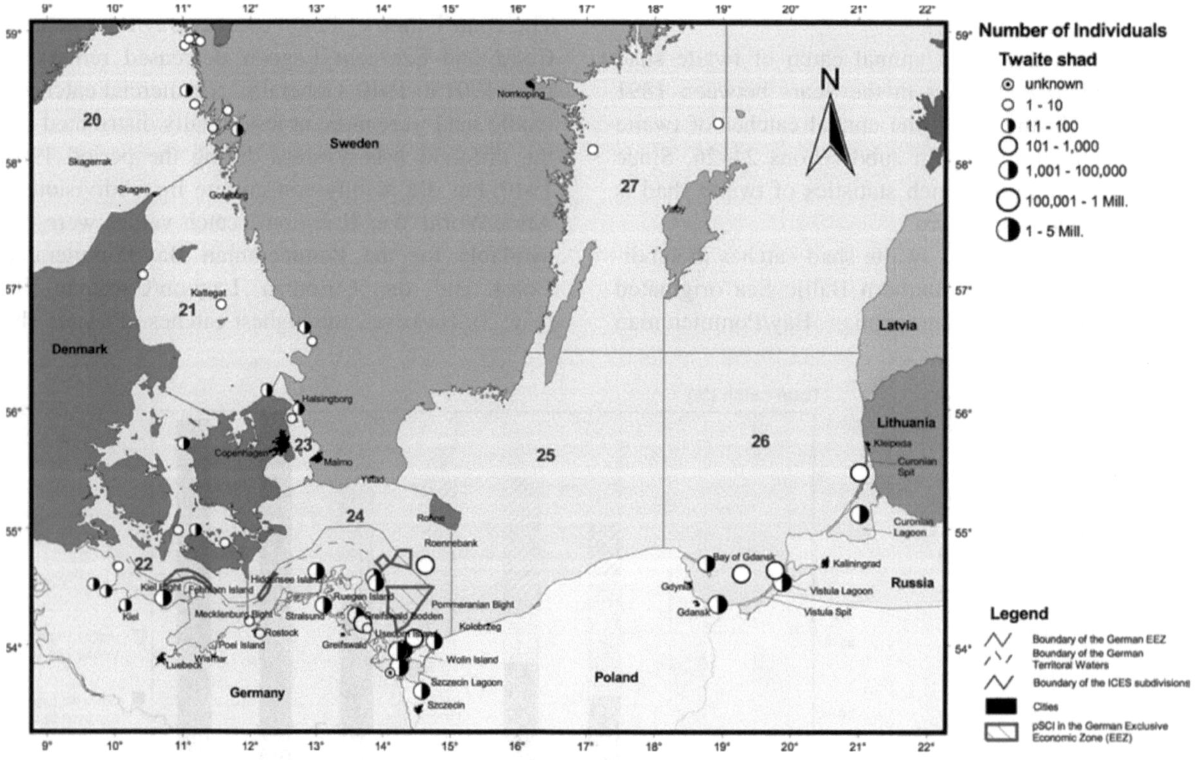

Fig. 5 Distribution of historical records of twaite shad in subdivisions 20–27 of the Baltic Sea until 1959

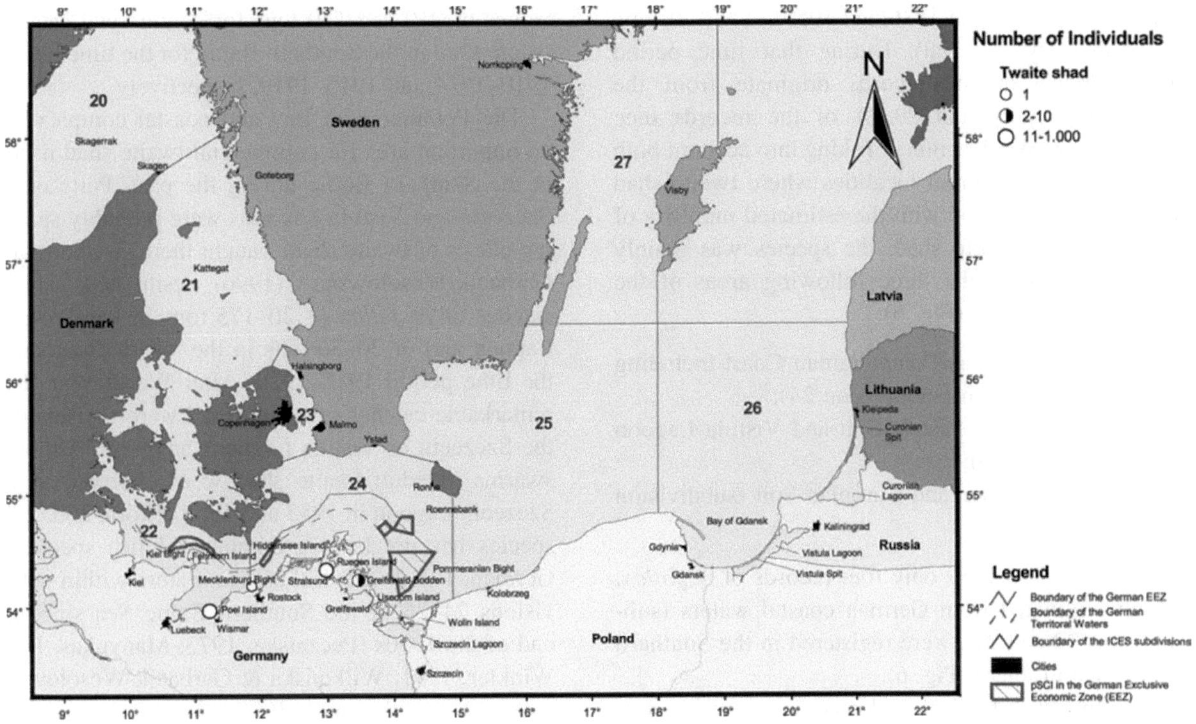

Fig. 6 Distribution of historical records of twaite shad in subdivisions 20–27 of the Baltic Sea from 1960 to 1989

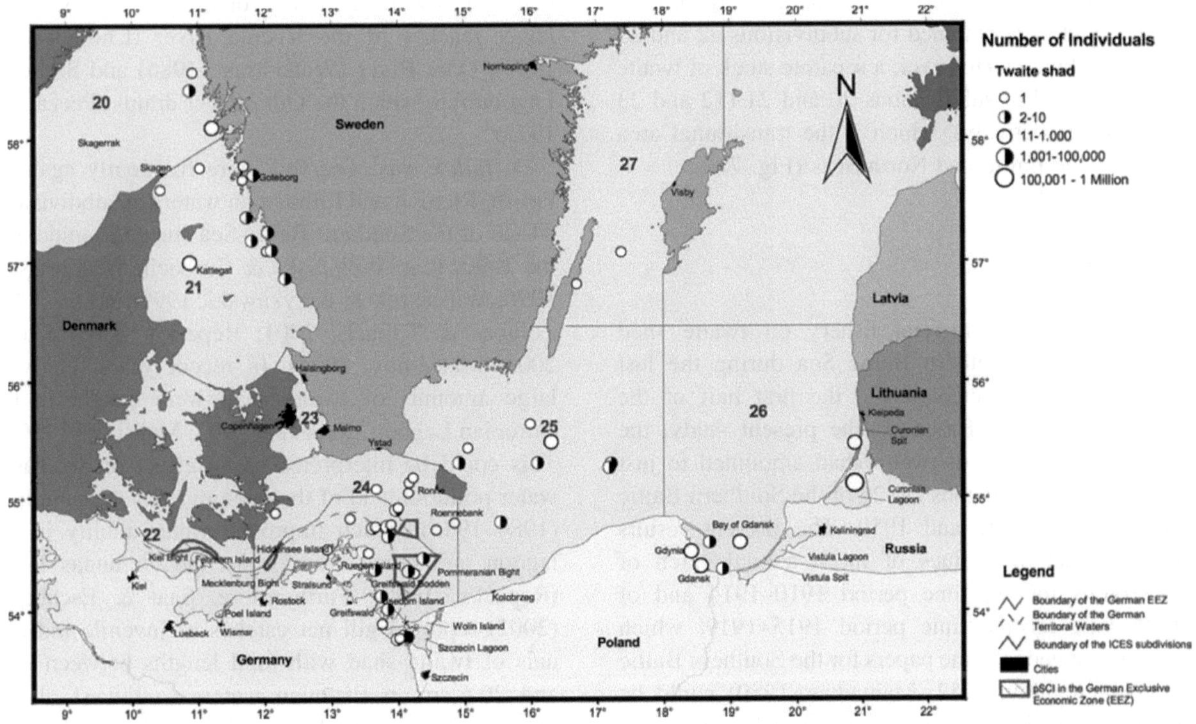

Fig. 7 Distribution of recent records of twaite shad in subdivisions 20–27 of the Baltic Sea from 1990 to 2005

time interval between 1836 and 1959 (see Electronic supplementary material). During that time period about 70.4% of the records originate from the twentieth century and 29.6% of the records date from the nineteenth century. Taking into account both the number of different localities where twaite shad was found in addition with the estimated numbers of individuals of twaite shad, the species was mainly distributed within the three following areas of the southern Baltic Sea (Fig. 5):

- Pommeranian Bay/Pommeranian Coast including Szczecin Lagoon (subdivision 24);
- Bay of Gdańsk, Vistula Spit and Vistula Lagoon (subdivision 26);
- Curonian Lagoon and Curonian Spit (subdivision 26).

From 1960 to 1989 only four records of *A. fallax*, which originated from German coastal waters (subdivisions 22 and 24), were registered in the Southern Baltic Sea (Table 2, Fig. 6).

In contrast, a total of 107 records of twaite shad was obtained in subdivisions 20–27 from 1990 to 2005 (Table 3, Fig. 7). Most of these recent records of twaite shad originate from a stock in subdivisions 24, 25 and 26 (30, 9 and 31 records). No records of twaite shad were obtained for subdivisions 22 and 23 (Table 3, Fig. 7). However, a separate stock of twaite shad occurred in subdivisions 20 and 21 (12 and 23 records, respectively), which is the transitional area between the Baltic and North Seas (Fig. 7).

Discussion

An important commercial fishery on twaite shad existed in the Southern Baltic Sea during the last quarter of the nineteenth and the first half of the twentieth century. Based on the present study, the mean annual catch of twaite shad amounted to just 90 tons for subdivisions 24–26 of the Southern Baltic Sea between 1891 and 1959. The present results indicate that the values of mean annual catch of 2,100 tons for the time period 1910–1914 and of 1,700 tons for the time period 1915–1919, which were presented in some papers for the Southern Baltic (e.g. Svetovidov, 1952; Manyukas, 1989) could be wrong. In their original papers, Ehrenbaum (1921) and Mohr (1941) already published the more likely values of 210 and 170 tons for commercial catch of twaite shad in the Southern Baltic for the time periods 1910–1914 and 1915–1919, respectively.

The Pommeranian Bay and coastal complex was an important area for commercial twaite shad fishery in the Southern Baltic during the past. Parts of the Szczecin and Vistula Lagoons were probably spawning places of twaite shads caught there. Wilkońska & Garbacik-Wesołowska (1996) estimated annual catches of *A. fallax* of 20–175 tons in the Szczecin Lagoon and of 20–56 tons in the Vistula Lagoon in the time period 1916–1920. After World War II no remarkable catches of twaite shad were registered in the Szczecin or Vistula Lagoons anymore. Although swarms of adult twaite shad were observed in the Szczecin Lagoon in 1953 and 1955 (Skóra, 2003), the species has not been a commercial fish species in German, Polish and Lithuanian waters within subdivisions 24–26 of the Southern Baltic Sea since the end of the 1950s (Peczalska, 1973; Manyukas, 1989; Winkler, 1991; Wilkońska & Garbacik-Wesołowska, 1996). Until the mid 1990s *A. fallax* was only occasionally caught in the Curonian Lagoon (Mileriene, 1997; Repečka, 1999) as well as over the Baltic Sea region (Wiktor, 1989; Winkler, 1991). During that time, twaite shad was found sporadically in the lower reaches of the Vistula River (Chmielewski, 1965), Odra River (Waterstraat, 1986) and Szczecin Lagoon into which the Odra River drains (Peczalska, 1973).

A. fallax was observed more frequently again in Polish, Russian and Lithuanian waters of subdivisions 24–26 of the Southern Baltic Sea since the middle of the 1990s (e.g. Wilkońska & Garbacik-Wesołowska, 1996; Więcaszek & Krzykawski, 1999; Hesse, 2000; Zolubas & Toliušis, 2001; Repečka, 2003; Skóra, 2003; Maksimov, 2004). In recent years relatively large amounts of twaite shad were caught in the Curonian Lagoon (Repečka, 2003; Maksimov, 2004). This could be interpreted as an effect of decreased water pollution and of the opening of a shipping lane (1984–1986), which improved water quality in the lagoon and created access to the Nemunas River (Repečka, 1999). Furthermore, Saat & Eschbaum (2002) reported gill net catches of juvenile individuals of twaite shad with total lengths between 20.2 and 20.6 cm in Estonian waters southwest of Hiiumaa Island (Saarnaki Island and Matsalu Bay) in 2000.

The increased number of records in Estonian, Lithuanian and Polish waters are in agreement with the results of the present study which clearly indicate that the number of records of twaite shad in the German EEZ and coastal waters of Germany in the Baltic Sea, especially northeast of the Ruegen Island, in the Pommeranian Bay and in the German part of the Szczecin Lagoon, increased remarkable since the turn of the millennium. Based on these observations, it must be assumed that the Baltic population of twaite shad has been increasing since the middle of the 1990s after about 40 years of very low population level.

This population increase could be originated by the increase of the eastern twaite shad stock of the Curonian Lagoon (Thiel et al., 2004). A migration of greater numbers of twaite shads from the North Sea into the Baltic Sea has not been observed so far. However, the existence of a separate stock of twaite shad within the transitional waters between the North and Baltic Seas (subdivisions 20–21) could also be an eastern part of the North Sea population. According to Kloppmann et al. (2003), high-density areas of twaite shad are occurring within the German coastal waters (12-nautical mile zone) of the North Sea.

The present study has shown that the commercial catches of twaite shad are characterized by a 20-year-cyclicity in subdivisions 24–26 of the Southern Baltic Sea from 1887 to 1959. Although there are not enough data available to relate the cyclicity of yields with the population dynamics of twaite shad in the Southern Baltic Sea, it seems that the amount of commercial catches in the Southern Baltic Sea and the number of records of twaite shad from other Baltic areas follow similar patterns. This is the case, for instance, for the time period 1960–1989, when no commercial catches were performed and the number of records of twaite shad from different sources was also very low. Additionally, there are indications that the cyclicity of commercial twaite shad catches could correspond with events of changing annual mean surface air temperatures (Anonymous, 2006) in the Baltic Sea basin south of the latitude of 60° N. During the end of the nineteenth and the early part of the twentieth centuries, periods of increasing annual mean temperatures are followed by periods of maximum yields. On the other hand, periods of minimum yields of twaite shad seem to correspond with previous decreasing annual mean temperatures. Especially the early twentieth century warming (Anonymous, 2006) until the 1930s seems to be followed by the maximum annual yield of twaite shad in the Southern Baltic in 1940. Furthermore, the marked warming in the Baltic Sea basin during the last decades of the twentieth century, which started around 1990 in Denmark and Estonia (Anonymous, 2006), seems to correspond with the following increase of the Baltic population of twaite shad since the middle of the 1990s.

Generally, the most northerly reported spawning population of *A. fallax* is located in the Nemunas River in Lithuania (Manyukas, 1989). However, in the 1930s, during the early twentieth century warming, numerous individuals of *A. fallax* were observed in the Gulf of Finland and Gulf of Riga and also in the Narva River in Estonia (Mikelsaar, 1984). Therefore, this area may be regarded as the present northern limit for a spawning population (Aprahamian, 2003b). However, the use of this region as spawning and nursery area is probably related to distinct climatic conditions. Drimmelen (1951) observed similar patterns of recruitment in disparate stocks of twaite shad which were mainly depending on water flow and temperature. Aprahamian & Aprahamian (2001) also suggested that a relationship may exist between the latitude of the Gulf Stream north wall and year class strength of twaite shad in the Severn Estuary.

However, the results of the present article clearly indicate that the German, Polish, Russian and Lithuanian waters within subdivisions 24–26 of the Southern Baltic Sea composed the main fishery areas as well as distribution range of twaite shad in the past. Around Scandinavian waters, *A. fallax* has been reported infrequently (Aprahamian, 2003a). Although juveniles were caught off the Swedish coast around the Skåne peninsula and adult individuals were found in the Swedish rivers Lagan and Nissan, it remains unclear whether a separate Scandinavian spawning population exists (Aprahamian, 2003b). Tributaries of the coastal lagoons within subdivisions 24–26, e.g. rivers Odra, Vistula and Nemunas, and probably parts of the lagoons were used as spawning areas (e.g. Chmielewski, 1965; Waterstraat, 1986; Aprahamian et al., 2003a).

Instead of climatic variation, the three main reasons for the decline of Baltic stocks of twaite shad from the 1950s to the 1990s are the construction of weirs in rivers with spawning sites of twaite shad, water pollution in the lagoons of the Southern Baltic and in their tributaries and commercial fishery (e.g. Repečka, 1999; Maksimov, 2004). The first two factors made it difficult or even impossible for *A. fallax* to reach the original spawning grounds to reproduce successfully (Gerkens & Thiel, 2001). For instance in the Nemunas River in Lithuania, the spawning stock of twaite shad originally ascended approximately 400 km up the Nemunas River. After the construction of the Kaunas hydroelectric dam in 1959, 224 km upstream of the mouth of the Nemunas River, the access to the upper reaches of the river was blocked (Maksimov & Toliušis, 1999; Repečka, 1999; Žiliukas & Žiliukienė 2002). Additionally, the discharge of industrial wastewater from paper mills into the lower streches of the Nemunas River and the Curonian Lagoon, and an increase in tonnage of marine shipping, including the Klaypedskiy railroad crossing, greatly impeded the migration of twaite shad and other anadromous fishes to their spawning grounds (Manyukas, 1989). In the 1950s and 1960s, all fish species inhabiting in the Curonian Lagoon were subjected to intensive fishing (Maksimov, 2004). Hence, the population declined (Maksimov & Toliušis, 1999; Repečka, 1999). All three of the mentioned factors contributed to the rapidly decreasing twaite shad population and catches in the Curonian Lagoon (Maksimov, 2004). However, the main reason for the decline of the species in Europe is probably the construction of dams (Aprahamian et al., 2003b). The effects of dams, weirs or other barriers may have been exacerbated by overfishing as fish congregate and become easier to capture below obstructions. According to Aprahamian et al. (2003b) poor water quality, especially in the lower reaches and estuaries has also been identified as a threat. Habitat destruction as consequence of gravel extraction and reengineering scheme to improve navigation and for flood defence purposes has also been implicated for the decline of *A. fallax* in European waters.

The decline of the Baltic twaite shad population is paralleled by the population decline of other anadromous species.

According to Imam et al. (1958), Wilkens & Köhler (1977), Möller (1984), Thiel & Salewski (2003), Krappe (2004) and Spratte (2004) a considerable decrease of anadromous lamprey populations during the mid 1950s has been observed in inland waters of Germany. Lelek (1987) assumed that pollution of estuaries and the building of weirs and dams prevented lampreys from reaching their original spawning grounds. Similarly, Kelly & King (2001) concluded that populations of anadromous lampreys have declined in European inland waters mainly due to pollution, to the construction of dams associated with hydroelectric power stations, weirs and other man-made barriers in rivers and channelisation and land-management practices that lead to increased siltation on spawning gravels. In Germany, no river lamprey fishery occurs anymore which must be attributed to the decreased population density of the species. In contrast, about 100 years ago, the river lamprey was an important commercial species for coastal fisheries in the Southern Baltic waters (Thiel et al., 2005b).

According to Thiel & Backhausen (2006) sturgeons were only caught occasionally in the Baltic Sea after 1950. The last two individuals of Atlantic sturgeon, *Acipenser oxyrinchus* Mitchill, 1815, from subdivisions 20–27 of the Baltic Sea were recorded closed to Gotland (in 1967) and to the Bay of Gdansk (in 1971; Thiel et al., 2006). Now anadromous sturgeons are considered extinct in Germany (Kirschbaum & Gessner, 2002). Similar to twaite shad, the extinction of sturgeons during the last century was caused mainly by river regulation and damming, pollution, and overfishing (Debus, 1995). Sturgeons exhibit an unusual combination of size, behaviour, and life history characteristics. These make them highly vulnerable to anthropogenic impact such as fisheries, hydroconstruction, and habitat degradation (Gessner, 2000; Mamcarcz, 2000). In a few river sections or rivers flowing into the Southern Baltic Sea, like the lower part of the Odra River, hydroconstruction presently is not a problem (Gessner & Bartel, 2000). Potential spawning habitats were found in the catchment area of the Odra River which is important in the framework of restoration programmes.

However, the restoration of twaite shad, lampreys, sturgeons and other species in the Baltic Sea will only be successful in the frame of close and continuous international cooperation (e.g. Kirschbaum & Gessner, 2000).

Conclusion

The results of this study demonstrate that the commercial catches of twaite shad are characterized by a

20-year-cyclicity in subdivisions 24–26 of the Southern Baltic Sea from 1887 to 1959. However, based on the available data it is not possible to confirm that the cyclicity of yields is also true for the population dynamics of twaite shad in the Southern Baltic Sea. On the other hand, there are some indications that the amount of commercial catches and the number of records of twaite shad follow similar patterns.

Furthermore, there is no clear evidence which is the key factor controlling disappearance and recovery of twaite shad stocks. Decreased pollution with nutrients, heavy metals and other pollutants in the lagoons of the southern Baltic and less strong winter periods in those areas during the last 15 years might have affected the recent population increase of twaite shad. Finally, it has to be pointed out that it seems to be necessary to start an international study about distribution, habitat use, genetics, morphology and conservation status of twaite shad from different regions of the Baltic Sea, North Sea and their transitional waters.

Such a study should also develop the scientific basis for the future conservation and fishery policy with regard to *A. fallax*. The twaite shad is listed in Annex II of the European Union Habitats Directive, which lists animal and plant species of interest to the European Community whose conservation requires the designation of SAC. *A. fallax* is also listed in Annex V of the Directive, which lists those animal and plant species of European Community interest whose exploitation and taking in the wild may be subject to management measures. Although it is unlikely that a commercial fishery for twaite shad will ever be proposed or start again in Germany, recent stock recoveries of *A. fallax* in Poland and Lithuania created problems in fishery and fisheries administration when twaite shad was landed as bycatch.

Acknowledgements We thank Bernd Bruns, Jens Heischkel, Volker Huckstorf, Dietmar Lill, Nardine Löser, Rimantas Repečka and Ronny Weigelt for their helpful contributions to this study. The study was assigned by the German Federal Agency for Nature Conservation and sponsored by the Federal Ministry for the Environment, Nature Conservation and Nuclear Safety under grant number 80385220.

References

Alexandrino, P., 1996. Estudo de populações de sável (*Alosa alosa* L.) e savelha (*Alosa fallax*, Lacépède). Análise da diferenciação interespecífica, subestruturação e hibridaçao. Doctoral dissertation. Universidade do Porto, Porto, Portugal: 185 pp.

Anonymous, 1887–1920. Jahresberichte über die deutsche See- und Küstenfischerei, Bd. V–XXXVII.

Anonymous, 2006. Climate Change in the Baltic Sea area. Draft HELCOM Thematic Assessment in 2006. HELCOM Stakeholder Conference on the Baltic Sea Action Plan, Helsinki, Finland, 7 March 2006: 48 pp.

Aprahamian, M. W. & C. D. Aprahamian, 2001. The influence of water temperature and flow on year class strength of twaite shad (*Alosa fallax fallax*) from the River Severn, England. Bulletin Français de la Pêche et de la Pisciculture 362/363: 953–972.

Aprahamian, M. W., C. D. Aprahamian, J. L. Baglinière, M. R. Sabatié & P. Alexandrino, 2003a. *Alosa alosa* and *Alosa fallax* spp. Literature Review and Bibliography. R&D Technical Report W1-014/TR: 349 pp.

Aprahamian, M. W., J. L. Baglinière, M. R. Sabatié, P. Alexandrino, R. Thiel & C. D. Aprahamian, 2003b. Biology, status, and conservation of the anadromous Atlantic twaite shad *Alosa fallax fallax*. American Fisheries Society Symposium 35: 103–124.

Chmielewski, S., 1965. From the history of fresh-water fisheries in Poland. In Backiel, T. (ed.), Fresh-water Fisheries of Poland, 5–13. XVI. Limnologorum Conventus in Polonia MCMLXV, Cracow: 93 pp.

Debus, L., 1995. Historic and recent distribution of Acipenser sturio in the North Sea and Baltic Sea. Proceedings of the International Sturgeon Symposium, VNIRO, Moscow: 189–203.

van Drimmelen, D. E., 1951. Beschouwingen over de fintvangsten. Visserij-Nieuws: 21–24.

Ehrenbaum, E., 1921. Mitteilungen über die Lebensverhältnisse unserer Fische. 17. Die Maifischarten. Der Fischerbote 17 and 18: 681–726 and 726–731.

Eichelbaum, E., 1926–1940. Jahresbericht über die deutsche Fischerei. Reichsministerium für Ernährung und Landwirtschaft. Verlag Gebr. Mann, Berlin.

Gerkens, M. & R. Thiel, 2001. Habitat use of age-0 twaite shad *Alosa fallax* (Lacépède, 1803) in the tidal freshwater region of the Elbe River, Germany. Bulletin Français de la Pêche et de la Pisciculture 362/363: 773–784.

Gessner, J., 2000. Reasons for the decline of *Acipenser sturio* L., 1759 in central Europe, and attempts at its restoration. Boletin Instituto España Oceanograph 16 (1–4): 117–126.

Gessner, J. & R. Bartel, 2000. Sturgeon spawning grounds in the Odra River tributaries: a first assessment. Boletin Instituto España Oceanograph 16 (1–4): 127–137.

Hesse, T., 2000. Parposz. In Brylińska, M. (ed.), Ryby slodkowodne Polski, Wydawnictwo Naukowe PWN, Warszawa 2000: 173–178.

Imam, A. K. E. S., M. Lühmann, & H. Mann, 1958. Über Neunaugen und Neunaugenfischerei in der Elbe. Fischwirt 9: 249–261.

Kartas, F., 1981. Les clupéidés de Tunisie. Caractéristiques biométriques et biologiques. Etude comparée des populations de l'Atlantique est de la Méditerranée. In The Clupeidae of Tunisia: Their biometrical and biological characteristics. A comparative study of eastern Atlantic and Mediterranean populations. Thèse de Doctorat d'Etat. Tus, Tunisia: Université de Tunis: 608 pp.

Kelly, F. L. & J. J. King, 2001. A review of the ecology and distribution of three lamprey species, *Lampetra fluviatilis* (L.), *Lampetra planeri* (Bloch) and *Petromyzon marinus* (L.): a context for conservation and biodiversity considerations in Ireland. Biology and Environment: Proceedings of the Royal Irish Academy 101B(3): 165–185.

Kirschbaum, F. & J. Gessner, 2000. Re-establishment programme for *Acipenser sturio* L., 1758: the German approach. Boletin Instituto España Oceanograph 16(1–4): 149–156.

Kirschbaum, F. & J. Gessner, 2002. Perspektiven der Wiedereinbürgerung des Europäischen Störs, *Acipenser sturio* L., im Einzugsgebiet der Elbe. Zeitschrift für Fischkunde, Suppl. 1: 217–232.

Kloppmann, M. H. F., U. Böttcher, U. Damm, S. Ehrich, B. Mieske, N. Schultz & K. Zumholz, 2003. Erfassung von FFH-Anhang II-Fischarten in der deutschen AWZ von Nord- und Ostsee. Final report for the German Federal Agency of Nature Conservation. (http://www.habitatmare.de/en/downloads.php): 82 pp.

Kottelat, M., 1997. European freshwater fishes. Biologia Bratislava 52, Supplement 5: 1–271.

Krappe, M., 2004. Quantitative Analysen populationsbiologischer Phänomene im Lebenszyklus des Bachneunauges *Lampetra planeri* (Bloch, 1784). PhD thesis, University of Rostock: 241 pp.

Lelek, A., 1987. Threatened fishes of Europe. In Holcik, J. (ed.), The Freshwater Fishes of Europe, Vol. 9. Aula Verlag, Wiesbaden: 343 pp.

Maksimov, Y., 2004. The "revival" of the twaite shad (Alosa fallax, Lacepede 1803) population in the Curonian Lagoon. Bulletin of the Sea Fisheries Institute 1(161): 61–62.

Maksimov, Y. & S. Toliušis, 1999. Water pollution effect on abundance of coastal herring (*Clupea harengus*) and twaite shad (*Alosa fallax* Lacépède) in the Curonian Lagoon (Lithuania). Proceedings of Symposium on Freshwater Fish and the Herring (*Clupea harengus*) Populations in the Coastal Lagoons. Environment and Fisheries. Sea Fisheries Institute, Gdynia (Poland): 153–160.

Mamcarcz, A., 2000. Decline of the Atlantic sturgeon *Acipenser sturio* L., 1758 in Poland: an outline of problems and prospects. Boletin Instituto España Oceanograph 16(1–4): 191–202.

Manyukas, Y. L., 1989. Biology of the Atlantic shad, *Alosa fallax fallax*, in Kurshskiy Bay. Journal of Ichthyology 29: 125–128.

Mikelsaar, N., 1984. Eesti NSV Kaland. In Fishes of the Estuanian SSR. Tallinn. Valgus: 432 pp.

Mileriene, E., 1997. Investigations on the fish fauna in the Lithuanian parts of Kuršhiu (Curonian) lagoon. FAO Fisheries Report No. 541 Supplement. Food and Agriculture Organization of the United Nations: 74–78.

Mohr, E.,1941. Maifische. In Demoll, R. & H. N. Maier (eds), Handbuch der Binnenfischerei Mitteleuropas: E. Schweizerbart'sche Verlagsbuchhandlung, Stuttgart: 527–551.

Möller, H., 1984. Daten zur Biologie der Elbfische. Verlag Heino Möller, Kiel.

O'Maoileideigh, N., S. Cawdery, J. J. Bracken & A. Ferguson, 1988. Morphometric, meristic character and electrophoretic analysis of two Irish populations of twaite shad, *Alosa fallax* (Lacépède). Journal of Fish Biology 32: 355–366.

Peczalska, A.,1973. Parposz, *Alosa fallax* (Lacépède, 1800) – ryba malo znana. Przeglad Zoologiczny 17: 195–200.

Quignard, J. P. & C. Douchement, 1991a. *Alosa alosa* (Linnaeus, 1758). In Hoestlandt, H. (ed.), The freshwater fishes of Europe. Volume 2: Clupeidae, Anguillidae 2. Aula-Verlag, Wiesbaden: 89–126.

Quignard, J. P. & C. Douchement, 1991b. *Alosa fallax fallax* (Lacépède, 1803). In Hoestlandt, H. (ed.), The freshwater fishes of Europe. Volume 2: Clupeidae, Anguillidae 2. Aula-Verlag, Wiesbaden: 225–253.

Repečka, R., 1999. Biology and resources of the main commercial fish species in the Lithuanian part of the Curonian Lagoon. Proceedings of Symposium on Freshwater Fish and the Herring (*Clupea harengus*) Populations in the Coastal Lagoons – Environment and Fisheries. Sea Fisheries Institute, Gdynia (Poland): 185–195.

Repečka, R., 2003. Changes in the biological indices and abundance of salmon, sea trout, smelt, vimba and twaite shad in the coastal zone of the Baltic Sea and the Curonian Lagoon at the beginning of spawning migration. Acta Zoologica Lituania 13(2): 195–216.

Saat, T. & R. Eschbaum, 2002. Fishes of the Väinameri; changes during recent decades. In Saat, T. (ed.), Fishes and Fisheries of the Väinameri. University of Tartu, Estonia: 9–45.

Sabatié, M. R., 1993. Recherches sur l'écologie et la biologie des aloses du Maroc (*Alosa alosa* Linné, 1758 et *Alosa fallax* Lacépède, 1803). Exploitation et taxinomie des populations atlantiques; bioécologie des aloses de l'oued Sebou. Thèse de Doctorat, U.B.O., Brest: 326 pp.

Saemundsson, B., 1949. Zoology of Iceland: marine pisces. Munksgaard, Copenhagen and Reykjavik 4: 72–124.

Schaarschmidt, T. & R. Lemcke, 2004. Quellendarstellungen zur historischen Verbreitung von Fischen und Rundmäulern in Binnengewässern des heutigen Mecklenburg-Vorpommerns. Mitteilungen der Landesforschungsanstalt für Landwirtschaft und Fischerei Mecklenburg-Vorpommern 32: 1–261.

Skóra, M. E., 2003. Charakterytyka populacji parposza *Alosa fallax fallax* (Lacépède, 1803) z rejonu Zatoki Gdanskiej. Magisterwork, Uniwersytet Gdanski: 85 pp (in Polish).

Spratte, S., 2004. Projektierungsstudie zur Verbesserung der Bestandsentwicklung von Neunaugen in Schleswig-Holstein. Landessportfischerverband Schleswig-Holstein e.V.: 36 pp.

Svetovidov, A. N., 1952. Fauna of the USSR, Fishes. Herrings (Clupeidae). Izd. Akad. Nauk SSSR, Moscow-Leningrad: 331 pp (in Russian).

Taverny, C., 1991. Pêche, biologie, écologie des aloses dans le système Grionde-Garonne-Dordogne. Contribution à la connaissance de la dynamique des populations d'aloses (*Alosa alosa* et *Alosa fallax*) dans le système fluvio-estuarien de la Gironde. Etudes CEMAGREF, Séries Réssources en eau 4: 375 pp.

Thiel, R. & I. Backhausen, 2006. Survey of NATURA 2000 fish species in the German North and Baltic Seas. In von Nordheim, H., D. Boedeker & J. C. Krause (eds), Progress in Marine Conservation in Europe: Springer-Verlag, Berlin, Heidelberg: 157–178.

Thiel, R. & V. Salewski, 2003. Verteilung und Wanderung von Neunaugen im Elbeästuar (Deutschland). Limnologica 33: 214–226.

Thiel, R., H. M. Winkler & R. Neumann, 2004. Erfassung von FFH-Anhang II-Fischarten in der deutschen AWZ von Nord- und Ostsee (ANFIOS). 1. Report for the German Federal Agency of Nature Conservation (http://www.habitatmare.de/en/downloads.php): 67 pp.

Thiel, R., H. M. Winkler & R. Neumann, 2005a. Erfassung von FFH-Anhang II-Fischarten in der deutschen AWZ von Nord- und Ostsee. 2. Report for the German Federal Agency of Nature Conservation (http://www.habitatmare.de/en/downloads.php): 106 pp.

Thiel, R., H. M. Winkler, P. Riel & R. Neumann, 2005b. Survey of river and sea lampreys in German waters of the Baltic Sea – basis of successful rebuilding programmes. ICES CM 2005/W:06: 33 pp.

Thiel, R., H. M. Winkler & R. Neumann, 2006. Erfassung von FFH-Anhang II-Fischarten in der deutschen AWZ von Nord- und Ostsee. Final report for the German Federal Agency of Nature Conservation. (http://www.habitatmare.de/en/downloads.php).

Waterstraat, A., 1986. Aktuelle Aufgaben zum Schutz gefährdeter Rundmäuler und Fische in Mecklenburg in Auswertung der Artenschutzbestimmung von 1984. Naturschutzarbeit in Mecklenburg 29: 87–92.

Wheeler, A., 1969. The Fishes of the British Isles and North West Europe. MacMillan, London: 613 pp.

Więcaszek, B. & S. Krzykawski, 1999. "Majowe śledzie" znów plywają w Baltyku. Magazyn Przemyslu Rybnego 3/99, 29–30.

Wiktor, J., 1989. Fish farming in Polish Baltic waters and its role in preserving the stocks of valuable species. Bulletin of the Sea Fisheries Institute, Gdynia 20: 50–54.

Wilkens, H. & A. Köhler, 1977. Die Fischfauna der mittleren und unteren Elbe: die genutzten Arten, 1950–1975. Verhandlungen Naturwissenschaftlichen Verein Hamburg 20: 185–222.

Wilkońska, H. & A. Garbacik-Wesolowska, 1996. Powrót parposza *Alosa fallax* (Lacépède, 1800). Komunikaty Rybackie 6/96: 13 pp.

Winkler, H. M., 1991. Changes of structure and stock in exploited fish communities in estuaries of the southern Baltic coast (Mecklenburg-Vorpommern, Germany). Internationale Revue der Gesamten Hydrobiologie 76: 413–422.

Winkler, H. M., K. Skóra, R. Repečka, M. Ploks, A. Neelov, L. Urho, A. Gushin & H. Jespersen, 2000. Checklist and status of fish species in the Baltic Sea. ICES CM 2000/ Mini 11: 15 pp.

Žiliukas, V. & V. Žiliukienė, 2002. Ichthyological evaluation of fish passes constructed in Lithuania. Acta Zoologica Lituanica 12(1): 47–57.

Zolubas, T. & S. Toliušis, 2001. Particular fish community changes in the Baltic Sea Lithuanian coastal zone in 1994–2000, Books of abstracts, Symposium on the Occasion of the 80th Anniversary of the Sea Fisheries Institute, Sea Fisheries Institute, Gdynia.

Hydrobiologia (2008) 602:179–186
DOI 10.1007/s10750-008-9291-6

FISH AND DIADROMY IN EUROPE

Does global warming impact on migration patterns and recruitment of Allis shad (*Alosa alosa* L.) young of the year in the Loire River, France?

C. Boisneau · F. Moatar · M. Bodin · Ph. Boisneau

Abstract The hydrological and thermal changes in the Loire River were investigated to test the influence of climatic changes on a short freshwater stage anadromous fish species, Allis shad, for the 1995–2004 period. The mean water temperatures during the adult migration and juvenile growth phases showed significant increase, and mean water flow during these two phases decreased significantly. The period below the threshold of 18°C shortened, and the period between 18°C and the maximum lengthened, while the temperature amounts (number of degree-days) received by aquatic organisms between 18 and 24°C showed an increase. The pattern of young-of-the-year downstream migration was modified. The first day when the juvenile catches reached 5% occurred 17 days earlier at the end of the 1995–2004 period than at the beginning. The first day when the juvenile catches reached 50% was related to the 18°C threshold (reproductive threshold) and the temperature amounts accumulated between the 18 and 20°C thresholds. The year-on-year levels of young-of-the-year abundance showed wide variations, which were not explained by environmental parameters, probably because of the long distance between the study site and the spawning grounds.

Keywords Water temperature · Hydrology · Trends · Migration pattern · Abundance · Allis shad

Guest editors: S. Dufour, E. Prévost, E. Rochard & P. Williot
Fish and diadromy in Europe (ecology, management, conservation)

C. Boisneau (✉) · F. Moatar
University of Tours, Parc de Grandmont, Tours 37200, France
e-mail: boisneau@univ-tours.fr

C. Boisneau
IMACOF Ingénierie des Milieux Aquatiques et des Corridors fluviaux, Tours, France

F. Moatar
EA2100, Laboratoire de Géologie des Environnements Aquatiques Continentaux, Tours, France

M. Bodin · Ph. Boisneau
Association Agréée Interdépartementale des Pêcheurs Professionnels du bassin Loire-Bretagne, Tours, France

Introduction

The oldest daily air temperature records in Europe date back to the nineteenth century, but this type of records is very rare for water temperatures (Webb & Nobilis, 1994). During the twentieth century, the water temperature in the Danube in Austria has shown a significant increase in monthly mean water temperatures of 0.8°C, varying from 0.66 to 2.0°C. Severe low water levels and human activities are accelerating this process (Webb & Nobilis, 1994, 1995). In the upper Rhône River (France), between 1979 and 1999, the average water temperature rose by about 1.5°C due to atmospheric warming, and the flow rates fluctuated around the annual mean without showing any particular trends (Daufresne et al., 2003). Linear regression models based on monthly air temperature and flow rate data were used to reconstruct the temperatures for the period 1881–2003 in the middle section of the Loire River (France). The rapid rise in water temperatures observed since 1976 is part of a general trend over the century, marked by other warm periods around 1900 and 1950 (Moatar & Gailhard, 2006). The mean annual water temperatures increased by 1.6°C between 1976 and 2003.

The Allis shad (*Alosa alosa*) is an anadromous Clupeid historically ranging from south Morocco to north Germany (Baglinière, 2000). This species frequents the Loire watershed where spawning grounds are located about 550–750 km from the sea (Boisneau et al., 1990). As there are no stocking phenomena, this species provides an interesting model for assessing the impacts of global changes. The freshwater stage is very short, ranging from 3 to 6 months. The juveniles migrate downstream to the sea during the summer and early autumn (Taverny et al., 2000).

The global warming of rivers in temperate zones is likely to affect ectothermic aquatic organisms. Flow and water temperature exercise fundamental control over these ecosystems and their communities (Ward, 1992). Although density-dependent processes interact with environmental factors to determine recruitment levels and variability, water temperature and flow are two important factors that affect freshwater fish dynamics (Schlosser & Angermeier, 1990; Cattaneo et al., 2001, 2002). Water temperature controls physiology and influences phenology, while survival of fish during their early life may be affected by water temperature and water flow rates. This can lead to earlier maturity, change the population age breakdown and generate year class fluctuations (Mann, 1991; Elliott et al., 2000; Davidson et al., 2006).

As climatic factors are known to influence freshwater fish abundance, our question in this article is whether these factors influence an anadromous species with a short freshwater stage. Especially, we address the following questions: (1) What are the hydrological and thermal changes in the middle section of the Loire River during Allis shad freshwater phases? What are the consequences of these changes? (2) For the migration pattern of young of the year, and (3) On their inter-annual abundance?

Methods

Study sites

The Loire River runs over a 1,000-km course from the centre of France and drains an area of about 117,000 km ^{2}. The study site where fish were caught is located 250 km from the sea and 300 km downstream from the spawning grounds (Fig. 1). The annual mean flow at the study site is 364 m^3/s. (1863–2004).

Temperature and flow data

Daily flow rate data were taken from the HYDRO national database managed by the French Department of the Environment for the nearest site. Daily temperature data were recorded by Electricité de France upstream from the Chinon-Avoine nuclear power plant, the nearest spot where water temperatures are recorded (Fig. 1). The levels of thermal waste flow into the Loire are very low, with a median rise of 0.1°C (Moatar et al., 2006) and they do not interfere with anadromous fish migration.

To evaluate the effects of hydrological and thermal changes on Allis shad, the freshwater stage was divided into two periods, adult upstream migration and spawning from 15/03 to 15/06, and the young of

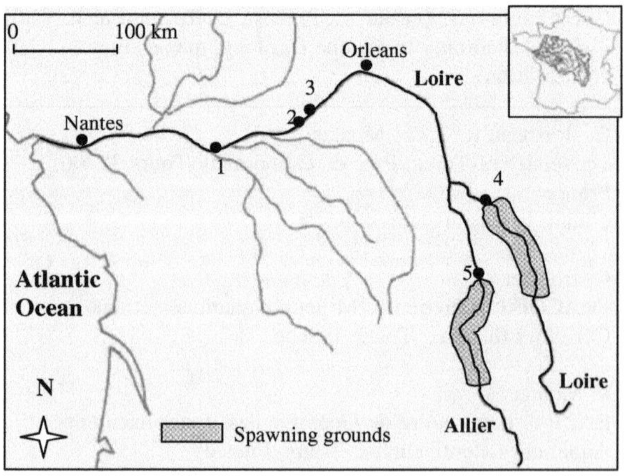

Fig. 1 Location of the temperature recording site, Avoine (1), the fish sampling site (2), the flow rate recording site (3) and Allis shad adult control stations (4–5)

the year (YOY) early growth and downstream migration, from 15/06 to 15/10. Several hydroclimatic variables were used (Carrel, unpublished data). They were calculated from daily data, for each year, from 1976 to 2004:

- The number of days (Julian calendar) between 1 January and the date of exceeding a given threshold. This shows whether the temperature rise phase occurs late or early. The thresholds adopted were 16, 18, 20, 24°C and the maximum. The temperature range chosen covers the adult migration period, the egg incubation period and the initial live stage (Roule, 1923; Hoestlandt, 1958). The variables are called C16, C18, C20, C24 and Cmax.
- The duration, in days, between the two thermal thresholds, between 16 and 18°C (D16–18), between 18 and 20°C (D18–20) and between 20 and 24°C (D20–24). This expresses the variations in length of the periods between two thresholds.
- The sum of daily thermal differences between the annual curve and an inter-annual smoothed curve (1976–2004) for the period between two thresholds, 16 and 18°C (S16–18), 18 and 20°C (S18–20), 20 and 24°C (S20–24). This is the equivalent of a degree-day amount as compared with the inter-annual curve. It quantifies temperature increases or decreases.
- The average conditions of seasonal water flow and temperature were estimated from the mean water temperature, the mean water flow rate and the median daily flow rate (Q50) over each migration season.
- The high and low flow level events were described with the 0.1, 0.2, 0.8, 0.9 percentiles of water flow expressed as Q10, Q20, Q80 and Q90 for the two periods of shad migration.
- Flow rate variability was expressed as the ratio Q10/Q90.

Fish data

Downstream migrating juvenile were intercepted with a beach seine, because Allis shad juveniles are near the bank (Taverny et al., 2000) (location 2, Fig. 1). The mesh size was 10 mm. The water depth ranged from 0.5 to 1.2 m and the water flow speed varied from 0.3 to 0.8 m/s. The substrate was sand and gravel. Sampling was carried out from Mondays to Fridays, from mid-June to mid-October, and from 1995 to 2004. The water flow rate ranged from 55 to 254 m^3/s. The fish were measured (fork length) to the nearest millimetre.

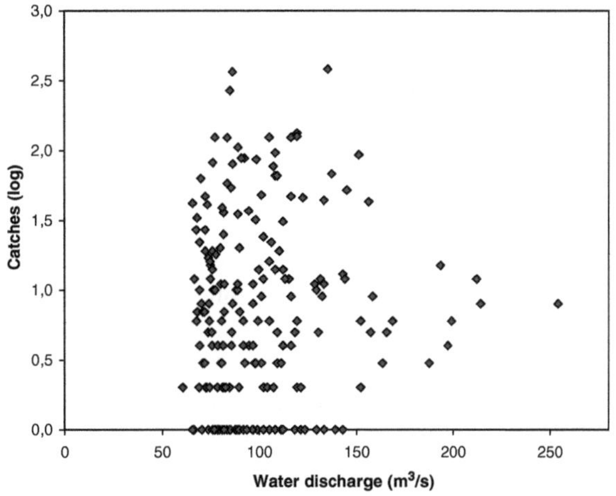

Fig. 2 Relationship between catches of Allis shad YOY and water flow rate during fish sampling

The links between YOY catches and effort were checked for each year. They showed no relationship except for 2001, which accounted for 6.3% of catches over the entire series. Catches and effort were still considered as independent. The links between catches and water flow were also checked for each year and did not show any relationship ($P >> 0.05$, Fig. 2). The YOY abundance was expressed as catch per unit effort (CPUE), calculated for each year, as a ratio of the total number of catches to the total effort.

The pattern of juvenile migration was determined by the days (Julian calendar) when the 0.05, 0.5 and 0.95 percentile of catches occurred. The durations, in days, between two of these dates (days when 0.05 and 0.5 of the catches occurred; days when 0.5 and 0.95 of the catches occurred) were calculated. This expresses the variations in the length of the migrating period.

The numbers of adults reaching the main spawning grounds in the watershed were obtained from fish control stations managed by LOGRAMI[1], since 1998, the starting date for data collection. They were expressed as the total number of adults getting over

[1] LOGRAMI is an NGO bringing together recreational and commercial fishermen from the Loire River watershed. It provides assistance for diadromous fish management.

the fish pass on the Loire and the Allier Rivers (Fig. 1, locations 4 and 5), over a period ranging from 14 April to 3 July.

Data analysis

In order to detect trend over time for thermal and hydrological parameters and in migration patterns, we used a non-parametric seasonal Kendall test and seasonal Kendall slope estimator. This non-parametric test (based on ranks) seeks trend once autocorrelation effects are removed (Webb & Nobilis, 1994).

The influence of flow and thermal changes on the migration pattern for YOY was tested with a Spearman rank correlation between the days of 0.05 and 0.5 percentile of catches and the environmental variables that showed significant prior trends.

Spearman rank correlations were used to test the potential relationship between YOY abundance and some selected environmental variables that showed significant prior trends, and between YOY abundance and the number of adults reaching the main spawning grounds.

Results

Temperature and flow rate data

The mean water temperature during the two periods of shad migration showed a significant increase of 2.5°C for the adult upstream migration period and 2.0°C for the YOY downstream migration period, between 1976 and 2004 (Table 1). The numbers of days before the thermal thresholds (Ci) showed significant decrease for the 18, 20 and 24°C levels, but not for the maximum (Table 1). The duration, in days, between the thresholds (Di) remained unchanged, but the increase in the heat amount (Si) between 18 and 20°C and 20 and 24°C was significant (Table 1).

The mean water flow rate during the two periods of shad migration showed a significant decrease (Table 2). These seasonal flow rate falls were part of a general trend over the entire year. The annual mean flow rate was 469 m^3/s in 1976 and 264 m^3/s in 2004 ($r_k = -0.547$, $P = 0.002$), the decrease is around 40%. The mean water flow rate during adult upstream migration was 626 m^3/s in 1976 and 298 m^3/s in 2004, and the mean water flow rate during YOY downstream migration was 218 m^3/s in 1976 and 98 m^3/s in 2004 (Table 2). The high flow levels for each migration season showed significant falls, though the low flow levels only dropped during the adult upstream migration (Table 2). Flow rate variability increased significantly during the juvenile downstream migration.

None of these trends were sensitive to extreme years; they remained significant at the 0.05 confidence level when we removed the hot years 1976 and 2003.

Fish data

Allis shad juveniles are very difficult to catch, and it was the first time in France that they has been successfully sampled over a 10-year period (1995–2004). 4,613 young Allis shad were sampled, with an absence of catches in 1996. The mean length was

Table 1 Trends in thermal variables of the median part of the Loire River between 1976 and 2004

Parameter	Slope of linear trend	Significance of trend (P)	Change over the study period Data for 1976 and 2004
C18 (Julian day for threshold 18°C)	−0.56	0.041	Day 140–day 124
C20 (Julian day for threshold 20°C)	−0.83	0.010	Day 159–day 135
C24 (Julian day for threshold 24°C)	−1.3	0.001	Day 198–day 161
S18–20 (temperature increase)	1.64	0.021	46°C. day
S20–24 (temperature increase)	5.31	0.000	102°C. day
Mean water temperature during adult upstream migration (15/03–15/06)	0.09	0.000	13.3–15.8°C
Mean water temperature during YOY downstream migration (15/06–15/10)	0.07	0.001	19.3–21.3°C

Only variables showing significant trends are shown

Table 2 Trends in the Loire River flow rate variables between 1976 and 2004

Parameter	Slope of trend	Significance of trend (P)	Change over the study period Data for 1976 and 2004
River flow rate during adult upstream migration			
Q20 (percentile 0.2)	−6.72	0.046	372–183 m^3/s
Q50 (percentile 0.5)	−9.94	0.048	538–260 m^3/s
Q80 (percentile 0.8)	−15.68	0.039	826–387 m^3/s
Q90 (percentile 0.9)	−21.37	0.036	1087–489 m^3/s
Mean river flow rate	−11.7	0.036	626–298 m^3/s
River flow rate during YOY downstream migration			
Q50 (percentile 0.5)	−3.36	0.038	172–78 m^3/s
Q80 (percentile 0.8)	−6.4	0.041	306–128 m^3/s
Q90 (percentile 0.9)	−9.36	0.047	417–155 m^3/s
Flow rate variability (Q10/90)	0.07	0.029	0.266–0.468
Mean river flow rate	−4.299	0.039	218–98 m^3/s

Only variables showing significant trends are shown

58 mm (standard deviation: 11 mm) with extremes ranging from 45 to 76 mm.

YOY were intercepted between 28 June and 15 October. The starting date for their downstream migration was 17 days earlier at the end of the time lapse than at the beginning (Fig. 3). The days when 50% and 95% of the catches occurred did not show any significant trend. This means that the duration of the first half of the migration period is lengthening.

No significant relationships were found between environmental variables and the day when 5% of catches of YOY occurred, but significant relationships were found between the day when 50% of catches occurred and the 18°C threshold ($r_s = 0.745$, $P = 0.012$), and the number of degree days between 18 and 20°C ($r_s = 0.745$, $P = 0.012$).

The year-on-year changes in abundance of YOY showed considerable variability, with very low levels in 1996, 1999, 2000 and 2003, and a very high level in 2004 (Fig. 4), but no links were found between juveniles abundance and adult abundance, thermal variables or river flow rates (Table 3).

Discussion

Temperature and flow rate data

The rise in water temperatures observed for the Loire River is consistent with the changes occurring in other European rivers, despite differences in their hydrological regime (Daufresne et al., 2003; Davidson et al., 2006; Mouthon & Daufresne, 2006; Webb & Nobilis, 1995). For the Loire, the water temperatures show significant increases in the spring, summer and winter (Moatar & Gailhard, 2006). During the spring and summer, the 18, 20 and 24°C thresholds are reached earlier, but the day when the maximum is reached does not show any significant trends. The period before the 18°C threshold has shortened, and the period between 18°C and the maximum has lengthened accordingly, while the temperature amount received by aquatic organisms between 18 and 24°C has increased.

In contrast with the trends noted concerning water flow rates in the Rhône and Saône Rivers (Daufresne et al 2003; Mouthon & Daufresne, 2006), the mean annual flow rate of the Loire is decreasing. This is also observed for the flow rate parameters, Q50, Q80 and Q90 for spring and summer. The maximum flow rate and flow rate variability (Q10/Q90) during the adult migration period do not show any trends. On the other hand, flow rate variability during the juveniles migration is increasing. We can assume that species spawning or/and having their larval development during spring and early summer probably emerge earlier and grow faster. This could be accelerated by lower water flow rates and lower levels of variability.

YOY of Allis shad

Climatic conditions during the larval period generally determine the survival rate of the fish during their

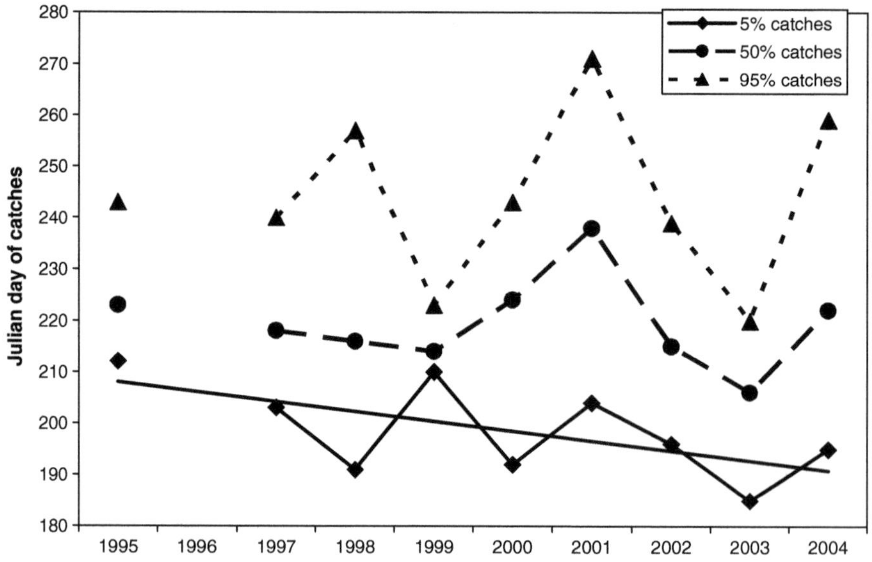

Fig. 3 Temporal evolution of the Julian days when 0.05, 05 and 0.95 percentile of YOY of Allis shad catches occurred. Only significant trends are shown. 5% of catches (seasonal Kendall test, $r_k = -0.700$, $P = 0.03$; 50% of catches, $r_k = -0.367$, $P = 0.09$; 95% of catches, $r_k = 0.112$, $P = 0.327$)

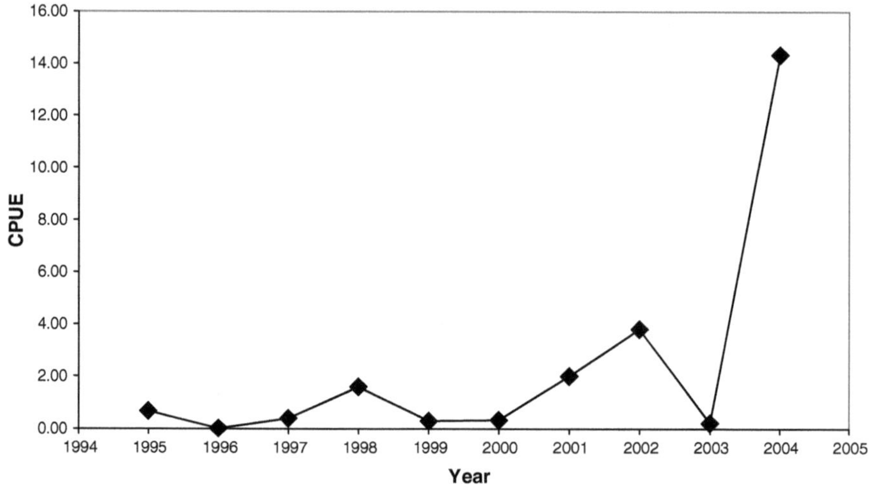

Fig. 4 Temporal evolution in abundance of Allis shad YOY

Table 3 Relationship between Allis shad YOY abundance and environmental parameters (1995–2004)

Parameter	Correlation (r_s)	Probability level (P)
Mean water temperature during adult upstream migration (15/03–15/06)	−0.479	0.16
Mean water temperature during YOY downstream migration (15/06–15/10)	−0.333	0.217
C18 (Julian day for threshold 18°C)	0.033	0.463
C20 (Julian day for threshold 20°C)	−490	0.15
C24 (Julian day for threshold 24°C)	0.382	0.138
S18–20 (temperature increase)	−0.261	0.191
S20–24 (temperature increase)	−0.127	0.36
Mean water flow during adult upstream migration (15/03–15/06)	0.406	0.122
Mean water flow during YOY downstream migration (15/06–15/10)	0.539	0.054
Number of adults reaching the main spawning grounds	0.464	0.147

initial life, and their abundance in small or large rivers (Cattaneo et al., 2001, 2002; Elliott et al., 2000; Daufresne, 2003; Grenouillet et al, 2001). Annual juveniles abundance shows strong fluctuations, without any apparent relationship with environmental variables. If the fishing effort is constant in terms of time and surface when the flow rate increases, the relative proportion of sampled habitat is reduced. This could account for the absence of links between YOY abundance and summer water flow rate. The distance between the study site and the spawning grounds is another possible explanation for the absence of links between YOY abundance and hydroclimatic variables. The spawning grounds are located 300 km upstream from the sampling site, and the environmental parameters are probably different. It was not possible to test the impact on the YOY survival rate of flows and temperatures during incubation and the early stages of life in the spawning grounds.

There are positive relationships between the day when 50% of the juveniles catches occurred and the 18°C threshold and the number of degree-days between 18 and 24°C. As is the case for other fish species, the durations of egg and larval development are related to temperatures, and especially to the temperature for adult reproduction, 18°C being the median temperature value when spawning occurs (Hoestlandt, 1958).

The day when the 0.05 percentile of catches was reached, occurred earlier (27 July in 1995 and 10 July in 2004). The downstream migration now starts earlier than the mid-August date given for the Garonne and Loire rivers (Taverny et al., 2000).

The absence of a relationship between the day for the 0.05 percentile of YOY catches and the environmental variables could be explained by the location of the sampling site; but according to Bernard & Larinier, (1988) water temperature and water flow rate are not at the origin of juveniles migration on the Garonne River. It could be the same on the Loire River; certain biological and physiological parameters (length and hormone concentration) could be involved in the downstream departure process, as is shown for twaite shad (*Alosa fallax*) (Aprahamian, 1988) and *Alosa sapidissima* (Limburg, 1996).

To conclude, our results indicate impacts of climatic change in the Loire River either in the thermal or the hydrological regime for the 1976–1995 period. This influences the departure of YOY Allis shad on their downstream migration, but not their annual abundance. This strongly suggests the need to take into account the impact of climatic change in anadromous fish monitoring programs.

Acknowledgements We thank HYDRO national database managed by the French department of the Environment for the flow rate data and Electricité de France at the Chinon-Avoine nuclear power plant for the temperature data. We are grateful to the three referees who helped to make considerable improvements to the quality of the manuscript, and to J. Casas, IRBI, University of Tours, for his helpful remarks.

References

Aprahamian, M., 1988. The biology of twaite shad, *Alosa fallax fallax* (Lacépède) in the Severn Estuary. Journal of Fish Biology 33(Supplement A): 141–152.

Baglinière, J. L., 2000. Le genre Alosa sp. In Baglinière J. L. & P. Elie (eds), les aloses (*Alosa alosa* et *Alosa fallax* spp.), écobiologie et variabilité des populations. INRA, Paris, 3–30.

Bernard, S. & M. Larinier, 1988. Echantillonnage d'alosons dans les chambres anti vortex de l'usine hydro électrique de Golfech. Rapport CEMAGREF, CSP, Ministère de l'Environnement 32.

Boisneau, P., C. Mennesson-Boisneau & J. L. Baglinière, 1990. Description d'une frayère et comportement de reproduction de la grande alose (*Alosa alosa)* dans le cours supérieur de la Loire. Bulletin Français de Pêche et Pisciculture 316: 15–23.

Cattaneo, F., G. Carrel, N. Lamouroux & P. Breil, 2001. Relationship between hydrology and cyprinid reproductive success in the Lower Rhône at Montélimar, France. Archiv für Hydrobiologie 151: 427–450.

Cattaneo, F., N. Lamouroux, P. Breil & H. Capra, 2002. The influence of hydrological and biotic processes on brown trout (*Salmo trutta*) population dynamics. Canadian Journal of Fisheries and Aquatic Sciences 59: 12–22.

Daufresne, M., M. C. Roger, H. Capra & N. Lamouroux, 2003. Long-term changes within the invertebrate and fish communities of the upper Rhône River: effects of climatic factors. Global Change Biology 10: 124–140.

Davidson, I. C., M. S. Hazlewood & R. J. Cove, 2006. Predicted growth of juvenile trout and salmon in four rivers in England, Wales based on past and possible future temperature regimes linked to climate change. In Harris, G. S. & N. J. Milner (eds), Sea Trout: Biology, Conservation & Management. Proceedings of the First International Sea Trout Symposium, Cardiff, Wales.

Elliott, J. M., M. A. Hurley & S. C. Maberly, 2000. The emergence period of sea trout fry in a Lake District stream correlates with the North Atlantic Oscillation. Journal of Fish Biology 56: 208–210.

Grenouillet, G., B. Hugueny, G. A. Carrel, J. M. Olivier & D. Pont, 2001. Large-scale synchrony and inter-annual variability in roach recruitment in the Rhône River: the

relative role of climatic factors and density-dependent processes. Freshwater Biology 46: 11–26.
Hoestlandt, D. H., 1958. Reproduction de l'alose atlantique (*Alosa alosa*, Linné) et transfert au bassin méditerranéen. Verhandlungen der Internationalen Vereinigung für Theoretische und Angewandte Limnologie 13: 736–742.
Limburg, K., 1996. Growth and migration of 0 year American shad (*Alosa sapidissima*) in the Hudson River estuary: otolith microstructural analysis. Canadian Journal of Fisheries and Aquatic Sciences 53: 220–238.
Mann, R. H. K., 1991. Growth and production. In Winfield, I. J. & J. S. Nelson (eds), Cyprinid Fishes. Systematics, biology and exploitation. Fish and fisheries Series 3. Chapman & Hall, London: 456–482.
Mouthon, J. & M. Daufresne, 2006. Effects of the 2003 heat wave and climatic warming on mollusc communities of the Saône: a large lowland river and of its two main tributaries. Global change Biology 11: 1–9.
Moatar, F. & J. Gailhard, 2006. Water temperature behaviour in the River Loire since 1976 and 1881. C. R. Geoscience 338: 319–328.
Roule, L., 1923. Notes sur les aloses de la Loire et de l'Aquitaine. Bulletin de la Société Centrale d'Agriculture et de Peche 30: 14–22.
Schlosser, I. J. & P. L. Angermeier, 1990. The influence of environmental variability, resource abundances and predation of juvenile cyprinid and centrarchid fishes. Polskie Archwum Hydrobiologii 37: 265–284.
Taverny, C., J. J. Cassou-Leins, F. Cassou-Leins & P. Elie, 2000. De l'œuf à l'adulte en mer. In Baglinière, J. L. & P. Elie (eds), les aloses (*Alosa alosa* et *Alosa fallax* spp.), écobiologie et variabilité des populations. INRA, Paris: 93–124.
Ward, J. V., 1992. Aquatic Insect Ecology, 1; Biology and Habitat. JWiley, New York, USA.
Webb, B. W. & F. Nobilis, 1994. Water temperature behaviour in the River Danube during the twentieth century. Hydrobiologia 291: 105–113.
Webb, B. W. & F. Nobilis, 1995. Long-term water temperature trends in Austrian rivers. Hydrological Sciences 40: 83–96.

MIX
Papier aus verantwortungsvollen Quellen
Paper from responsible sources
FSC® C105338

If you have any concerns about our products,
you can contact us on
ProductSafety@springernature.com

In case Publisher is established outside the EU,
the EU authorized representative is:
Springer Nature Customer Service Center GmbH
Europaplatz 3, 69115 Heidelberg, Germany

Printed by Libri Plureos GmbH
in Hamburg, Germany